# LIZARDS
## AND SNAKES
## OF ALABAMA

Philip Henry Gosse as a young man of twenty-nine, the year of his return to England from Alabama, painted by his brother, William Gosse. (1839, watercolor on ivory, courtesy of the National Portrait Gallery—London)

**Philip Henry Gosse** (1810–1888) was an English naturalist and illustrator who spent eight months of 1838 on the Alabama frontier, teaching planters' children in Dallas County and studying the native flora and fauna. Years after returning to England, he published the now-classic *Letters from Alabama: Chiefly Relating to Natural History,* with twenty-nine important black-and-white illustrations included. He also produced, during his Alabama sojourn, forty-nine remarkable watercolor plates of various plant and animal species, mainly insects, now available in *Philip Henry Gosse: Science and Art in "Letters from Alabama" and "Entomologia Alabamensis."*

**The Gosse Nature Guides** are a series of natural history guidebooks prepared by experts on the plants and animals of Alabama and designed for the outdoor enthusiast and ecology layman. Because Alabama is one of the nation's most biodiverse states, its residents and visitors require accurate. accessible field guides to interpret the wealth of life that thrives within the state's borders. The Gosse Nature Guides are named to honor Philip Henry Gosse's early appreciation of Alabama's natural wealth and to highlight the valuable legacy of his recorded observations. Look for other volumes in the Gosse Nature Guides series at http://uapress.ua.edu.

THE UNIVERSITY OF ALABAMA PRESS  TUSCALOOSA

# LIZARDS AND SNAKES OF ALABAMA

**CRAIG GUYER, MARK A. BAILEY**

**AND ROBERT H. MOUNT**

WITH LINE DRAWINGS BY **CLAIRE C. FLOYD**

The University of Alabama Press
Tuscaloosa, Alabama 35487-0380
uapress.ua.edu

Typeface: Scala Pro

Manufactured in China
Cover image: *Nerodia erythrogaster flavigaster,*
Elmore County, AL; photo courtesy of James C. Godwin
Cover and interior design: Michele Myatt Quinn

Publication is supported in part by the
ALABAMA MUSEUM OF NATURAL HISTORY,
Tuscaloosa, Alabama

Cataloging-in-Publication data is available from the Library of Congress.
ISBN: 978-0-8173-5916-4
E-ISBN: 978-0-8173-9192-8

I see that you are surprised at the rustling noise and motion that occurs among the dry leaves on either side, at almost every step. It is caused by the nimble feet of little lizards, which dart along like lightning as we approach, to the shelter of the nearest log or stone, under which they may hide: they move so quickly that it is very seldom we can catch a glance of their bodies; we trace them only by their motion and their sound. There are three or four species, the most common of which is called, by a strange misnomer, the Scorpion (*Agama undulata*); and it is this species which so rapidly scuttles along under the crisped leaves. It is about six inches long, of which half is tail: above, it is greyish, with darker bands; underneath it is palish, with a patch of bright blue under the throat, larger in some (I think, males) than in others. It is covered with prominent scales, each having a sharp ridge, which gives it a rough appearance. They are very abundant, and may be often seen chasing each other about some old log, running by little starts, now on the top, now on the sides, and now on the bottom, it being all the same whether the back be upward or downward.

—Philip Henry Gosse, *Letters from Alabama*, Letter III, Dallas County, June 1, 1838

# Contents

# Skinks, Whiptails, and Their Relatives

# Glass Lizards and Their Relatives

# Snakes

# Vipers

# Abbreviations

Throughout this book, various agencies, programs, and legislation are frequently represented as acronyms with which the reader should become familiar.

ADCNR Alabama Department of Conservation and Natural Resources
AUM    Auburn University Museum of Natural History
ESA    Endangered Species Act of 1973, as amended
SWAP   Alabama's State Wildlife Action Plan (2005, revised 2015)
USFS   US Forest Service
USFWS  US Fish and Wildlife Service

# Introduction

This book is designed to update the squamate (lizard and snake) fauna described in Mount's (1975) comprehensive volume on the reptiles and amphibians of Alabama. Our treatment represents the second in a series of volumes that will cover each major taxonomic group described in Mount's seminal work. Alabama possesses one of the most species-rich biotas in north temperate areas, and this richness is reflected in some groups of lizards, such as skinks, and especially in snakes. Here, we provide a modern description of that diversity.

Our summary of the lizards and snakes of Alabama centers on describing their biodiversity. This approach examines all species known from the state, describes important regional variation within each species, and describes changes in species across the many habitats that comprise the state. Significant field studies, especially of Alabama's threatened and endangered species, have been performed, and we use these to guide our discussion of each species. The field of systematics has re-emerged as a primary goal of biological sciences, and this has been coupled with a healthy debate on species concepts (e.g., Frost and Hillis 1990). This debate has expanded the focus of studies of speciation from tests of reproductive isolation (e.g., Shine et al. 2002) to discovery of diagnostic features indicative of unique lineages on phylogenetic trees (e.g., Zaher et al. 2009). These changes have increased the known diversity of the state and have pointed to new directions for research that are likely to continue to expand Alabama's known squamate fauna. We provide diagnostic features and summarize key life history variables of each species and then indicate conservation efforts and management tools designed to maintain each of them. To reach this goal, we first list the taxa currently known from the state and then present climatic, geologic, and geographic features that shape squamate diversity. We end our introductory material by outlining the information selected to characterize each species account.

Our target audience remains the same as that for Mount (1975). We aim to enlighten people who are interested in the natural history of their local biota because we know these people will develop responsible attitudes toward the role that humans play in sustaining the Earth's ecosystems. Moreover, those with knowledge of natural history and a willingness to experience nature have a vast new world full of opportunities for soul-enriching experiences that we have had as biologists and hope to generate for others. This publication is a compromise of sorts in that it was prepared for use by the layman as well as the serious student of southeastern herpetology. The life history accounts are focused on providing information of interest to the lay public, and detailed information regarding taxonomy as well as recent citations of key field research are presented for serious students of herpetology.

## THE SQUAMATE FAUNA OF ALABAMA

### Indigenous Species

The classification scheme that follows organizes the native lizards and snakes of Alabama. These are taxa that are thought to have evolved within the state or to have dispersed to it without the assistance of humans. Changes to systematic biology since the publication of Mount (1975) have generated a growing number of taxonomic problems. Generally, these are associated with a desire for taxonomic groupings that are monophyletic (groups in which members are all more closely related to each other than any member is to a species outside of the group) and a desire for restricting a proliferation of named groups associated with monophyletic taxonomies. In order to reach these goals we adopt some of the philosophy argued by de Queiroz and Gautier (1992), who advocate reducing a reliance on taxonomic levels of the Linnean hierarchy in favor of generating indented lists of increasingly more-restricted monophyletic groups. Even in such taxonomies, species are identified as binomials, with a genus name identifying a group of closely related species and the specific epithet identifying a particular one of those species. The species name includes both the genus and the specific epithet, simultaneously generating a unique name for each species and identifying it as part of a more-inclusive taxonomic group. In addition to this convention, we retain the level of family as a useful taxonomic category because this level is so heavily entrenched in the taxonomic literature and because the content of reptile families has remained relatively consistent. We have avoided use of terms associated with levels of the Linnean hierarchy above the level of the family because these vary substantially among schemes and the choice of a term for these levels (e.g., superfamily versus suborder) is a

matter of personal choice rather than providing any increased understanding of biology.

Our classification scheme uses the format of an indented list, starting with the group Squamata, the raciation that contains all living and
fossil lizards and snakes. At the first level of indentation are five major
squamate lineages known from Alabama, which are listed as a phyletic sequence from the oldest evolutionary division among them to
the most recent division. Within each of these lineages, families (and
subfamilies of Colubridae) are listed as a phyletic sequence with species and subspecies listed in alphabetical order.

Squamata
  Iguania
    Dactyloidae
      *Anolis carolinensis carolinensis*—Northern Green Anole
    Phrynosomatidae
      *Sceloporus undulatus*—Eastern Fence Lizard
  Scincomorpha
    Scincidae
      *Plestiodon anthracinus anthracinus*—Northern Coal Skink
      *P. anthracinus pluvialis*—Southern Coal Skink
      *P. egregius similis*—Northern Mole Skink
      *P. fasciatus*—Five-lined Skink
      *P. inexpectatus*—Southeastern Five-lined Skink
      *P. laticeps*—Broad-headed Skink
      *Scincella lateralis* (three genetic lineages)—Ground Skink
    Teiidae
      *Aspidoscelis sexlineata sexlineata*—Eastern Six-lined Racerunner
  Anguimorpha
    Anguidae
      *Ophisaurus attenuatus longicaudus*—Eastern Slender Glass Lizard
      *O. mimicus*—Mimic Glass Lizard
      *O. ventralis*—Eastern Glass Lizard
  Serpentes
    Viperidae
      *Agkistrodon contortrix*—Copperhead
      *A. piscivorus piscivorus*—Northern Cottonmouth
      *A. piscivorus conanti*—Florida Cottonmouth
      *Crotalus adamanteus*—Eastern Diamondback Rattlesnake

*C. horridus*—Timber Rattlesnake

*Sistrurus miliarius barbouri*—Dusky Pygmy Rattlesnake

*S. miliarius miliarius*—Carolina Pygmy Rattlesnake

*S. miliarius streckeri*—Western Pygmy Rattlesnake

Elapidae

*Micrurus fulvius*—Eastern Coralsnake

Colubridae

Colubrinae

*Cemophora coccinea copei*—Northern Scarletsnake

*Coluber constrictor constrictor*—Northern Black Racer

*C. constrictor helvigularis*—Southern Black Racer

*C. flagellum flagellum*—Eastern Coachwhip

*Drymarchon couperi*—Eastern Indigo Snake

*Lampropeltis calligaster calligaster*—Prairie Kingsnake

*L. calligaster rhombomaculata*—Mole Kingsnake

*L. elapsoides*—Scarlet Kingsnake

*L. getula getula*—Eastern Kingsnake

*L. getula nigra*—Black Kingsnake

*L. getula holbrooki*—Speckled Kingsnake

*L. triangulum syspila*—Red Milksnake

*L. triangulum triangulum*—Eastern Milksnake

*Opheodrys aestivus aestivus*—Northern Rough Greensnake

*Pantherophis guttatus*—Red Cornsnake

*P. obsoletus*—Gray Ratsnake

*Pituophis melanoleucus lodingi*—Black Pinesnake

*P. melanoleucus melanoleucus*—Northern Pinesnake

*P. melanoleucus mugitus*—Florida Pinesnake

*Tantilla coronata*—Southeastern Crowned Snake

Dipsadinae

*Carphophis amoenus amoenus*—Eastern Wormsnake

*C. amoenus helenae*—Midwestern Wormsnake

*Diadophis punctatus edwardsii*—Northern Ring-necked Snake

*D. punctatus punctatus*—Southern Ring-necked Snake

*D. punctatus stictogenys*—Mississippi Ring-necked Snake

*Farancia abacura abacura*—Eastern Mudsnake

*F. abacura reinwardtii*—Western Mudsnake

*F. erytrogramma*—Rainbow Snake

*Heterodon platirhinos*—Eastern Hog-nosed Snake

*H. simus*—Southern Hog-nosed Snake

*Rhadinaea flavilata*—Pinewoods Littersnake

Natricinae

*Haldea striatula*—Rough Earthsnake

*Liodytes pygaea pygaea*—Northern Florida Swampsnake

*L. rigida sinicola*—Gulf Glossy Swampsnake

*Nerodia clarkii*—Gulf Saltmarsh Watersnake

*N. cyclopion*—Green Watersnake

*N. erythrogaster erythrogaster*—Red-bellied Watersnake

*N. erythrogaster flavigaster*—Yellow-bellied Watersnake

*N. fasciata confluens*—Broad-banded Watersnake

*N. fasciata fasciata*—Banded Watersnake

*N. fasciata pictiventris*—Florida Watersnake

*N. floridana*—Florida Green Watersnake

*N. rhombifer*—Diamond-backed Watersnake

*N. sipedon pleuralis*—Midland Watersnake

*N. taxispilota*—Brown Watersnake

*Regina septemvittata*—Queensnake

*Storeria dekayi limnetes*—Marsh Brownsnake

*S. dekayi wrightorum*—Midland Brownsnake

*S. occipitomaculata occipitomaculata*—Northern Red-bellied Snake

*Thamnophis sauritus sauritus*—Eastern Ribbonsnake

*T. sirtalis sirtalis*—Eastern Gartersnake

*Virginia valeriae elegans*—Western Smooth Earthsnake

*V. valeriae valeriae*—Eastern Smooth Earthsnake

## Introduced Species

Because of increased trade in vertebrates, establishment and expansion of non-indigenous species has become an increasing problem in maintaining native North American faunas (Romagosa et al. 2009). Much of the trade in lizards and snakes passes through south Florida, where individuals that escape captivity, or are intentionally released, have established populations of a growing number of species. The majority of these have not spread from the areas occupied by the founding populations (Meshaka et al. 2004). However, in a few cases, invasive species have expanded rapidly and extensively from the founding population, and concern has emerged that these taxa might disrupt native communities. For example, the Burmese Python (*Python molurus*) is now established in Everglades National Park, and distribution models have suggested that it may be capable of spreading throughout much of the southeastern United States, including Alabama (Rodda et al. 2009). Because this species is a voracious predator that appears to be altering assemblages of native mammalian predators, spread of the Burmese Python might disrupt predator-prey systems of native habitats (Dorcas et al. 2012).

Two established non-indigenous species were listed by Mount (1975) for the state of Alabama: the Turkish House Gecko (*Hemidactylus turcicus*) and the Texas Horned Lizard (*Phrynosoma cornutum*). The Turkish House Gecko has expanded its range in the state, occupying most major cosmopolitan areas by invading large buildings that maintain warmth during winter months. This invasion has inserted a sixth major lineage of squamates, Gekkota, into Alabama's herpetofauna. Of the four places previously known to have populations with reproducing individuals of the Texas Horned Lizard (Mount 1975), at least one retained the species as late as 1994. However, the continued viability of these populations deserves monitoring. To these invasive taxa, we add one more reptile species, the Cuban Brown Anole (*Norops sagrei*), which appeared first in South Florida and has expanded across the Southeast via transport in potted plants. Because of the mild climate along the Gulf Coast of Alabama, this species has become established, from which we infer other non-indigenous squamates may do the same.

The path to establishment of non-indigenous species is not an easy one. Most aliens are introduced to the state as individuals that escape from captivity. This process makes it all but impossible for the species to become established because founding populations are too small. Nevertheless, records of species that failed to become established are important to the development of predictive models for understanding which species will invade and why. Mount (1975) indicated one apparently unsuccessful invasion by the Panama Least Gecko, *Sphaerodactylus lineolatus*, which was observed in Mobile, Alabama, likely having been transported in shipping trade. To these we add eight species (appendix 1), all of which have been brought to us or were observed by competent staff at parks within Alabama. Undoubtedly, all represent escaped or released pets that, even if they were to survive a winter season, are unlikely to find a mate, thereby failing to become established.

Of the currently established species, the Turkish House Gecko clearly has expanded its distribution from the restricted range (north Eufaula, Barbour County) described by Mount (1975). Because this expansion is associated with invasion of buildings and does not involve invasion into native habitats, no ecological consequence to this range expansion appears likely to occur. Additionally, those forms transported in plant trade, like the Brahminy Blind Snake (*Rhamphotyphlops braminus*), which appear destined to invade Alabama's snake fauna, seem unlikely to cause problems of conservation concern because they are likely to fill roles not currently filled by indigenous taxa.

Taxonomic Changes and Problems

Our taxonomic list includes seventy-seven lineages (species, subspecies or genetic clades) of indigenous squamates. At first glance, this represents a modest change from the seventy-eight lineages listed in Mount (1975). However, these numbers mask significant taxonomic changes to the squamate fauna. All taxa on our list have valid scientific names, with the exception of the three molecular lineages of Ground Skinks known from Alabama, based on Jackson and Austin (2012), and the Gulf Coast clade of Southern Black Racers known from Alabama, based on Burbrink et al. (2008). We suspect that future evaluation will document these lineages to deserve scientific names. We add two taxa to those listed by Mount (1975): the Mimic Glass Lizard, a cryptic species previously included with the Eastern Slender Glass Lizard, and the Prairie Kingsnake, a subspecies noted by Mount (1975) as likely being present in the state. We eliminate the Northern Brownsnake from the state based on conclusions in Christman (1982). Four taxa are no longer recognized as being distinct enough to warrant recognition and six lineages have been elevated to species status (appendix 2). Most notably, six generic designations are changed (*Aspidoscelis* for *Cnemidophorus*, *Plestiodon* for *Eumeces*, *Coluber* for *Masticophis*, *Pantherophis* for *Elaphe*, *Liodytes* for *Regina rigida*, and *Liodytes* for *Seminatrix pygaea*).

Our taxonomy also reflects important changes to the higher taxonomy of lizards and snakes. In particular, we do not recognize a single taxonomic group for lizards, instead recognizing several independent groups that represent phylogenetic lineages of these organisms (Anguimorpha, Gekkota, Iguania, and Scincomorpha). Additionally, we recognize two families, Dactyloidae and Phrynosomatidae, placed by Mount (1975) into the single large family Iguanidae.

## CLIMATE OF ALABAMA

Because of its location, with a southern border along the Gulf Coast and a northern border along the southern extent of the Appalachian Mountains, the climate of Alabama is classified as humid subtropical (McKnight and Hess 2000). This climate is characterized by mild winters and hot, humid summers. Mean temperatures are warmer and more constant in the southern portion of the state than the northern portion, both because of lower southern topography and a stronger influence of Gulf breezes in the south. Rainfall is distributed throughout the year because of cold, dry, polar fronts moving against warm, moist, coastal air during autumn and winter, yielding intense thunderstorms, and moist, warm Gulf air moving north during spring and summer, rising over terrestrial areas, and generating afternoon rains. Rainfall is slightly increased along the coast of Alabama because of the increased moisture content of air associated with the Gulf of Mexico. Measurable snowfalls are exceedingly rare in the southern half of the state, and annual totals of more than 6 inches (150 mm) are seldom recorded even for the northernmost stations.

These typical patterns of weather are broken by annual occurrences of violent weather associated with tornadoes, mostly during spring, and hurricanes during late summer and fall. These storms can cause periods of intense rains that saturate soils and flood extensive areas. Such occurrences cause many lizards and snakes to move to upland areas to avoid advancing waters. These storms also kill trees by tipping them up from the roots, snapping them off at the trunk, or severely stressing them from storm surge of salt water. Each of these add fallen logs and leaves to the floor of Alabama's forested habitats, thus providing sites used as refuges during extreme temperatures, as well as nests for many egg-laying species. Intense breeding aggregations of amphibians that can occur during these storms also concentrate food resources that snakes exploit. In short, Alabama's climate is mild and moist enough, the geography diverse enough, and disturbances associated with storms frequent enough to support an unusually diverse mix of squamates.

## ALABAMA GEOGRAPHY

Alabama covers 52,419 mi$^2$ (135,765 km$^2$) of land and water in the southeastern United States that is divided into sixty-seven counties. Essentially, all of this area is habitable by lizards and snakes.

Distributions of these organisms are affected most strongly by the major river systems and physiographic regions. Here, we summarize these geographic features of the state and discuss their effects on Alabama's snake and lizard fauna.

## River Basins

One remarkable feature of Alabama's geography is the diversity of rivers that drain its surface. Eight river systems are found in the state, all of which drain to the Gulf of Mexico. Three of these are major rivers, being wide enough to present a challenge to most terrestrial organisms attempting to cross them. The Tennessee River enters the northeast corner of the state, flows east to west, and exits at the northwest corner. This river drains the northern one-eighth of the state and joins the Ohio River, eventually exiting via the Mississippi River into the Gulf of Mexico. However, the tortured pathway of the Tennessee suggests that it changed its course from a Mesozoic path that took it southwestward down the Ridge and Valley formation to the Alabama River system (Appalachian River). In the Paleocene (56–65 million years ago), erosion and tectonic events altered this ancient path by opening a channel from the Ridge and Valley formation into the Sequatchie Valley west of Chattanooga, following that valley southwestward to Guntersville, and then turning abruptly northwestward at Guntersville by cutting through the Pottsville sandstone formation into the Tuscumbia limestone. The Tennessee River takes a final turn at the Alabama-Mississippi-Tennessee border, flowing northward from there (Mills and Kaye 2001). Among Alabama's squamate fauna, the Prairie Kingsnake is the only taxon that appears to have its distribution limited by this river.

Two additional major river basins are present in the state. One, the Chattahoochee, forms the southern half of Alabama's eastern border (draining 6 percent of the state), and the other, the Mobile, drains most of the state (64 percent), entering the Gulf near the southwestern corner. These are ancient rivers that have remained in close association with upland regions of the southern Appalachians. However, their connections to the Gulf of Mexico have lengthened or shortened because of marine inundation or subsidence that altered terrestrial lowlands during the 300 million years during which reptiles have occupied this region. Aquatic taxa tied to these waters have been isolated from other rivers for exceptionally long time periods, allowing

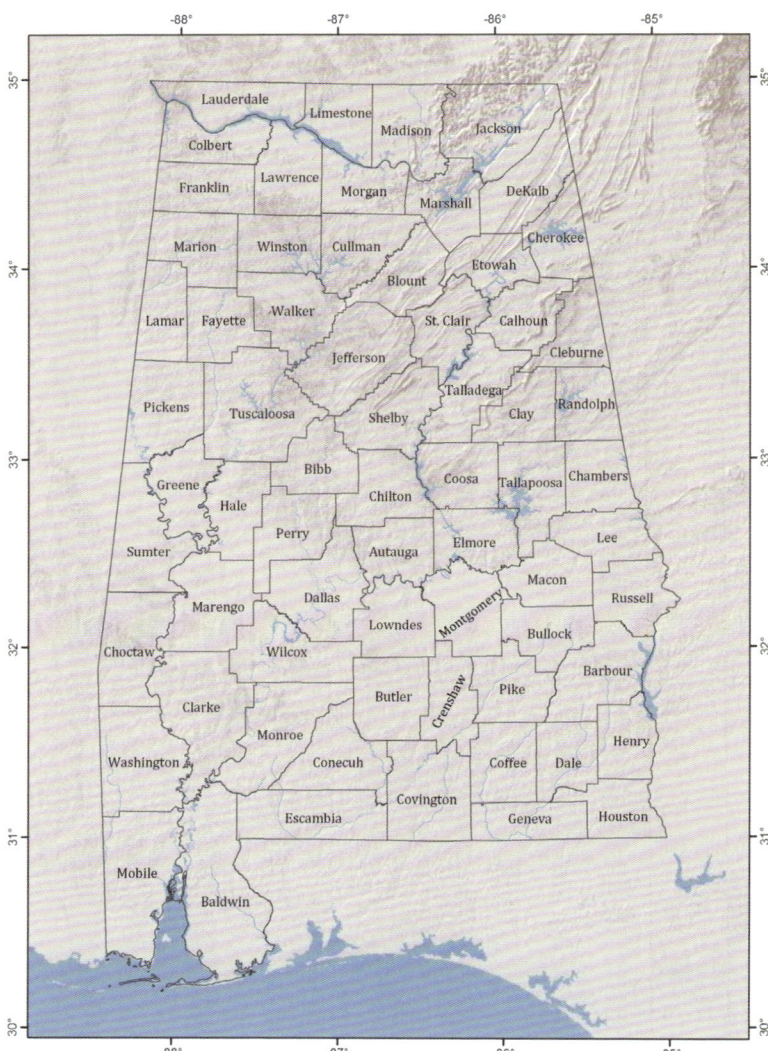

Counties
of Alabama

for the evolution of unique taxa (e.g., *Nerodia taxispilota* in the Chatta-hoochee; *N. rhombifer* in the Mobile).

Five smaller rivers are also found in the state. One of these, the Escatawpa, originates in southwestern Alabama, draining the western portions of Mobile and Washington Counties, about 3 percent of Alabama's surface area, and exits the state into Mississippi, where it opens to the Gulf in Pascagoula Bay. The Perdido River drains Escambia and Baldwin Counties, opening into Perdido Bay. This short river drains only about 1 percent of Alabama. Two rivers, the Conecuh and

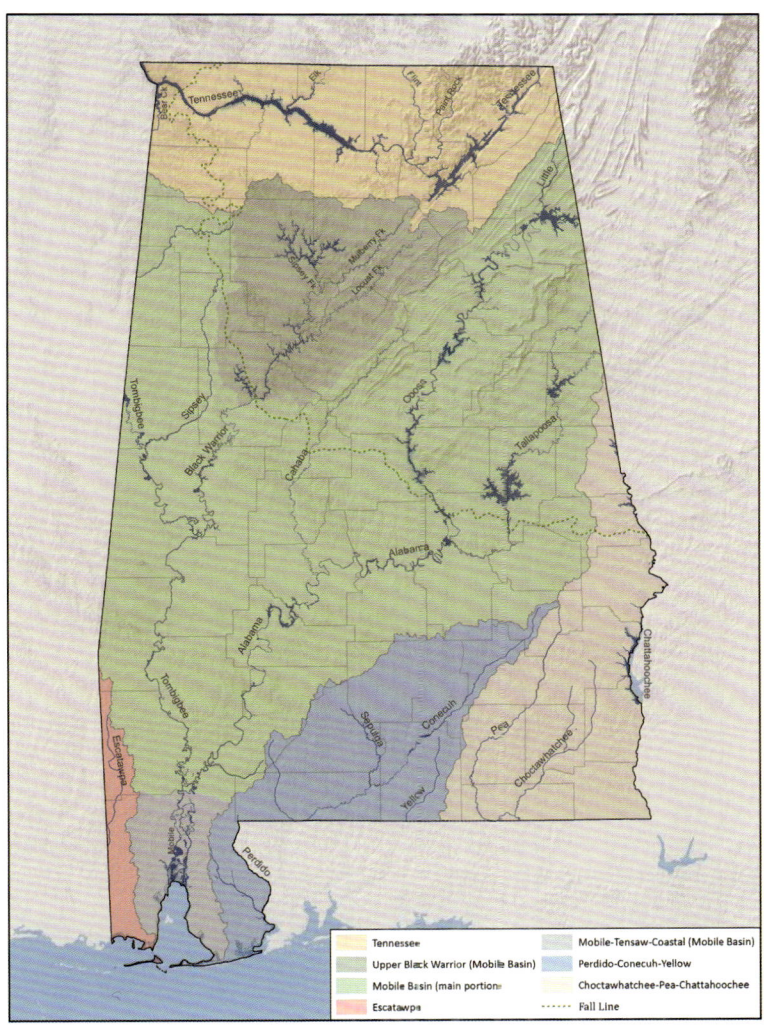

River basins
of Alabama

Yellow, drain 8 percent of Alabama in the south central counties of the state. These rivers eventually enter Pensacola Bay in Florida. Finally, the Choctawhatchee and Pea Rivers drain about 6 percent of south-eastern Alabama and unite before entering Florida and, eventually, Choctawhatchee Bay. In general, these smaller rivers have not affected distributions of squamates within Alabama.

Taxa attempting to enter Alabama from the west face a significant dispersal barrier from the Mobile drainage (appendix 3); taxa entering Alabama from the east face an apparently weaker barrier from the

Chattahoochee and Choctawhatchee/Pea Rivers (appendix 4). Similarly, squamate taxa found within Alabama between these rivers have had considerable opportunities to evolve within this region and face a challenge when attempting to disperse out of the state and adjacent areas of the Panhandle of Florida.

## Important Geographic Units

From the perspective of understanding squamate diversity, Alabama is divisible into ten meaningful physiographic units, which Mount (1975) termed herpetofaunal regions. These units fall into two natural groupings, Coastal Plain units of the southern part of the state that were created by ancient sea shores and upland units that were formed by the southern end of the Appalachian Mountains. These two groupings have a distinct boundary, the Fall Line, where streams change from rocky, fast-flowing waterways of upland units to sandy or muddy, slow-moving waterways of the Coastal Plain.

## The Coastal Plain

The Coastal Plain in Alabama is distinctly belted and physiographically more variable than it is in either Georgia or Mississippi. Topographically, it varies from flat to almost montane. The soils vary from acid sands and sandy loams, the dominant soils of the southern and northern belts, to clay sands, silts, and heavy, calcareous, alkaline types, soils that dominate regions of the central belts. The sandy soils are covered with pine forests, dominated by longleaf (*Pinus palustris*), in areas retaining native forests, and loblolly (*P. taeda*) and slash (*P. elliottii*), in areas affected by forestry operations. Those soils that are not primarily sandy allow for the formation of steep ravines that support dense hardwood forests and include upland areas with heavy soils that are covered by grasslands lacking a pine overstory.

Rocks and rock outcrops occur in some portions of the Coastal Plain but seldom to the extent that they do in provinces above the Fall Line. Most streams of the Coastal Plain are fairly sluggish and have sand, silt, or gravel bottoms, but some flow over bedrock. Several of the streams have broad, low floodplains with sloughs and oxbows. Swampy habitats are fairly common, many of them having been created by beaver dams.

Thirty-five lineages (species, subspecies, or genetic clades) of squamates, constituting about 45 percent of the state's lineages, occur

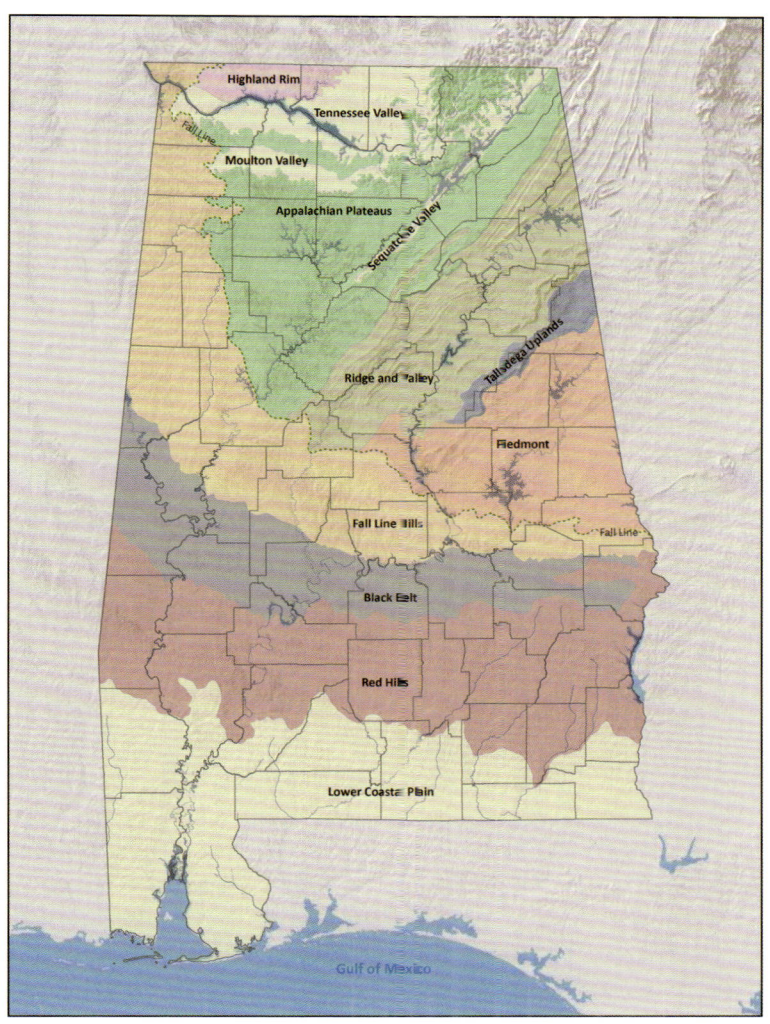

Herpetofaunal
regions of Alabama

exclusively or almost exclusively within the Coastal Plain. Of these, sixteen lineages are widespread in the Coastal Plain (appendix 5). From a herpetofaunal perspective, the Coastal Plain is separable into four major regions: Lower Coastal Plain, Red Hills, Black Belt, and Fall Line Hills.

*Lower Coastal Plain*

The Lower Coastal Plain extends completely across the state and includes the southernmost tier of counties and the lower portions of

Longleaf pine stand in Conecuh National Forest, Covington County, AL

some of those in the second tier. Relief varies from flat to gently rolling. This region encompasses about 20 percent of Alabama's total surface area, and its soils are sands, sandy loams, and sandy clays with occasional gravelly phases. The Lower Coastal Plain is delimited to the north by its boundary with the Red Hills region. The transition is rather abrupt west of the Conecuh River. Eastward the distinction is not well defined, and in eastern Alabama, where creeks drain into the Chattahoochee River, the Red Hills region tends to lose its biogeographic and physiographic integrity relative to that of the Lower Coastal Plain. This creates a corridor connecting the biota of the Lower Coastal Plain to the Fall Line Hills (described below).

The dominant forest communities originally occurring on the upland sites are fire-maintained ones, such as sandhills dominated by longleaf pine and turkey oak (*Quercus laevis*) and pine flatwoods, although a variety of other types are also represented. Intensive forestry is practiced in much of the Lower Coastal Plain, with a recent trend

toward replacing existing off-site loblolly (*Pinus taeda*) and slash pine (*P. elliottii*) stands with longleaf pines. The intensive mechanical and chemical site preparation that often precedes replanting, however, tends to eliminate a number of native plant species, including wiregrasses (*Aristida beyrichiana* and *Sporobolus junceus*) and bluestem (*Andropogon gerardii*), the dominant herbs in many natural communities within the Lower Coastal Plain. Because they require regular prescribed fires to persist in modern times, extensive intact areas of these natural communities have become quite scarce, and a large percentage of the squamate fauna is of conservation concern because of this habitat loss (Guyer and Bailey 1993).

Streams of the Lower Coastal Plain are highly variable in character. The major rivers flow generally southward and along some stretches have broad, low floodplains. The smaller streams are typically sand bottomed and tea colored. Some stretches of the latter have typical floodplain forest development along them, while others have high

Longleaf pine flatwoods, Covington County, AL

banks forested with pine or hardwoods. Swampy places along the smaller streams may be dominated by thickets of titi (*Cliftonia monophylla* and *Cyrilla racemiflora*). Most natural ponds and lakes are of sinkhole origin with clear, shallow water and abundant aquatic vegetation. The largest of these is Lake Jackson, which is situated on the Alabama-Florida boundary at Florala, Covington County.

The dominant feature of Alabama's coast is Mobile Bay. To the southeast, the bay is delimited by the Fort Morgan peninsula, a narrow strip of land extending westward from the body of the mainland for about eighteen miles. Fort Morgan peninsula, as well as the remainder of the coast of Baldwin County, with which it is contiguous, is dominated by dune communities and xeric pine flatwoods with some salt marsh habitat on the leeward side. West of Mobile Bay, in Mobile County, the coast of the mainland is low with extensive salt marsh communities. Several barrier islands lie offshore of Mobile County, the largest of which is Dauphin Island. Dauphin Island is

about 5 miles (8 km) long and about 1.5 miles (2.4 km) wide at the widest point. The island has beach and dune habitat, as well as pine flatwoods and salt marshes. There is one small freshwater pond. Virtually unchecked coastal development, in the form of condominiums, beach houses, commercial districts, and roads, has significantly altered much of Alabama's coastline to the detriment of many lizard and snake taxa.

Nineteen lineages of squamates are restricted to the Lower Coastal Plain (appendix 6), some of which are encountered only rarely. This diversity appears to have been generated by aquatic snakes that became isolated in specific drainage systems (Soltis et al. 2006) and by repeated expansion and contraction of coastal regions during periods of marine inundation (VanZant and Wooten 2007). Two snakes are adapted to brackish coastal waters (Gulf Saltmarsh Watersnake) and coastal marshes and dunes (Marsh Brownsnake). Of particular importance to the distribution of coastal squamates is Dauphin Island on

Sand dunes at Fort Morgan peninsula, Baldwin County, AL

the western side of the mouth of Mobile Bay and, therefore, part of Mobile County. Two snake species from that island indicate a shared history with the mainland region to the east (Baldwin County) rather than with that to the north (Mobile County). These are the Eastern Kingsnake *(Lampropeltis getula getula)* and the Eastern Mudsnake *(Farancia abacura abacura)*. The counterparts of these forms on the mainland of Mobile County are the Speckled Kingsnake (*L. getula holbrooki*) and Western Mudsnake (*F. abacura reinwardtii*). It appears, on the basis of this evidence, that Dauphin Island was the western extremity of what is now the Fort Morgan peninsula and that, in the relatively recent past, this island became isolated via altered water flow into and out of Mobile Bay, perhaps associated with hurricane storm surges of the past. Finally, the diversity of squamates of the Lower Coastal Plain involves taxa (Florida Cottonmouth, Red-bellied Watersnake, Florida Watersnake) that evolved in peninsular Florida and invaded the southeast corner of Alabama.

### Red Hills

The Red Hills region is a belt of Eocene age (34–56 million years ago), 30 to 40 miles (48–64 km) in width, and cutting a semicircular swath from the Mississippi border to the Georgia border. This region encompasses about 10 percent of Alabama's total surface area, much of which has deeply dissected topography, relatively fertile soils, and frequent outcrops of siltstone, claystone, or limestone. In its western extent in Alabama (Choctaw, Clarke, and Monroe Counties), the Red Hills is almost montane in character with rocky bluffs, deep ravines, and clear, rock-bottomed brooks. Here, the Red Hills is abruptly differentiated from the Lower Coastal Plain to the south and the Black Belt to the north. Some of the ridge tops rise as high as three hundred feet above the floors of the intervening valleys. The ridge tops and upper ravine slopes tend to support communities with mixtures of pine and hardwood, while the coves and lower slopes often have luxuriant hardwood forest stands, with oaks, hickories, beeches, and magnolias predominating. At its eastern extent in Alabama, the Red Hills topography is characterized by rolling hills that grade less abruptly and interdigitate with the Lower Coastal Plain and Black Belt, a distinction that is masked by the scale of our map.

Some elements of the biota of the Red Hills are usually associated with regions above the Fall Line, and others are found in no other

Coastal Plain province. For example the steep forested slopes of this region include disjunct populations of trees and shrubs typically found in the Appalachian Plateaus and Piedmont regions farther north. This feature is not strongly reflected in the squamate fauna. However, west of the Alabama River, the Speckled Kingsnake is found in the Red Hills region and the Black Belt, giving way to the Black Kingsnake in the Upland regions. Much of the Red Hills is in private ownership, but recent purchases of land by the state of Alabama have expanded public holdings on which conservation actions can be planned.

Red Hills slope in Monroe County, AL

### Black Belt

The Black Belt is roughly crescent-shaped and extends almost continuously from western Tennessee southward through Mississippi and generally east–southeastward across central Alabama. It is about 35 or 40 miles (56–64 km) wide at its widest point in extreme western

Black Belt prairie in Dallas County, AL

Alabama. This region covers 8 percent of the total surface area of the state. In Alabama, the western portion of this formation is distinctive and easy to differentiate from the Fall Line Hills to the north and the Red Hills to the south but becomes poorly defined at its eastern terminus well west of the Chattahoochee River in Russell County. The Alabama River forms the northern boundary of the Black Belt for the eastern half of the state and the Cahaba, Black Warrior, and Tombigbee Rivers dissect the Black Belt in the western half. These low-gradient rivers have extensive floodplains that are forested with hardwoods, cypress, or bamboo or have been converted to agriculture. Ponds along these rivers are common, forming breeding habitats for amphibians, and the banks are often steep and covered with lush vegetation, features that are attractive to many snakes and lizards.

A predominance of heavy calcareous soils of Cretaceous (66–145 million years ago) origin characterizes the Black Belt and creates gently rolling topography on which prairie vegetation, outcrops of Selma chalk, and forests of hardwood and mixed pine and hardwoods occurred naturally (Schotz and Barbour 2009). Fire maintained the tall grasses that carpeted natural prairie vegetation, and the area was rich in flowering plants. However, a near-complete conversion of this region to agriculture reduced the ancestral biota to very small parcels of land, most of which have become heavily encroached by trees and shrubs that invaded due to a lack of fire.

Scattered through the Black Belt are inclusions of acid soils, some of which are sandy in texture and appear suitable for squamates of the Lower Coastal Plain. Despite this, several species of the Coastal Plain appear to be scarce or absent from much, if not all, of the Black Belt. These include the Mole Skink, Florida Pinesnake, and Southern Hognosed Snake. The lack of extensive upland forests also contributes to the scarcity of some forms. The numerous farm ponds and lakes, however, often support an abundance, though not a great variety, of watersnakes. Among the larger terrestrial snakes, the Copperhead is common in the Black Belt; the Speckled Kingsnake also characterizes the Black Belt, but this subspecies has declined in number.

*Fall Line Hills*
The Fall Line Hills, called Upper Coastal Plain by some authorities and Central Pine Belt by others, lies between the Black Belt and the Fall Line. The region has a crescent shape that is widest along Alabama's western border with Mississippi and narrows to about 5 miles (8 km) along the Georgia border in Lee and Russell Counties. Topographically this formation varies from moderately hilly in the west to

Fall Line Hills, Tuskegee National Forest, Macon County, AL

gently rolling in the east and, in surface area, represents about 20 percent of the state.

The soils of the Fall Line Hills are Cretaceous in origin, mostly well drained, and vary from clay to sand. Gravelly phases are common. The sandy, well-drained sites often support communities dominated by longleaf pine and turkey oak, the best developed and most extensive of these being in Autauga and Russell Counties. These communities are generally similar to those that occur on dry, sandy sites in the Lower Coastal Plain. Diffuse boundaries between the Lower Coastal Plain, Red Hills, and Black Belt along the Chattahoochee River appear to allow passage of some taxa (Mole Skink, Southern Hog-nosed Snake, and Eastern Coral Snake) that are otherwise limited to the Lower Coastal Plain into the sandy soils of the Fall Line Hills.

## Upland Regions

The upland regions, considered collectively to encompass the faunal provinces above the Fall Line, include the Piedmont, Talladega Upland, Ridge and Valley, Appalachian Plateaus, Tennessee Valley, and Highland Rim. The ranges of ten species, or approximately 14 percent of Alabama's squamates, are limited essentially to one or more of these regions. Three of these (Northern Black Racer, Northern Pinesnake, and Midland Watersnake) are widespread throughout the uplands with the rest being restricted to its components.

### Piedmont

The Piedmont extends into Alabama from the east and occupies a triangular area representing about 10 percent of the state's surface area in the eastern central portion. Along its lower margin, where it makes contact with the Coastal Plain, the transition is relatively abrupt, biotically and physiographically. To the north, the Piedmont borders the Talladega Upland. The transition there is not abrupt, and the northern portion of the Piedmont and the Talladega Upland have several features in common.

The Piedmont is hilly, for the most part, with clay soils that tend to be rocky. Much of the land is forested with shortleaf (*Pinus echinata*) and loblolly (*P. taeda*) pines on the ridge tops and hardwoods along the lower slopes and in the bottomlands. Granite outcrops are common throughout the region. Because the soil is so poor, extensive areas of agricultural lands have reverted to forested tracts.

No squamate lineage is endemic to this region, but dense lizard and snake populations are maintained here. Loss of forested zones along streams in this region creates open habitats that are invaded by aquatic snakes, some invading from the Coastal Plain (Barrett and Guyer 2008).

Piedmont roadcut in Lee County, AL

### Talladega Upland

Above the Piedmont is a series of ridges, extending approximately 100 miles (160 km) southwestward. that we term Talladega Upland, following Griffith et al. (2001), but which Mount (1975) designated as Blue Ridge. The region covers about 1 percent of the surface of Alabama and has been considered by most recent accounts to be a subunit of the Piedmont. We treat it separately because of its apparent role in limiting distributions of some squamate species.

The Talladega Upland gives way abruptly to the Coosa Valley, the main body of which lies to the northeast. The highest point in Alabama, Mt. Cheaha (elevation 2,413 feet [730 m]), is a component of the Talladega Upland, as are a number of other peaks in the region that exceed 2,000 feet (600 m) in elevation.

The soil tends to be rocky and fairly friable, with some sandy phases. A great majority of the region is devoted to forestry, and a variety of forest habitat types are represented. Longleaf pine is one of the most abundant trees on the drier sites. Although the Talladega Upland has no endemic squamate lineage, at least one, the Mole Skink, reaches its northern limit in the Talladega Upland of Calhoun County.

### Ridge and Valley

The Ridge and Valley region lies between the Talladega Upland and the Appalachian Plateaus. It extends southwestward from DeKalb County, near the northeast corner of Alabama, to the Fall Line, and covers about 9 percent of the surface of the state. The region is considered here to consist of the Coosa Valley, the Cahaba Valley, and the uplands arising within these valleys. The soils range from gravelly loams to clay. The rocks at the lower elevations are mostly limestone and shale,

with sandstone and chert at higher elevations. The most prominent ridge is Double Oak Mountain, most of which lies in Shelby County, near the region's southwestern terminus. Springs are common in this region and include some of the largest in the state.

No species of squamate is endemic to the Ridge and Valley. However, the lizard and snake fauna of this region is unique for upland subdivisions in having at least one species, the Eastern Glass Lizard, that is typically associated with the Coastal Plain and is not known to occur elsewhere above the Fall Line. Similarly, the Southern Hog-nosed Snake was rare in this region, common in the Lower Coastal Plain, and now is likely extirpated in both. The presence of these Coastal Plain taxa may be related to the occurrence of sizable longleaf pine flatwoods with sandy or gravelly soil in the historical landscape of Cherokee and Etowah Counties. Harper (1943) mentions these and cites earlier references to them.

*Appalachian Plateaus*

The Appalachian Plateaus lie immediately above the Ridge and Valley region and cover about 15 percent of the surface of Alabama. Physiographically, these plateaus are considered subdivisions of the Cumberland Plateau. They include Lookout Mountain and Sand Mountain south of the Tennessee River and extensive mountains in Jackson County north of that same river. Lookout Mountain originates in Tennessee, extends southwestward across the northwestern corner of Georgia and into Alabama, where it narrows and gradually loses its identity around Gadsden in Etowah County. Sand Mountain lies to the west and northwest, across the valley of Big Wills Creek. At the Alabama-Georgia boundary, Sand Mountain is about 15 miles (24 km) wide. Southwestward, it expands to occupy much of the north-central portion of Alabama above Birmingham. The integrity of Sand Mountain as a plateau is maintained westward to within about 20 miles (32 km) of the Fall Line before it breaks up into an area of deeply dissected terrain in Lawrence and Winston Counties. Much of this rugged terrain is now included within the Bankhead National Forest.

Lying north of Sand Mountain, from Morgan County westward, is Little Mountain, a portion of the Appalachian Plateaus. This narrow, somewhat irregular ridge has biogeographical characteristics similar to those of Sand Mountain from which it is separated by Moulton Valley, a narrow, low-lying intrusion of the Tennessee Valley.

The soils of the Appalachian Plateaus are mostly sandy loams, although drier soil types are not uncommon in the southern portion and at some of the lower elevations in the north. Rocky phases are found at many sites. Throughout most of the region, sandstone is the dominant exposed rock. Shale, limestone, and chert are other common rock formations in the southern portion and at lower elevations along the northern edge.

The gently rolling tops of the plateaus, especially portions of Sand Mountain, are often intensively farmed, but the edges and sides have much of the natural habitat remaining. The streams draining protected watersheds are clear with rock and sand bottoms and frequently descend over waterfalls from plateau overhangs to valleys below. Similarly, caves are common in this region and streams frequently fall into them or emerge from their mouths. Most of the coal deposits in Alabama are situated within this region, and strip mining has drastically altered the natural habitats at many localities.

Three squamates are confined to the Appalachian Plateaus. The entire ranges of the Red and Eastern Milksnakes in Alabama are found only in rocky exposures of DeKalb, Jackson, Lawrence, Madison,

*Right:*
Little River Canyon, Cherokee County, AL

*Below:*
Sipsey Fork in Winston County, AL

Morgan, and Winston Counties. Additionally, the Northern Coal Skink has been recorded largely from higher elevation sites in Lawrence and Winston Counties.

*Tennessee Valley*

The main body of the Tennessee Valley is a broad expanse of relatively level, fertile land that lies along the Tennessee River and covers about 4 percent of the surface area of Alabama. On the east and south, where it meets the Appalachian Plateaus, there are bluffs and steep slopes, the lower reaches of which have extensive limestone outcrops. The soil of the valley is mostly red clay of limestone origin, and the few remaining forests are composed mostly of hardwoods and scattered patches of eastern red cedar (*Juniperus virginiana*).

The Tennessee River is impounded throughout its course in Alabama and bears little resemblance to its former state. The shoals of this river, which once contributed to a diverse aquatic fauna, are now all under deep water. This region has a limited effect on squamates and no species is endemic to it.

Tennessee River and Valley, Marshall County, AL

*Highland Rim*

Above the Tennessee Valley, in northwestern Alabama, is the Highland Rim, a region termed the Chert Belt by Mount (1975). Physiographically, the Highland Rim is the southernmost subdivision of a vast province whose components are termed collectively the Interior Low Plateaus. In Alabama, the Highland Rim covers about 3 percent of Alabama's total land area in Lauderdale and Limestone Counties and portions of uplands in Madison, Colbert, and Lawrence Counties.

The Highland Rim is a moderately elevated region, with topography varying from hilly to nearly flat. The region lies entirely within the Tennessee River drainage, and the greatest relief is typically found near streams. The soils are mostly heavy and fairly fertile and once supported extensive hardwood forests. In the current landscape, the areas that remain forested are mostly in Lauderdale County.

The squamate fauna of the Highland Rim has no taxon that is endemic to it, but when distribution patterns for this region become clearer, the Prairie Kingsnake is likely to reach its southern limit in this region. Additionally, the reptile fauna of this area is unusual in that Northern Green Anoles and Coachwhips are scarce.

Highland Rim from County Road 20 at Baker Hill Road Bridge, Lauderdale County, AL

# Species Accounts

The remainder of this book describes squamates as a major radiation of amniotes (terrestrial, egg-laying or live-bearing vertebrates), each family found in Alabama, and each of the state's species within each family. Presentation of each family, subfamily (for colubrid snakes), genus, species, and subspecies is in the order of appearance within the keys provided rather than listed alphabetically. Important genetic variation supported by published analyses is discussed within each species account. Each species or subspecies account has distinct sections that we describe below.

## Keys

Keys are tools designed to aid in identification of organisms. These tools present paired descriptions, one of which will conform to an individual organism of interest and the other description will not. At the end of each consistent description is a number indicating the next couplet to be considered. This process of making dichotomous choices is followed until a final description identifying the organism of interest is reached. We include taxonomic keys for the squamates of Alabama and take the unusual step of dispersing these keys throughout the accounts rather than including a single key. We do this to place information close to sections of text for which the keys are most useful. Keys to the major lineages of squamates appear at the end of the description of the group Squamata, and keys to the families of squamates appear at the end of the description of each major squamate lineage. When necessary, keys to genera are placed at the end of the description of each family, keys to species appear at the end of descriptions of each genus, and keys to subspecies are placed at the end of the taxonomy section of the appropriate species account.

## Names

The generic, specific, subspecific, and common names applied are, in most cases, those listed in Crother (2017). However, a few designations, such as our treatment of *Lampropeltis getula*, represent personal choices that are our recommendations for taxonomic allocations given uncertainty in phylogenetic estimations.

## Photographs

We have benefitted from the talents of a large number of photographers. Where possible, we have selected images that show key features rather than those that have the best background or artistic composition. When the location of the specimen photographed is known, we identify it.

## Descriptions

Our descriptions are intended to provide sufficient information to enable the reader to distinguish a particular taxon from all others occurring within the state. Each description is based on a composite of specimens representing variation within Alabama and surrounding states. Because nature is variable, it should be kept in mind that occasional individuals belonging to the described taxon will not conform to the descriptions presented here.

## Alabama Distribution

In addition to a general statement describing the distribution of each species or subspecies occurring within Alabama, a map is included depicting its range in the state. Solid dots on the maps indicate localities of a) specimens the authors have examined, b) photo-vouchered specimens submitted to the Alabama Herp Atlas Project, c) occurrences documented in the databases of the Alabama Natural Heritage Program and/or ADCNR State Lands Division, Natural Heritage Section, and d) literature records believed valid. Each record has been georeferenced and plotted to the greatest possible precision. For taxa that are not found throughout the state, we include a shaded region indicating the likely limit to its distribution within Alabama. The state distribution maps also include black and white insets (lower right corner) showing the approximate distribution of each taxon within North America. These are derived from public data developed through two sources: the International Union for the Conservation of Nature (IUCN 2014) and the USGS Gap Analysis Program (GAP). In some cases, these maps were further modified to better reflect current information.

## Habits

Here, we provide description of habitat specialization, seasonal patterns of activity and reproduction, mating strategies of males and females, and major diet items. In general, this section is designed to summarize where and when each species is likely to be active and what activities make the species detectable by humans. Additionally, we describe the timing and duration of each major growth stage in the life cycle of the species.

## Conservation and Management

In this section, we describe the current conservation status of each species or subspecies in Alabama. Alabama's squamates are generally threatened by habitat loss and fragmentation, loss of natural community integrity, and, in the case of large snakes, direct persecution. Because conservation issues are likely to increase in the future, we summarize human activities that might imperil each species or subspecies as well as those activities that are likely to enhance populations. Similar data are provided for taxa that have conservation status within the state, and for these, we provide information on key public properties that will play crucial roles in long-term maintenance of Alabama's imperiled herpetofauna. In developing the State Wildlife Action Plan (Division of Wildlife and Freshwater Fisheries, Alabama Department of Conservation and Natural Resources 2005), the state of Alabama used the findings from its 2002 Nongame Symposium, which assembled scientific experts to compile the best data on Alabama's wildlife and used those data to identify those species most in need of conservation action. The Nongame Symposium's Amphibian and Reptile Subcommittee reconvened in 2012, identifying five lizards and ten snakes as being of immediate conservation need (Priority 1 or 2, on a scale of 1 to 5; Mirarchi, Bailey, Haggerty, Best 2004), and we summarize the subcommittee's recommendation in each species account.

## Taxonomy

We accept the concept that species are lineages that are discovered through careful analysis of variation in the characteristics of organisms. These discoveries arise from creation of phylogenetic trees built from character data. Under this species concept, any diagnosable terminal branch is sufficient to discover a new species. Additionally, we accept the concept that taxonomic groups at any level of classification should be monophyletic so that such groups will carry the additional evolutionary information that members of the group are more closely related to other group members than they are to any organism that does not belong to the group. In practice, ancestral species might survive through the branching process, generating some lineages that are not monophyletic (de Queiroz 1998). Such species present challenges for determining species boundaries, and the decisions that we make for the boundaries of Alabama's species undoubtedly will suffer from this challenge.

For squamates, color patterns, conformation of scales, counts of scales, and shapes of appendages are external features that traditionally have been used to diagnose species. To a lesser extent, features of the skeleton and muscles are used as well. However, use of these morphological characteristics requires collection of large series of specimens because differences between some forms are based on differences in modes of scale counts. Such counts have been provided by Mount (1975), and we do not repeat them here. Instead, we focus on diagnostic characteristics that allow identification of animals in the field. A knowledge of these characteristics and of general anatomical directions will be needed to use the keys that we provide.

To these traditional characters, we have added information from publications describing the mitochondrial and nuclear genomes. These sequence data have the advantage of allowing rapid development of data sets that are much larger than those based on morphology. Offspring inherit the mitochondrial genome entirely from the female side of the family tree, while the nuclear genome captures information about gene flow associated with both parents. For this reason, phylogenetic trees based on the mitochondrial genome are not guaranteed to be concordant with those based on the nuclear genome. Data on the mitochondrial genome are particularly voluminous because they are cheap and easy to procure. These data are particularly important in phylogeographic studies, a field of biogeography that uses patterns of evolution within species that are discernible by analysis of molecular data. Now, such studies are common for Alabama's squamates, and the intraspecific lineages generated by such studies likely will allow us to discover new taxa that were not evident from analysis of traditional morphological data. This creates an exciting environment for taxonomists because so many new lineages may

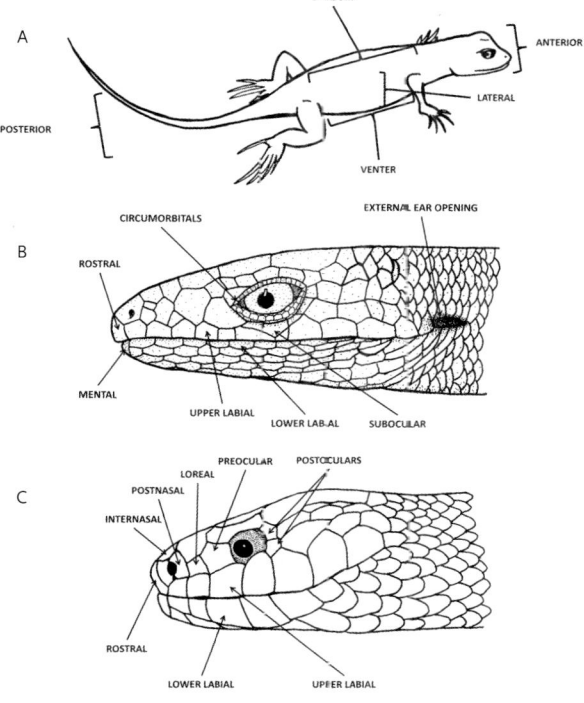

A

DORSUM

ANTERIOR

LATERAL

POSTERIOR

VENTER

EXTERNAL EAR OPENING

CIRCUMORBITALS

B

ROSTRAL

MENTAL

UPPER LABIAL

LOWER LABIAL

SUBOCULAR

PREOCULAR

POSTOCULARS

LOREAL

POSTNASAL

C

INTERNASAL

ROSTRAL

LOWER LABIAL

UPPER LABIAL

General morphological areas and features of a squamate (lizard or snake)

A: Broad areas of a squamate body

B: Head scales and structures of a typical lizard

C: Head scales and structures of a typical snake

be available for discovery, and such discoveries tell important stories about how Alabama's rich biodiversity was generated (e.g., Soltis et al. 2006). It also creates a scary environment for authors of field guides, such as this one, because of the likelihood that the guide will be obsolete before it is published. Because of this, we attempt to describe all lineages supported by character data, including the mitochondrial genome, and, therefore, that might indicate speciation events awaiting taxonomic recognition.

Mount (1975) was exemplary in recognizing important subspecific variation within Alabama. That these taxonomic distinctions are important is supported by the elevation of a large number of subspecies to species status, largely based on accumulating molecular data but clearly supported by traditional characters. Because of this, we retain subspecies that are based on characters that show apparently significant geographic discontinuities. Phylogeographic studies frequently find imperfect concordances between subspecies boundaries based on morphology and boundaries of mitochondrial lineages; this discordance frequently is used to argue against traditional subspecific boundaries (e.g., Burbrink et al. 2000). In fact, this trend is so strong that large numbers of lineages have emerged that represent numbered or lettered clades on phylogenetic trees (e.g., Jackson and Austin 2010). Where possible, we attempt to align subspecific names with numbered or lettered clades and use such alignment to retain most subspecific categories. In some cases, molecular data, in association with subspecific designations that appear to be based on clinal variation, are used to reject previously recognized subspecies. If no taxonomic names are available for clades discovered within phylogenetic studies, then we retain named, numbered, or lettered clades described in such studies.

# Lizards and Snakes—Squamata

Squamates, the living lizards and snakes, comprise 70 families containing about 9,900 species, making this the second-most species-rich radiation of land vertebrates. Only birds, with about 10,000 species, exceed the squamate radiation in species richness. As might be expected of such a diverse lineage, squamates exploit an amazing variety of habitats and display divergent foraging and reproductive modes. Many squamates are essentially ground-surface dwellers (e.g., Horned Lizards of the genus *Phrynosoma*), but some spend virtually their entire lives underground (e.g., Wide-snouted Wormlizards of the genus *Rhineura*), while others are almost completely arboreal (e.g., anoles of the genus *Dactyloa*), and a few are strongly aquatic with no need to return to land (e.g., Yellow-bellied Seasnakes of the genus *Pelamis*).

Organizing the taxonomy for such a large group has been a challenge in herpetology. Fortunately, some clarity is beginning to emerge in that all squamates, with the exception of the enigmatic family Dibamidae, consistently cluster into one of four major monophyletic clades. Here, we follow the parsimonious analysis of morphological data summarized by Gautier et al. (2012) in classifying squamates. We prefer this treatment because of its extensive coverage of both taxa and characters and because it provides essential data from fossils. However, this treatment is challenged by extensive molecular data that recover different numbers of major monophyletic groups and find different phylogenetic histories among these groups (e.g., Wiens et al. 2010; Vidal and Hedges 2009; Pyron et al. 2013). However, because morphological data, coupled with fossil information, overturned similar previous challenges from molecular data (e.g., Gautier et al. 1988), we place our support on the morphological data by recognizing four major squamate groups: Iguania (iguanians), Gekkota (gekkotans), Scincomorpha, and Anguimorpha with snakes (Serpentes) evolving as the sister group to Anguimorpha.

A basal divergence separates the group Iguania—a lineage of squamates with relatively short bodies, relatively long legs, and tongues used in prey capture—from all others, which have tongues that are

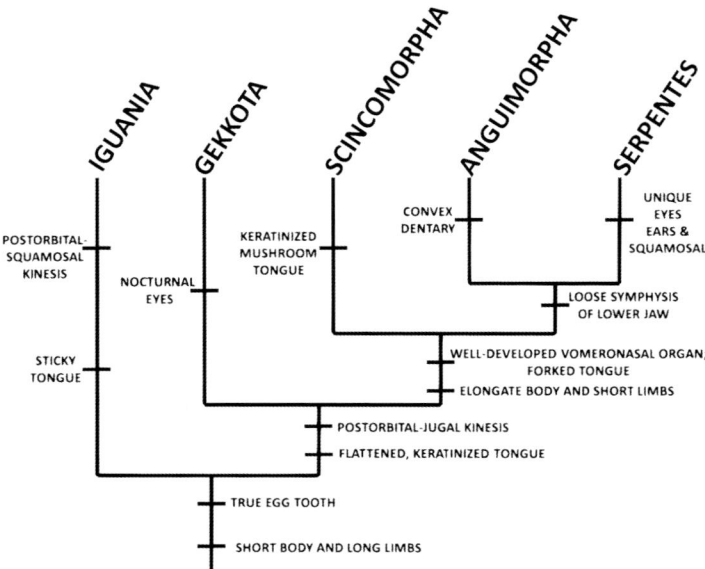

IGUANIA — GEKKOTA — SCINCOMORPHA — ANGUIMORPHA — SERPENTES

POSTORBITAL-
SQUAMOSAL
KINESIS

STICKY
TONGUE

NOCTURNAL
EYES

KERATINIZED
MUSHROOM
TONGUE

CONVEX
DENTARY

UNIQUE
EYES
EARS &
SQUAMOSAL

LOOSE SYMPHYSIS
OF LOWER JAW

WELL-DEVELOPED VOMERONASAL ORGAN;
FORKED TONGUE
ELONGATE BODY AND SHORT LIMBS

POSTORBITAL-JUGAL KINESIS

FLATTENED, KERATINIZED TONGUE

TRUE EGG TOOTH

SHORT BODY AND LONG LIMBS

Phylogeny of major groups of squamates

used to locate prey. The next divergence separates the group Gekkota, a lineage with specialized eyes for night vision, from all other squamates, which are characterized by relatively long bodies, relatively short limbs, and a forked tongue designed for detecting chemicals via a well-developed vomeronasal (Jacobsen's) organ. Among derived squamates, a divergence event occurs between those forms that have tongues covered with large plate-like structures (Scincomorpha) and those forms that retain a deeply forked projectile tongue (all others). Finally, a split occurs separating Serpentes, a species-rich group of squamates with highly modified vision and hearing, developed because the ancestral form was a burrower (Conrad 2008) or was aquatic (Lee 2005b), from Anguimorpha, a group of terrestrial predators retaining typical reptile eyes and ears but with lower jaws modified for prey capture. These major taxa will guide our presentation of Alabama's squamate fauna.

Limbs of many squamates are well developed, especially within iguanians. However, each of the other major lineages has representatives that have lost limbs in association with burrowing into loose soil, swimming, or crawling through grasslands. In forms that burrow, modifications of the eyes and ears occur because vision becomes practically useless within burrows and high-frequency sounds are not transmitted easily through soil. So, these radiations all contain lineages that look and act like snakes, including the lineage that eventually becomes true snakes (Gautier et al. 2012).

Food habits of lizards and snakes are equally diverse. Although rare, some forms, such as Green Iguanas (*Iguana*), are herbivorous and have specialized structures, such as slicing teeth, a muscular proventriculus (gizzard), elongate small intestine, and enlarged caecum, designed to digest foliage. Others are frugivorous, such as Chuckwallas (*Sauromalus*), which are famous for consuming the fruits of cacti. However, most squamates are predators, consuming animals as small as collembolans (Vitt et al. 2005) to as large as sun bears (Fredriksson 2005). Additionally, squamates may specialize on an extremely reduced array of prey. For example, some consume centipedes almost exclusively (e.g., Black-headed Snakes of the genus *Tantilla*) or consume any item small enough to fit in the mouth (e.g., Cottonmouths of the genus *Agkistrodon*), including road-killed animals (DeVault and Krochmal 2002).

Finally, reproductive modes of lizards and snakes are diverse. All forms have internal fertilization that is effected by insertion of an everted hemipenis, a unique copulatory structure of male squamates, into the cloacal opening of a female. This process frequently involves elaborate visual and olfactory cues that allow males to find females and females to select from available mates. Many forms, such as anoles (Dactyloidae), lay eggs without modifying the environment when creating a nest and provide no further parental care. In other forms, such as Glass Lizards (*Ophisaurus*), the female constructs a nest and then attends the eggs, likely driving away predators. Still other forms retain fertilized eggs in the uterine tract and develop a placenta, an anatomical feature that allows the female parent to provide protection and nutrients for the developing offspring. Such viviparous forms have developed independently at least 108 times within squamates (Blackburn 2006), suggesting strong selective pressure to protect the offspring during their early development.

Many lizards and snakes have developed modified salivary glands that produce venoms used in defense and prey capture. Precursors to venom systems are known from iguanians and complete systems are found in the groups Anguimorpha and Serpentes. The most famous venomous anguimorphans are the Gila Monsters and Beaded

Lizards (*Heloderma*) of the southwestern United States, Mexico, and Guatemala. In these forms, venoms are produced by glands of the lower jaw and are delivered by specialized elongate and grooved teeth. These venoms, thought to be defensive in function, are delivered by biting the potential predator, flipping over, and chewing. The other classic cases of venomous squamates occur within snakes, many of which have developed specialized upper teeth to deliver venom produced in modified salivary glands of the upper jaws. Some forms have enlarged, grooved teeth at the posterior end of the maxilla that serve to deliver venoms. These rear-fanged snakes include dipsadine colubrids, like Hog-nosed Snakes (*Heterodon*). A second lineage includes members of the family Elapidae that have short, immobile fangs at the front of the maxilla. Finally, the family Viperidae includes snakes with large, mobile fangs at the front of the maxilla. These venom systems are designed to subdue prey and, in some forms, begin the digestive process before a prey item is consumed.

In describing the group Squamata, we minimize use of the term "lizard." In previous classification schemes (e.g., Mount 1975), squamates were divided into two major groups, one containing all snakes and one containing the remaining forms, which were referred to as lizards. Unfortunately, this classification scheme is misinformative because it implies that lizards are an evolutionarily meaningful group (all lizards are more closely related to each other than they are to organisms outside of their group). This is not true. For example, skinks and geckos are both lizards, but skinks are more closely related to snakes than they are to geckos. Because scientists prefer group names that carry evolutionary meaning, as well as allowing for effective communication, the term "lizard" is slowly being purged from scientific terminology. However, the term is so entrenched in vernacular use that we do not eliminate the term entirely.

## Key to the Major Lineages of Squamata of Alabama

**1a** Tongue rounded, covered in mucus, and sticky.

    **Iguania—Iguanians . . . . page 43.**

**1b** Tongue flat and keratinized (dry and hard). Go to **2**.


**2a** Eyes designed for nocturnal activity pattern; tongue capable of wiping eyes; small velvety scales;.

    **Gekkota—Gekkotans . . . . page 55.**

**2b** Eyes designed for diurnal activity pattern; tongue incapable of wiping eyes; scales not small and velvety; Go to **3**.

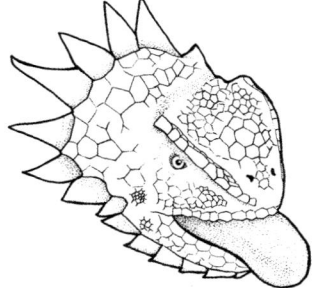 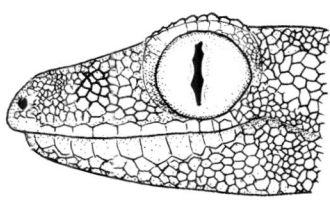

*From left to right:*

Tongue morphology of a Horned Lizard (*Phrynosoma*)

Lateral view of head of a House Gecko (*Hemidactylus*)

**3a** Tongue indented and with sclerotinized papillae; scales smooth and cycloid on dorsum and venter or velvety on dorsum and quadrangular on venter.

    **Scincomorpha—Skinks, Whiptails, and Their Relatives . . . . page 71.**

**3b** Tongue deeply forked; scales variable, but if cycloid, then with large plate-like scutes on venter. Go to **4**.

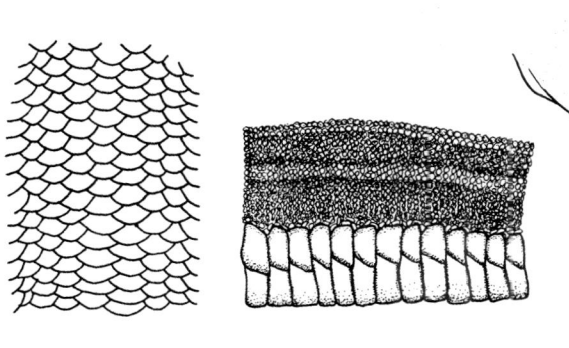

*From left to right:*

Smooth cycloid scales of a skink (*Scincidae*)

Velvety scales on dorsum and quadrangular on venter of a teiid (*Teiidae*)

Tongue morphology of a Watersnake (*Nerodia*)

**4a** External ear openings typically present; eyelids present; squamosal not mobile.

> **Anguimorpha—Alligator, Beaded, Glass, and Monitor Lizards . . . . page 101.**

**4b** External ear openings absent; eyelids absent; squamosal mobile.

> **Serpentes—Snakes . . . . page 113.**

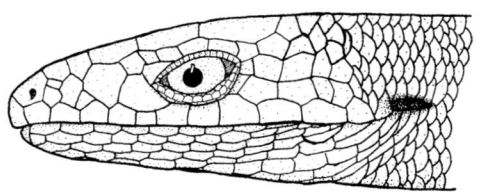

*From top to bottom:*

Lateral view of head of a Glass Lizard (*Ophisaurus*)

Lateral view of head of a Kingsnake (*Lampropeltis*)

# Iguanas, Anoles, and Their Relatives

## Group Iguania

This taxon includes squamates that generally have tongues that are simple in shape, covered in mucus, and used mainly to apprehend insect prey rather than to follow chemical trails. These features are associated with the fact that these animals are active, visually oriented predators. The group Iguania comprises a species-rich radiation that contains 14 families with about 1,600 species that are distributed across all major continents except Antarctica. Iguanians can be quite large and herbivorous, such as the Green Iguana (*Iguana*), or tiny and insectivorous, such as some chameleons (*Brookesia*). These squamates occupy equatorial rainforest areas (e.g., anoles of the genus *Dactyloa*) and range as far north as southern Canada (e.g., horned lizards of the genus *Phrynosoma*). There is no convenient external characteristic unique to iguanians. Perhaps for this reason no consistent placement of iguanians on the squamate phylogenetic tree has emerged. Conrad (2008) and Gautier et al. (2012) place them as sister to all other squamates, the traditional placement, while molecular data place them as the sister taxon to Anguimorpha + snakes (Townsend et al. 2004; Wiens et al. 2010). Regardless of this uncertainty, iguanians, unlike all other squamate radiations, have no member that has evolved limbless crawling. The state of Alabama has four species of iguanians belonging to two families. Half of these species have invaded the state by human activities.

### KEY TO THE FAMILIES OF IGUANIA OF ALABAMA

**1a** Scales granular; males with colorful dewlap.

Family **Dactyloidae** . . . . **page 44.**

**1b** Scales keeled; no colorful dewlap in males.

Family **Phrynosomatidae** . . . . **page 55.**

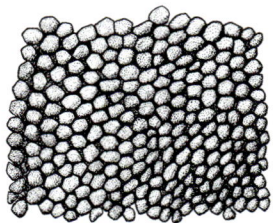 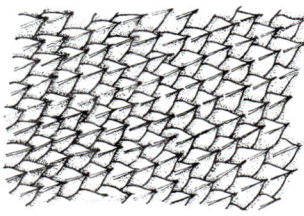

*From left to right:*

Scales of a Green Anole (*Anolis*)

Scales of a Fence Lizard (*Sceloporus*)

# Family Dactyloidae

This family, with eight genera and more than four hundred species, is widespread in the New World. Known as anoles, these animals are almost exclusively arboreal, diurnal, and largely insectivorous. Many forms can alter their color pattern to match the background of their habitat or in response to hormones released during social interactions. Because of this trait, anoles are frequently (and incorrectly) called chameleons. The family is diagnosed by an elongate cartilage located in the throats of males that can be extended from the body to display the dewlap, a brightly colored patch of skin that is used by males in territorial and courtship displays. A second diagnostic feature of the family is the presence of expanded toe pads on digits that have unique hairlike structures that cause the toe pads to adhere to smooth surfaces, such as leaves and tree bark. The radiation is tropical in origin, being particularly species rich in Central and South America as well as in the Greater Antillean Islands of the Caribbean. One species is native to the United States, and it is found in Alabama. Four genera and eight other species of non-indigenous anoles have become established in the United States, and one of these has become established in Alabama. The family Dactyloidae has uncertain systematic affinities, being found to be sister to *Enyalioides* by Townsend et al. (2004), sister to Polychrotidae + Leiosaurinae + Anisolepinae by Conrad (2008), and sister to Leiocephalidae by Wiens et al. (2010).

### Key to the Genera of Dactyloidae of Alabama

1a Caudal vertebrae with transverse processes posterior to autotomic septa.

> Genus *Norops*—Beta Anoles . . . . page 45.

1b Caudal vertebrae lacking transverse processes.

> Genus *Anolis*—Anole . . . . page 49.

*From top to bottom:*

Caudal vertebrae of a Cuban Brown Anole (*Norops sagrei*)

Caudal vertebrae of a Green Anole (*Anolis carolinensis*)

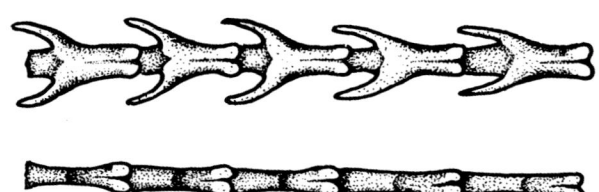

# Beta Anoles
## Genus *Norops* (Wagler, 1830)

We place all anoles with the transverse processes posterior to the autotomic septa of the caudal vertebrae (beta condition) in the genus *Norops* (Nicholson et al. 2012). It contains about 180 species that are distributed in Cuba, Jamaica, the Cayman Islands, and Central and South America. The radiation takes on a variety of shapes and sizes, features that appear to allow as many as seven species to inhabit the same area. These anoles occupy the arboreal habitat from the tops of canopy trees to the base of tree trunks and the leaf litter below. However, the lineage appears to come from a relatively small ancestor, whose niche focused on the ground, as did its sister genus *Ctenonotus* of Hispaniola, Puerto Rico, and the northern Lesser Antilles (Nicholson et al. 2012). No member of the genus *Norops* is thought to be native to the United States, but two have invaded south Florida, and one of these has become established in Alabama.

Adult male Cuban Brown Anole, *Norops sagrei sagrei*, Franklin County, FL

## Cuban Brown Anole
*Norops sagrei sagrei* (Duméril and Bibron, 1837)

DESCRIPTION This is a small to mid-sized anole that attains a maximum snout-vent length of about 60 mm. As in all anoles, the scales are small, weakly keeled, and non-overlapping, features that give the skin a velvety appearance. Lamellae, wide plate-like scales on the bottom of the next-to-last joint of each toe, expand to form an elongate

Adult female Cuban Brown Anole, *Norops sagrei sagrei*, Monroe County, FL

pad. Males have an extensible dewlap that is red in the center and yellow along its outer border. Adult males also possess a mid-dorsal crest on the nape of the neck and the center of the back that may be erected in social encounters. The head is pointed but not elongate. Dorsal color of each individual may be brown or grayish and, in males, may have faint light stripes, especially along each side; these may be disrupted by light flecks and dark V-shaped marks down the middle of the back. In females, a bold yellowish mid-dorsal stripe or series of diamond-shaped light markings may be present in some individuals.

**ALABAMA DISTRIBUTION** This subspecies is known from Baldwin, Mobile, and Houston Counties, where it was probably introduced in shipments of potted plants from Florida. It has been recorded at the Interstate 10 and Highway 231 welcome centers at the Alabama-Florida border, from gardens in the cities of Mobile and Fairhope, and from Dauphin Island. An apparently viable population is known from a large Lee County greenhouse, but the subspecies has not spread from the warm confines of the buildings.

**HABITS** This subspecies, which has relatively long hind legs, perches at the base of tree trunks or on low shrubs and scampers across the leaf litter to capture arthropods or engage in social interactions. Cuban Brown Anoles rarely climb more than 3 m up a tree, spending the vast majority of their time low to the ground or on leaf litter. These anoles move rapidly and often, frequently using their hind legs to

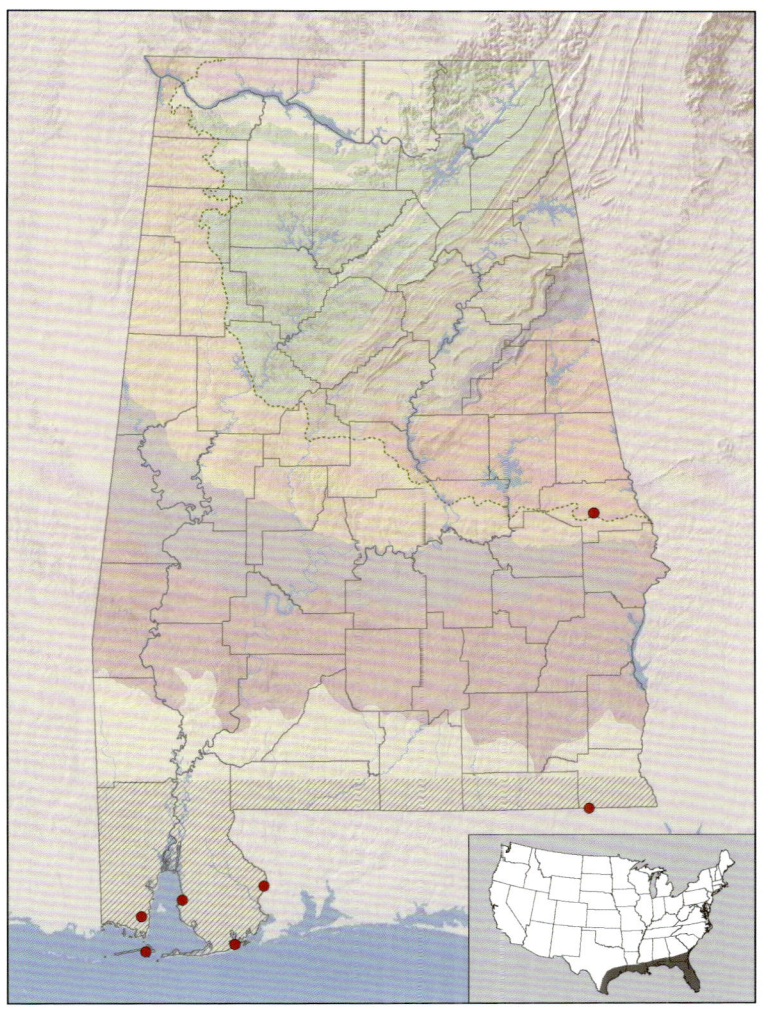

Distribution of Cuban Brown Anole, *Norops sagrei sagrei*. The presumed range in Alabama is indicated by hatching. Inset map depicts approximate range in the United States.

jump from perch to perch. Males are territorial and use their dewlaps and mid-dorsal crests as visual cues in territorial interactions with intruding males. Males use these same features in courtship displays toward females. Females produce a single egg at a time but may produce an egg as rapidly as every twelve days or so throughout the season of activity. These eggs are light gray, have soft, striated shells, and are deposited on or under leaf litter or in holes dug by females in loose soil. Like all trunk-ground anoles, Cuban Brown Anoles eat arthropods.

**CONSERVATION AND MANAGEMENT** This subspecies has expanded its geographic range through several states in the southeastern United States, generally after becoming established in southern Florida nursery businesses and then hitching rides to new areas in shipments of potted plants. Although some have expressed concern that this subspecies outcompetes the native Green Anole, the two lineages coevolved on Cuba and appear to avoid competitive exclusion when they are found together (e.g., Schoener 1968). Cuban Brown Anoles always occupy habitat structures close to the ground, regardless of whether or not Green Anoles are present. When Green Anoles occur in areas that lack Cuban Brown Anoles, the Green Anoles may be found anywhere from the base of tree trunks to the tops of trees. However, when the two subspecies occur together, the Green Anole is found only at the tops of trees, where its short limbs are better adapted for life in the canopy. For this reason, the invasion of Cuban Brown Anoles into Alabama is unlikely to cause a conservation crisis for native Green Anoles.

**TAXONOMY** The Cuban Brown Anole is sister to *N. bremeri* + *N. quadriocellifer*, two additional Cuban species (Nicholson et al. 2012). Three subspecies are recognized within *N. sagrei* and one of them has invaded Alabama. Previous authors have placed this subspecies in the genera *Anolis*, *Dactyloa*, and *Dracontura*.

# Anoles
## Genus *Anolis* (Daudin, 1802)

We restrict the content of this genus to that recommended by Nicholson et al. (2012). This concept of the genus includes only the monophyletic radiation of anoles centered on Cuba that are small, green, and with short hind legs. Dispersal of this radiation from Cuba has allowed them to become established on the Bahamas and the mainland of the United States, undoubtedly through peninsular Florida (Glor et al. 2005). This genus contains about fifty species, one of which occurs in Alabama. The genus is sister to *Ctenonotus + Norops* (Nicholson et al. 2012).

Adult male Northern Green Anole, *Anolis carolinensis carolinensis*, Lawrence County, AL

## Northern Green Anole
*Anolis carolinensis carolinensis* (Voigt, 1832)

**DESCRIPTION** This is a small to mid-sized anole that attains a maximum snout-vent length of about 70 mm. As in all anoles, the scales are small, weakly keeled, and non-overlapping, features that give the skin a velvety appearance. Lamellae, wide plate-like scales on the bottom of the next-to-last joint of each toe, expand to form an elongate pad. Males have an extensible reddish dewlap or throat fan. The head is long and pointed and becomes noticeably enlarged in adult males. The dorsal color of each individual may be green, brown, or grayish,

Adult female Northern Green Anole, *Anolis carolinensis carolinensis*, Clarke County, AL

depending on the background of the environment and the mood of the animal. Females, under some conditions, display an irregular, mid-dorsal light stripe.

**ALABAMA DISTRIBUTION** This subspecies is found throughout the state. It is uncommon to rare in extreme northern Alabama, becoming increasingly common southward.

**HABITS** This squamate, called chameleon by many who recognize it, is best known for its ability to change color. It may be green, brown, or gray, depending on how pigments are dispersed in the dermis of the skin. In general, these animals change color to match their background. However, territorial males perched on tree bark may become green to enhance their visibility to intruder males. Color changes from green to brown often are associated with initiation of social interactions by resident males (Jenssen et al. 1995). Frightened or angered individuals, if not green initially, will usually turn green. At night, presumably when asleep, these squamates are pale grayish in color making them easy to distinguish in light produced by headlamps.

The Northern Green Anole is largely arboreal and is not choosy about its habitat as long as vegetation and shade are abundant. In southern Alabama, Northern Green Anoles often abound in shady residential areas of cities and towns. During the active season, adults conform to temperatures available in shaded areas, which may be slightly

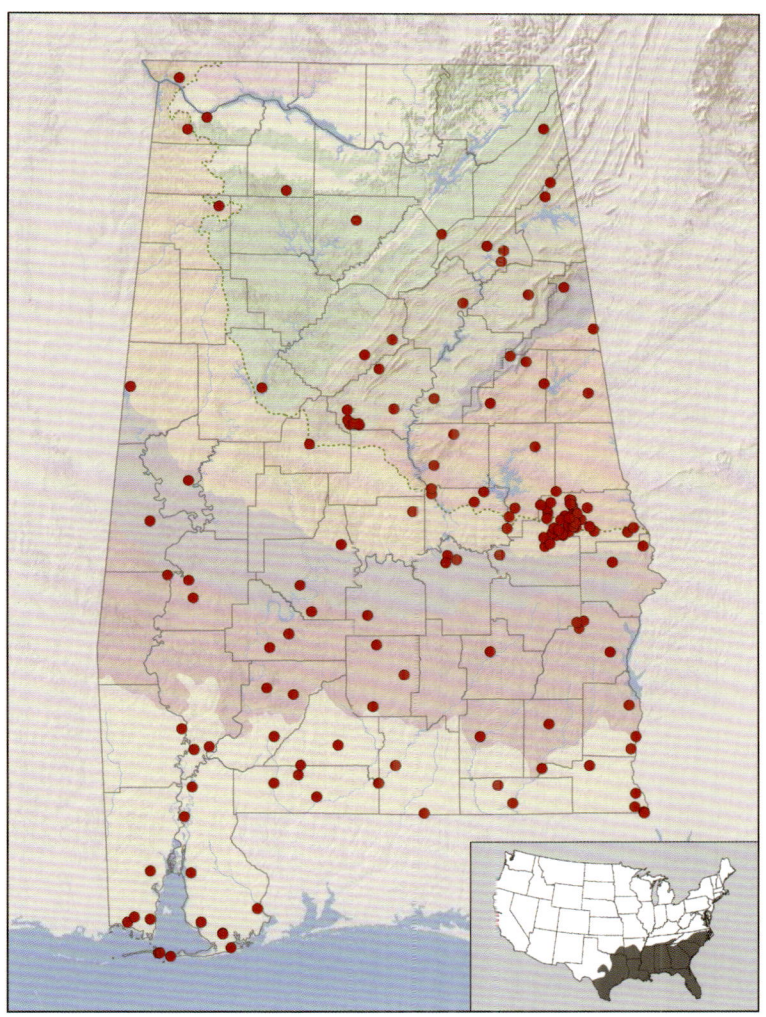

Distribution of Northen Green Anole, *Anolis carolinensis*. This species is assumed to occur throughout Alabama. Inset map depicts approximate range in the United States.

cooler than temperatures selected in laboratory thermal gradients and at which performance is maximized (Lailvaux and Irschick 2007). During winter, anoles do not cease surface activity, as most other reptiles do, but instead bask during sunny weather. Northern populations are more tolerant of cold temperatures than are southern populations (Wilson and Echternacht 1987). In areas where freezing temperatures are common, Northern Green Anoles aggregate on south-facing exposures, hiding in crevices that serve as overwinter sites. Individuals

emerge on sunny days and bask but do not wander far from their crevices (Bishop and Echternacht 2004) and are capable of growing during
these winter months (Bishop and Echternacht 2003). Individuals may
be found under wood chips, pieces of bark, facing boards of buildings,
and rock crevices during cold or cloudy periods.

The activity season for *A. carolinensis* is extensive. Male Northern
Green Anoles are strongly territorial, especially in spring. When approached by another male, or otherwise intimidated, a resident male
extends his conspicuous red dewlap. This behavior is accompanied by
a few bobs of the head. If the intruder male continues his trespass, a
heated battle likely ensues in which males bite one another and attempt to dislodge the rival from his perch. In these battles, two morphs
of males are recognizable. One is of large males with proportionately
enlarged head sizes. In these males, individuals with the strongest
bite forces usually win aggressive interactions. The second morph is of
smaller males with proportionately smaller head sizes. In these males,
individuals with the strongest jumping ability usually win aggressive
interactions (Lailvaux et al. 2004). However, neighboring males eventually establish territorial borders that allow them to recognize known
rivals with whom they have settled border disputes from novel males
with whom border disputes need to be settled (Qualls and Jaeger 1991).

Mating occurs from spring through summer, and males of the
large morph have greatest access to females (Ruby 1984). In Virginia,
male territories overlap home ranges of about three females (Jenssen
et al. 1995). Courtship involves head bobs and extensions of the red
dewlap by the male (colloquially called "showing his money purse"),
behaviors that are directed toward the female. Males bite receptive females in the neck and shoulder region, after which the male wraps his
tail around the female's to place his cloaca next to her cloacal opening. This allows the males to evert one hemipenis into the cloaca of
the female, effecting internal fertilization. Gravid females ovulate a
single egg at a time, and an egg may be produced as rapidly as every
seven to fourteen days (Lovern et al. 2004). Thus, males that success-
fully defend territories can mate with several females and may mate
with each female several times each season. Females typically deposit
eggs on soil, sphagnum, leaf litter, or rotting wood. Female anoles in
the northern part of the geographic range are larger in body size and
produce larger eggs than the southern populations. This is thought
to result from selection for larger egg size in northern populations,

a selective pressure constrained by the size of the pelvic aperture through which eggs much pass (Michaud and Echternacht 1995).

Northern Green Anoles have relatively short hind legs that are designed for creeping along twigs and branches, allowing them to glean arthropods, especially spiders and small insects, as they move. Mobile insects can also be captured by anoles that jump from branch to branch. Northern Green Anoles survive well in captivity if they are fed a variety of small insects and are sprayed with water on a daily basis.

**CONSERVATION AND MANAGEMENT** This subspecies is so common and survives so well in garden settings that it is hard to imagine that it would ever be in jeopardy of extirpation. For that reason, it receives no special protection by state law. Although Cuban Brown Anoles (*Norops sagrei*) outcompete Northern Green Anoles in urban areas (Gerber and Echternacht 2000), these two species co-occur on islands (e.g., Schoener 1968), and it is unlikely that this competition will lead to extirpation of Northern Green Anoles.

**TAXONOMY** Green anoles are sister to *A. porcatus* (Nicholson et al. 2012), a Cuban species with identical external appearance. Two subspecies of *A. carolinensis* are recognized, one of which occurs in Alabama. Previous authors have placed this subspecies in the genera *Dactyloa* and *Lacerta*.

# Horned and Spiny Lizards

## Family Phrynosomatidae

This family contains the spiny lizards of North America, Mexico, and Central America. These squamates have heavily keeled scales that allow them to live in open habitats, such as rock outcrops, grasslands, and deserts. The family contains 9 genera and 136 species, one of which is native to Alabama and a second of which has become established in the state through release by humans.

### KEY TO THE GENERA OF PHRYNOSOMATIDAE OF ALABAMA

**1a** Body greatly flattened; head with conspicuously enlarged spines.

Genus *Phrynosoma*—**Horned Lizards** . . . . page 56.

**1b** Body not greatly flattened; head lacking conspicuously enlarged spines.

Genus *Sceloporus*—**Fence Lizards** . . . . page 60.

 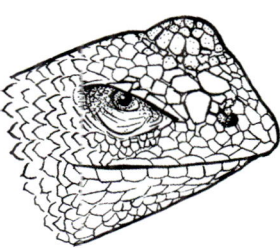

*From left to right:*

Head of a Horned Lizard (*Phrynosoma*)

Head of a Fence Lizard (*Sceloporus*)

## Horned Lizards
### Genus *Phrynosoma* (Wiegmann, 1828)

This genus contains the horned lizards of the deserts of North America, a group distributed from extreme southern Canada to Guatemala. Their flattened, pancake-like body plan makes these squamates easy to identify. Additionally, the head is short with a posterior margin that is expanded and that has short or elongate spines designed to deter some types of predation. These cryptic lizards typically remain motionless when approached, which allows them to escape detection. All species are insectivorous, but the genus is particularly famous for consuming ants as a primary dietary item. The genus *Phrynosoma* is sister to *Uma* + *Cophosaurus* + *Callisaurus* + *Holbrookia*, genera of lizards adapted to open sandy and rocky areas of the desert southwest (Reeder and Wiens 1996). There are sixteen species in the genus *Phrynosoma*, one of which has been introduced to Alabama's herpetofauna.

Texas Horned Lizard, *Phrynosoma cornutum*, Luma County, NM

## Texas Horned Lizard
*Phrynosoma cornutum* (Harlan, 1825)

**DESCRIPTION** This is a mid-sized iguanian that attains a maximum snout-vent length of about 100 mm. The body is flattened dorsoventrally, and the head is broad with eight spines that project laterally and caudally; the central two are much longer than the rest. The tail

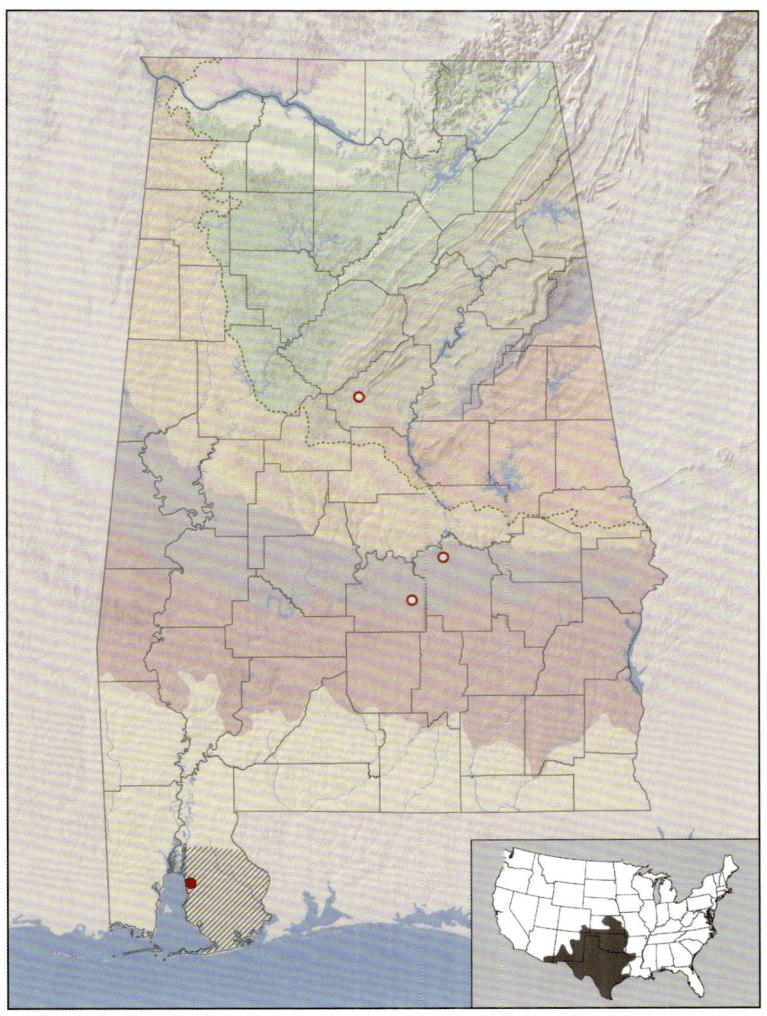

Distribution of Texas Horned Lizard, *Phrynosoma cornutum*. The presumed range of the species in Alabama is indicated by hatching. Open circles represent historical records not expected to represent viable populations. Inset map depicts approximate range in the United States.

is short, wide at the base, and flattened dorsoventrally. Scales of the dorsum include many that are noticeably enlarged and heavily keeled while the rest are small and velvety. Along the edge of the flattened body are two series of keeled scales that create a spiny border. The dorsal body color is tan to reddish brown with dark brown blotches. The venter is white to faint yellow with small gray spots. Adult males have a broader base to the tail than do females, and males have a pair of enlarged post-anal scales not present in females.

**Alabama Distribution** This species has been introduced through the pet trade into at least four places in Alabama: Daphne (Baldwin County), Letohatchee (Lowndes County), Montgomery (Montgomery County), and Siluria (Shelby County).

**Habits** Texas Horned Lizards specialize in eating harvester ants (*Pogonomyrmex*). Because *P. badius* is found in sandhill habitats of the Coastal Plain of Alabama, an appropriate prey base is available in some areas of the state. The lizards are diurnal and remain motionless on the tops of active ant mounds during morning and late afternoon hours, where they lap up ants along their trails. As the temperature rises, Texas Horned Lizards shuttle between shade and sun to maintain a constant body temperature. In extreme heat and cold (night or winter months), they bury themselves under loose soil. The color pattern makes these animals extremely cryptic so they do not move away from approaching humans unless nearly stepped upon. Males and females mate in the spring, after which females deposit an average of twenty-nine eggs in a nest dug in sandy soil. Perhaps the most bizarre feature of Texas Horned Lizards is their ability to squirt blood from the conjunctival sac associated with the lower eyelid. This occasionally occurs when humans handle these squamates, but it is almost assured to happen when an individual is nudged by a dog, coyote, or fox (Middendorf and Sherbrooke 1992). The blood contains a substance that causes an allergic response in the canid, suggesting that this is an anti-predatory mechanism.

**Conservation and Management** Two sites (Shelby and Baldwin Counties) have had several individuals discovered, suggesting that free-ranging individuals reproduced within the state. However, neither site has been sampled recently to determine whether populations persist. Nevertheless, Texas Horned Lizards appear to occupy an available ant-eating niche within the state and this niche does not appear to put Texas Horned Lizards in competition with any native species. Therefore, no attempt to extirpate the species seems warranted, if it remains established in the state. Texas Horned Lizards are now listed as threatened in the states of Texas and Oklahoma, where the species was common. The Red Imported Fire Ant (*Solenopsis invicta*) is thought to be the primary factor causing declines in these two states because of direct attacks of ants on lizards or their eggs and because of indirect effects of fire ants on harvester ants, the major dietary item of Texas Horned Lizards (Allen et al. 2004). Because Red Imported Fire Ants are now common in Alabama, any established populations may have been extirpated by the imported ants.

**Taxonomy** This widespread species is found in Texas, Arizona, and Oklahoma. It is one of the two basal species within the genus (Hodges and Zamudio 2004). No subspecific variation is recognized. Other authors have placed this species in the genera *Agama*, *Lacerta*, *Tapaya*, and *Tropidogaster*.

## Spiny Lizards
### Genus *Sceloporus* (Wiegmann, 1828)

This genus contains the fence lizards of North America, Mexico, and Central America. Its sister genus is *Sator* of Baja California (Reeder and Wiens 1996). These squamates all have heavily keeled scales and tend to occur on rocky areas or open grasslands, where they frequent tree trunks, fence posts, or rocks. The genus contains about ninety species, one of which is native to Alabama.

Adult male Eastern Fence Lizard, *Sceloporus undulatus*, Covington County, AL

## Eastern Fence Lizard
*Sceloporus undulatus* (Bosc and Daudin in Sonnini and Latreille, 1801)

**DESCRIPTION** Eastern Fence Lizards are medium-sized squamates attaining a maximum snout-vent length of around 85 mm. The stout body is covered by heavily keeled, overlapping scales that have spiny tips. Color varies with age and sex. Females and young are gray to brown or yellowish brown with dark wavy crossbars; their venters are white or light gray with small dark flecks that may have traces of light blue. Adult males are dark gray to dark brown or bronze with dark crossbars that become obscure with age. The venter of adult males has conspicuous dark metallic blue or blue-green patches edged in black. Males also have a pair of enlarged post-anal scales that are visible in individuals of all sizes and are absent in females.

Adult female Eastern Fence Lizard, *Sceloporus undulatus*, Winston County, AL

**ALABAMA DISTRIBUTION** These lizards are found throughout the state.

**HABITS** Eastern Fence Lizards prefer dry, open woodlands, abandoned farm buildings, rock outcrops, and piles of old lumber. Called rusty-backed lizard by most rural citizens of Alabama, these squamates are usually conspicuous because of their tendency to bask in exposed places and to move readily when approached. These lizards usually try to escape enemies by climbing the nearest tree or wall. However, this habit also allows their capture by approaching the tree from the opposite side selected by the lizard, peering around the trunk carefully to determine the lizard's position, and then grabbing at a point level with or slightly above the lizard's head. These lizards can also be captured by placing a noose made of dental floss or fine gauge copper wire at the tip of a fishing pole and placing the noose around the neck of the lizard. Most individuals perceive the noose as food and remain stationary while it is slipped into place.

Male Eastern Fence Lizards are strongly territorial, advertising their presence to other members of their species with an elaborate ceremony involving head bobbing and push-ups. Another male may challenge the territory owner but more frequently will move away and avoid direct confrontation. Resident males generally maintain the same territory from year to year. Females have home ranges that are smaller than male territories so that each male overlaps the ranges

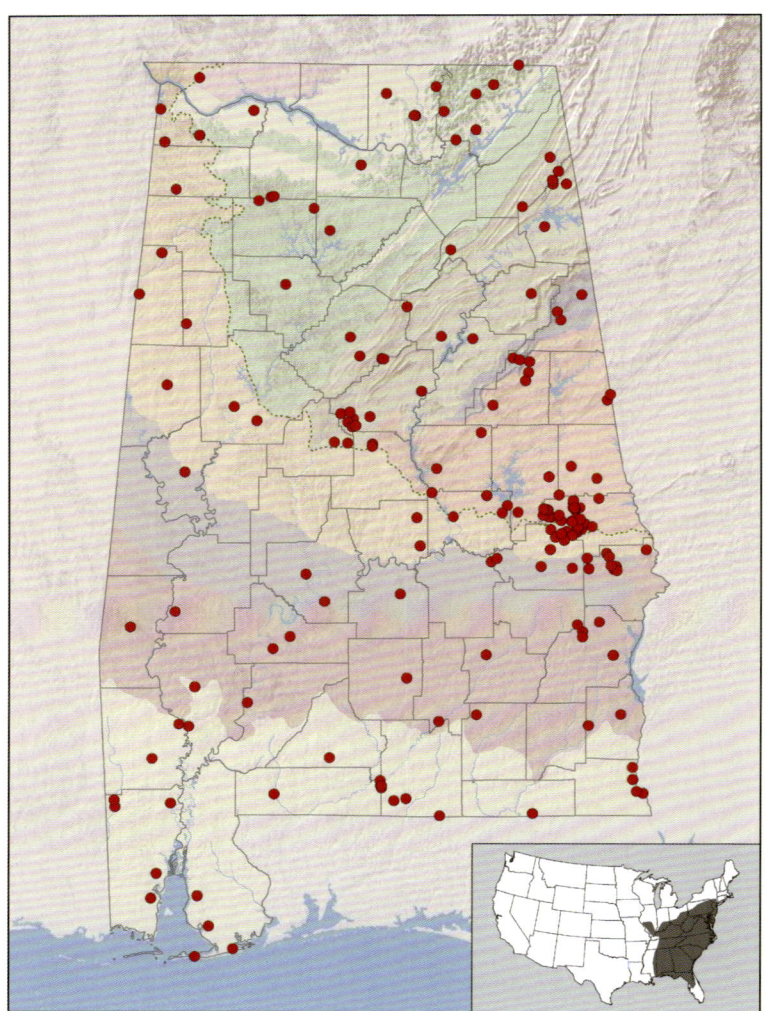

Distribution of Eastern Fence Lizard, *Sceloporus undulatus*. This species is assumed to occur throughout Alabama. Inset map depicts approximate range in the United States.

of several females. Mating occurs within a male's territory and involves courtship displays presented by the male to the female. These displays consist of the push-ups, head bobs, and body flattening used in male-male aggression. If the female is receptive, then the male bites the female in the neck or shoulder region and then wraps his tail around hers so that the cloacal openings of the mates are in close association. After insemination, the female lays six to fifteen eggs in a cavity she digs in soil. After covering the eggs, she pays no further attention to the nest. In Alabama, nesting activity usually begins in May. Two or more clutches of eggs may be laid each season.

Eastern Fence Lizards consume a wide variety of insects and other arthropods. These typically are selected visually and captured on a sticky tongue that is first projected from the mouth to contact the prey and then retracted into the mouth where the prey item is chewed before being swallowed. For prey possessing long legs, the lizard may wipe its jaws on a hard substrate in an attempt to break off prey legs, making the item easier to swallow. Fence lizards in northern Alabama burrow into loose soil or under rocks during winter months. In south Alabama, individuals may emerge on warm days throughout winter.

**CONSERVATION AND MANAGEMENT** Eastern Fence Lizards are widely distributed across Alabama and interact well with humans, taking advantage of gardens, even in urbanized areas. For these reasons, the species receives no special protection under state law. However, this species is not as abundant as it once was, perhaps as a result of predatory pressure by the Red Imported Fire Ant. Lizards in areas with fire ants have longer hind limbs and use these longer limbs to remove attacking fire ants more rapidly and effectively than lizards from areas lacking fire ants (Langkilde 2009; Freidenfelds et al. 2012). Lizards prefer fire ants to native ants as dietary items, in areas where both are present (Robbins et al. 2013), but a fire ant diet leads to higher juvenile mortality and smaller adult body size (Langkilde and Freidenfelds 2010). These data suggest that Eastern Fence Lizards have evolved to deal with fire ants, but that this may have reduced their carrying capacity. Maintenance of rock piles and outcrops, as well as retaining snags and large fallen logs, likely enhance habitat for this species by providing appropriate basking sites and refuges.

**TAXONOMY** We follow Leaché and Reeder (2002) in separating *S. consobrinus* as a distinct species and eliminating subspecies previously recognized within *S. undulatus*. Mitochondrial and morphological data do not support the previously accepted (Mount 1975) northern and southern subspecies within *S. undulatus*. Instead, monophyletic lineages east of the Appalachian Mountains and along the backbone of the Appalachian Mountains are evident. However, Leaché and Reeder (2002) do not give these subspecific status and we follow that convention. Specimens from Alabama are nested within the Appalachian lineage. Previous authors have placed this species in the genera *Agama*, *Lacerta*, *Stellio*, and *Tropidolepis*.

# Geckos and Their Relatives

## Group Gekkota

Members of the group Gekkota have simple, rounded tongues, which they extend frequently from the mouth and touch to the substrate in order to detect chemicals in the environment. This represents an initial step in a progression of events that develops in squamates characterized by increasingly complex tongues, which are used to detect chemicals from the environment and then interpreted by use of the vomeronasal organ, a structure on the roof of the mouth. Gekkotans also tend to be nocturnal, with large eyes designed to capture as much light as possible in low-light environments. Uniquely in this radiation, the tongue can be used to clean the surface of the head, especially the eyes. The group also is characterized by a unique positioning of the cartilaginous septum of the caudal vertebrae (Gautier et al. 2012) that allows these squamates to easily lose their tails during encounters with predators. We follow Vidal and Hedges (2009) in recognizing 7 families within Gekkota, a group that contains about 1,500 species and that is cosmopolitan in distribution. Most species are terrestrial, but some forms are limbless, burrowing in loose soil, and other forms have invaded arboreal habitats. No gekkotan is native to Alabama, but one species has invaded the state.

# Family Gekkonidae

This family contains the true geckos, a radiation of lizards that is so strongly tied to nocturnal activity that their eyes have become enlarged, the irises have become modified to focus numerous weak bands of light on to the retina, and the eyelids have been lost, being replaced by a transparent scale (the spectacle) that covers and protects the eye. Although the tongue is used to sample chemicals in the environment, geckos tend to be visually oriented and can extrude the tongue far enough from the mouth cavity to use it to wipe clean the spectacle. The family contains forty-nine genera that are found in Africa, Asia, and Australia. Of the nine hundred species in the family, none is native to Alabama, but one has become established in the state. The family Gekkonidae is sister to the family Phyllodactylidae, a radiation of leaf-toed geckos of the Middle East, Africa, and South and Central America (Pyron et al. 2013).

## House Geckos
### Genus *Hemidactylus* (Gray, 1825)

This genus of House Geckos is so named because of their propensity to invade human homes, frequently providing a benefit by consuming household arthropod pests. The genus has a center of diversity in southeastern Asia, where it occupies arid, rocky areas. Like many geckos, members of the genus communicate by vocalizing. It is a successful lineage including over one hundred species, several of which have invaded areas outside their native ranges because they do so well in homes and gardens of large cities.

A single species, the Turkish House Gecko (*Hemidactylus turcicus turcicus*), occurs in Alabama. Turkish House Geckos are the sister species to *H. robustus* + *H. macropholis* + *H. oxyrhinus* + *H. homoeolepis* + *H. forbesi*, a portion of an arid Asian and African clade (Carranza and Arnold 2006). Previous authors have placed this species in the genera *Gecko*, *Gecus*, and *Lacerta*.

## Turkish House Gecko
*Hemidactylus turcicus turcicus* (Linnaeus, 1758)

**DESCRIPTION** These small, delicate lizards attain a maximum total length of about 150 mm. The head is relatively large, pointed anteriorly, and expanded posteriorly to accommodate the large, dorsally oriented eyes. The skin is composed of extremely small velvety scales interspersed with large, beaded scales. The toes have expanded pads that are created by a series of wide, overlapping scales called lamellae, which cover the ventral surface of each digit. As in similar structures found in anoles, these lamellae allow *H. turcicus* to adhere to smooth surfaces, such as walls and glass. The ground color is light gray with darker gray blotches in adults and has a pinkish cast in juveniles. Juveniles also have a tail that is boldly marked with white and dark gray rings that become faded in adults.

**ALABAMA DISTRIBUTION** This subspecies has become established in several large cities within the southern half of the state. They have been documented from Birmingham and Huntsville, but it is not clear whether they are established in these cities (Marion and Bosworth 1982).

**HABITS** Any places with buildings that are large enough to have warm basements or heating pipes are capable of harboring these prolific invaders. During spring and summer, temperatures are warm enough to allow these geckos to move to the outside of buildings, where they capture insects attracted to lights at night. During winter months,

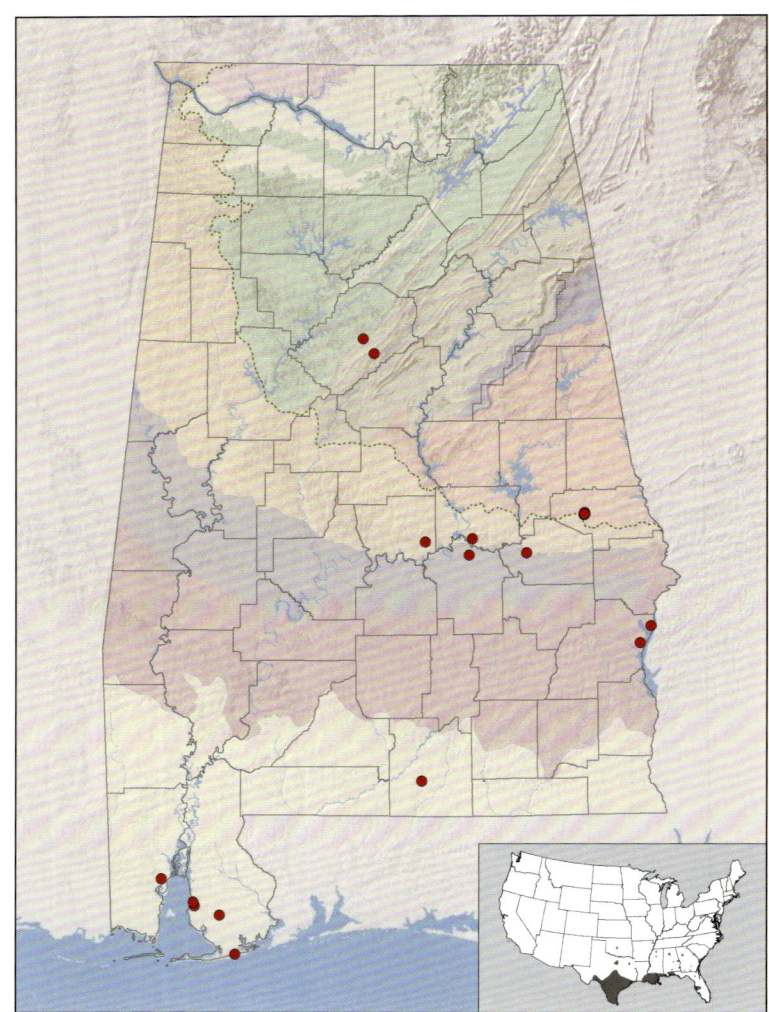

Distribution of Turkish House Gecko, *Hemidactylus turcicus*. This species is assumed to occur throughout Alabama. Inset map depicts approximate range in the United States.

they move indoors and consume insect pests associated with heating ducts. They cannot survive freezing weather and cannot establish themselves more widely than large buildings. Females produce two eggs per clutch; the eggs are nearly round in shape, white in color, and adhere with a glue-like substance to hard surfaces. The skin of these lizards is extremely delicate and tears easily upon capture, a mechanism used to escape predators, after which the skin is repaired and replaced with the next shed cycle. As in many other squamates, the tail detaches easily and twitches reflexively, another mechanism used by these lizards to escape predation. The tail can regenerate but will not be as long or as colorful as the original tail.

**CONSERVATION AND MANAGEMENT** This is an invasive subspecies that is viewed by some to be a nuisance. However, it does not compete with any native species and consumes arthropod pests and so is perhaps best referred to as a commensal of humans in cities. These geckos can be captured on sticky traps used for roaches; however, it is doubtful that this method could be used to eradicate them from a building.

# Skinks, Whiptails, and Their Relatives

## Group Scincomorpha

This lineage of elongate squamates is characterized by a tongue that is keratinized, indented at its anterior end, and covered by flat, overlapping papillae. These lizards use their tongues to sample chemicals in the environment, especially chemical trails laid down by conspecifics. These trails allow males to establish territorial boundaries and to find mates. The group contains 7 families, 2 of which are found in Alabama, and about 2,430 species, 7 of which are found in Alabama. These organisms are cosmopolitan in distribution, and the group tends to be terrestrial in habitat selection but includes some forms that are limbless burrowers and others that have invaded arboreal habitats.

### KEY TO THE FAMILIES OF SCINCOMORPHS OF ALABAMA

**1a** Dorsal and ventral scales smooth and cycloid.

Family **Scincidae—Skinks** . . . . page 72.

**1b** Dorsal scales small and velvety, ventral scales large and rectangular.

Family **Teiidae—Whiptails and Racerunners** . . . . **page 96.**

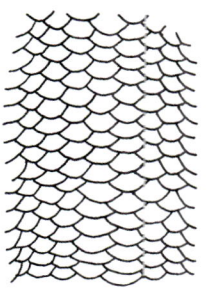

Smooth cycloid scales of a skink (*Scincidae*)

# Family Scincidae

This is a large cosmopolitan family containing forms that possess smooth, cycloid scales that give the animal a shiny, moist appearance. They represent the basal radiation within scincoids. There are about 128 genera with over 1,500 species, only a relatively few of which occur in the New World. Two genera are represented in the United States, both of which occur in Alabama. These are terrestrial organisms that do well in a variety of habitats, including garden areas within cities.

### KEY TO THE GENERA OF SCINCIDAE OF ALABAMA

**1a** Lower eyelid with a transparent window-like structure; maximum size small (50 mm snout-vent length); dorsum lacking conspicuous light stripes as juveniles and adults.

Genus *Scincella*—**Ground Skinks . . . . this page.**

**1b** Lower eyelid lacking transparent window-like structure; maximum size larger than above; dorsum plain or with light stripes of varying intensity as juveniles.

Genus *Plestiodon*—**Toothy Skinks . . . . page 76.**

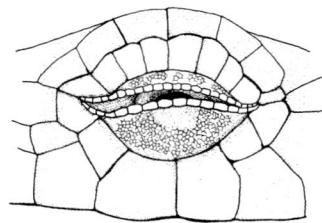

Lateral view of scales of lower eyelid of a Ground Skink (*Scincella lateralis*)

## Ground Skinks
### Genus *Scincella* (Mittleman, 1950)

This genus includes small skinks that are found on leaf litter of hardwood forests. The presence of a transparent disc on the lower eyelid is diagnostic of the genus and gives it a secondary common name, Window-eyed Skink. There are twenty-six species in the genus that are distributed in Asia, Mexico, and the eastern United States. Only five species are found in the New World, and of these, only one is found in Alabama.

# Ground Skink
*Scincella lateralis* (Say, 1823)

**DESCRIPTION** This is a small, elusive lizard attaining a maximum snout-vent length of around 50 mm. The presence of a transparent window on the lower eyelid characterizes this genus and species. The scales are smooth, shiny, overlapping, and cycloid. The dorsal color is tan to dark brown with a pair of dark dorsolateral stripes evident in relatively light-colored specimens, becoming obscure in dark-colored ones. The venter is yellowish to dull white.

**ALABAMA DISTRIBUTION** Ground Skinks are common and found throughout the state. However, each of the three genetic lineages within Alabama (Jackson and Austin 2010) is restricted in its distribution. The Tombigbee lineage is largely distributed west of the Alabama River, including specimens from the western border of the state to Tuscaloosa; the East Coast lineage is east of the Alabama River except for the southern tier of counties east of Mobile Bay, which belong to the Florida Panhandle lineage.

**HABITS** The Ground Skink, one of our most abundant reptiles, is usually seen scurrying about among leaves on the forest floor, and occasionally even on residential lawns. It is surprisingly difficult to catch, and the would-be captor seldom gets more than one opportunity. Ground Skinks exploit most forested, terrestrial habitat types and prefer mesic and dry sites over damp ones. Both sexes deposit chemical cues as they move through the environment. Adult males are

Ground Skink,
*Scincella lateralis*,
Bibb County, AL

attracted to the chemical cues of adult females, suggesting that males trail females during the mating season (Duval et al. 1980). Adult lizards move an average of 10.5 m when they move (Fitch and von Achen 1977). Females lay from one to seven, but usually two or three, eggs under a rock or in rotting wood or humus. Several clutches may be produced each season; unlike in other skinks, female Ground Skinks do not protect their nests. These skinks consume a variety of small arthropods in the leaf litter.

**CONSERVATION AND MANAGEMENT** Ground Skinks are common throughout the state. As currently defined, the species receives no special protection under state law. These skinks do well in garden areas and in hardwood forests, habitats that seem unlimited. Therefore, no special management tools are required for maintaining them across Alabama's landscape. This species is more abundant in mature forests than in clear-cuts that have been burned, roller chopped, and seeded on mechanically prepared mounds (Greenberg et al. 1994). Stand thinning and implementation of frequent fire are detrimental to Ground Skinks (Steen, McGee, et al. 2010).

**TAXONOMY** No subspecies of this wide-ranging skink has been described based on morphology. The three genetic lineages recovered by Jackson and Austin (2012) have been given no official status. Previous authors have placed this species in the genera *Mocoa, Leiolopisma, Lygosoma,* and *Scincus.*

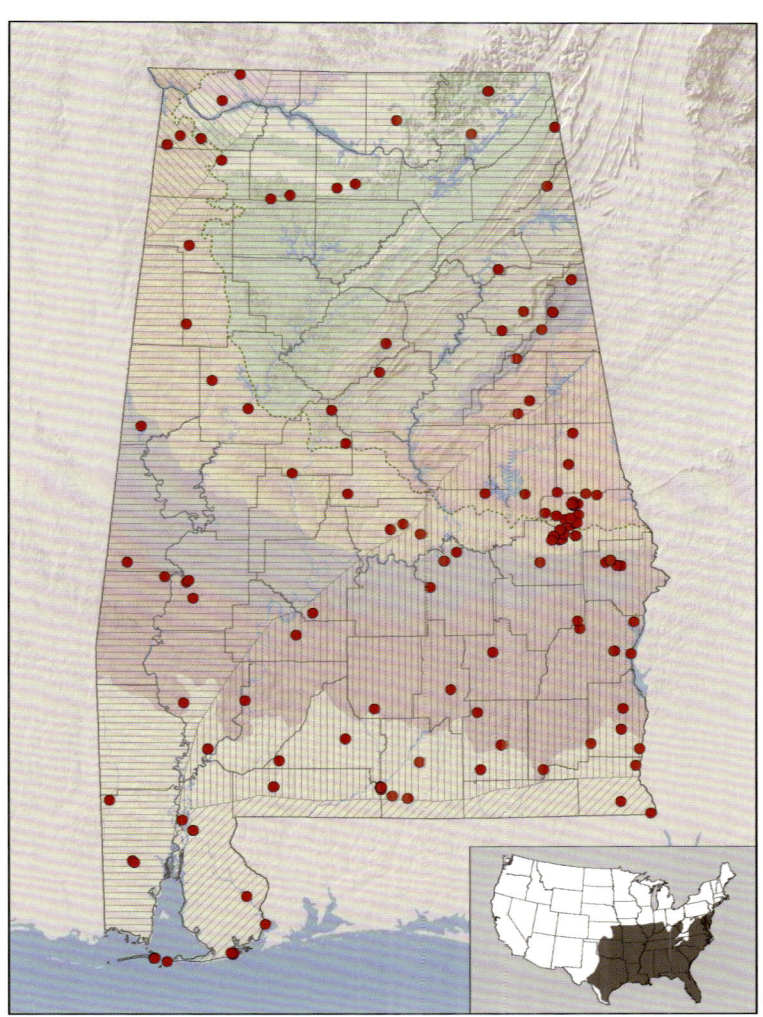

Distribution of Ground Skink, *Scincella lateralis.* This species is assumed to occur throughout Alabama. Inset map depicts approximate range in the United States. Horizontal hatching indicates Tombigbee River east lineage; vertical hatching indicates East Coast lineage; right-slant hatching indicates Florida Panhandle lineage.

## Toothy Skinks
### Genus *Plestiodon* (Griffith, 1991)

We follow Smith (2005) in using this genus name for the radiation of large skinks of Asia and North America. It is the sister genus to *Brachymeles* of Thailand, Malaysia, and the Philippines. The genus *Plestiodon* contains forty-five species, with the North American species forming a monophyletic lineage that likely dispersed from Asia across the Bering land bridge. Most *Plestiodon* have robust bodies and live in forested openings where they focus their activities on snags and fallen logs. Five species are found in Alabama. Four of these, *P. anthracinus*, *P. fasciatus*, *P. inexpectatus*, and *P. laticeps*, are closely related members of the five-lined skink group and, in the juvenile stage, are remarkably similar in general appearance, possessing a bright blue tail and five light stripes on the dorsum.

### Key to the Species of *Plestiodon* of Alabama

**1a** Postmental single.

*Plestiodon anthracinus* spp. . . . . . page 78.

**1b** Postmentals two in number. Go to **2**.

*From left to right:*

Chin scales of a Coal Skink (*Plestiodon anthracinus*); based on AUM 9363

Chin scales of a Five-lined Skink (*Plestiodon fasciatus*); based on AUM 5705

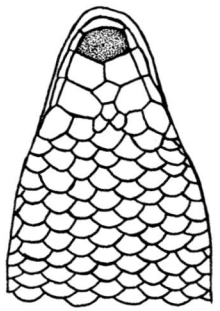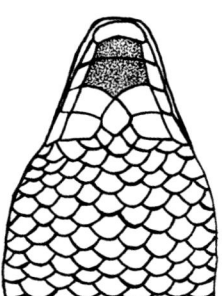

**2a** Body slender; limbs reduced in size; tail reddish (occasionally orange); head not noticeably enlarged and orange in males; maximum size around 55 mm in snout-vent length.

*Plestiodon egregius similis*—**Northern Mole Skink** . . . . page 84.

**2b** Body not slender; limbs not reduced in size; tail not red; head noticeably enlarged and reddish or orange in males; maximum size greater than 55 mm snout-vent length. Go to **3**.

**3a** Median subcaudal scales about the same width as those in adjacent rows.

    *Plestiodon inexpectatus*—Southeastern Five-lined Skink . . . . **page 87.**

**3b** Median subcaudal scales noticeably wider than those in adjacent rows. Go to **4.**

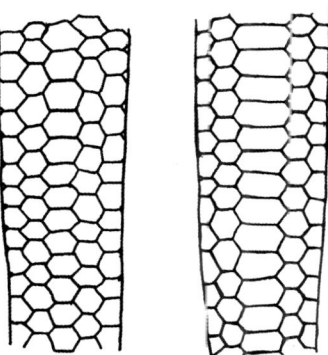

From left to right:

Scalation of ventral aspect of tail of a Southeastern Five-lined Skink (*Plestiodon inexpectatus*); based on AUM 13783

Scalation of ventral aspect of tail of a Broad-headed Skink (*Plestiodon laticeps*); based on AUM 3607

**4a** Upper labials usually six to center of eye; enlarged postlabials usually lacking; mid-dorsal stripe on juveniles very narrow; adult males and females large, up to around 130 mm in snout-vent length; adult males patternless, or nearly so, with conspicuously swollen heads.

    *Plestiodon laticeps*—Broad-headed Skink . . . . **page 90.**

**4b** Upper labials usually five to center of eye; postlabials usually evident and two in number; mid-dorsal stripe on juveniles relatively wide; adult males and females seldom, if ever, exceeding 85 mm in snout-vent length; adult males usually showing at least traces of striped pattern; head of adult male usually somewhat swollen but not conspicuously so.

    *Plestiodon fasciatus*—Five-lined Skink . . . . **page 93.**

From left to right:

Upper labial scales of a Broad-headed Skink (*Plestiodon laticeps*); based on AUM 3607

Upper labial scales of a Southeastern Five-lined Skink (*Plestiodon inexpectatus*); based on AUM 13783

# Coal Skink

*Plestiodon anthracinus* (Baird, 1850)

**TAXONOMY** This species is the basal member of the five-lined species of *Plestiodon* in North America (Pyron et al. 2013) but differs from those taxa in having a more robust body (Richmond 2006). Two subspecies of this lizard are currently recognized. One of these occurs in Alabama, and the other exerts its influence in intergradient populations. Previous authors have placed this species in the genus *Eumeces*.

### KEY TO THE SUBSPECIES OF *PLESTIODON ANTHRACINUS* OF ALABAMA

**1a** Light pigment on posterior upper labials continuous, scales at midbody usually twenty-five or fewer, dorsum usually lacking dark markings between dorsolateral light stripes.

> *Plestiodon anthracinus anthracinus*—Northern Coal Skink . . . . this page.

**1b** Light pigment on posterior upper labials broken by dark pigment along the sutures, producing a spotted appearance; scales at midbody usually more than twenty-five; dorsum often with dark stripes or rows of dark spots between dorsolateral light stripes.

> *Plestiodon anthracinus pluvialis*—Southern Coal Skink . . . . page 81.

Northern Coal Skink, *Plestiodon anthracinus anthracinus*, Winston County, AL

# Northern Coal Skink

*Plestiodon anthracinus anthracinus* (Baird, 1850)

**DESCRIPTION** This is a fairly small skink attaining a maximum snout-vent length of around 65 mm. It has scales that are smooth, shiny, overlapping, and cycloid. Scutellation of the head lacks postnasals and

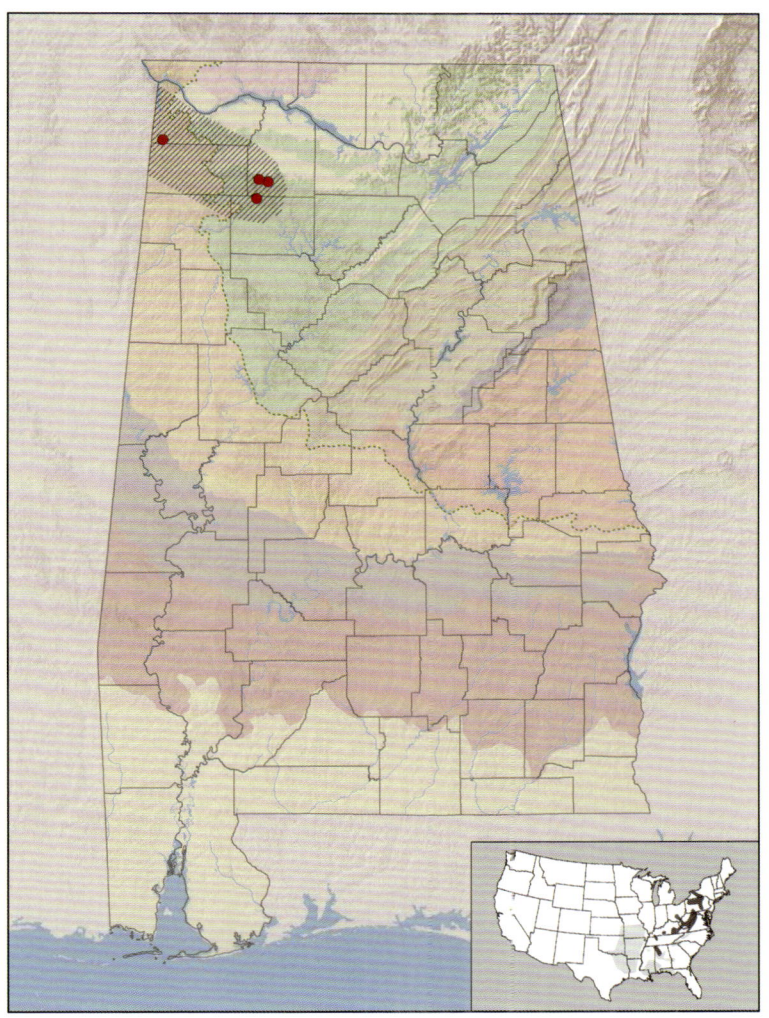

Distribution of Northern Coal Skink, *Plestiodon anthracinus anthracinus*. The presumed range of the subspecies in Alabama is indicated by hatching. Inset map depicts approximate range in the United States, with dark shading indicating the range of *P. a. anthracinus* and light shading indicating the range of the other subspecies.

includes a single postmental (two postmentals in *P. fasciatus* and *P. laticeps*). The median subcaudals are wider than those in adjacent rows. Each side of the body has a dark lateral band bordered below by a faded light stripe and above by a bold dorsolateral light stripe. Many individuals have an unbranched mid-dorsal light stripe that, when present, is less conspicuous than the dorsolateral stripes. In this subspecies, the head has a light stripe passing through the posterior upper labials and continuing to the ear (absent in *P. a. pluvialis*) Additionally, the scale rows at midbody usually are twenty-five or fewer (twenty-six or more in

*P. a. pluvialis*). Juveniles are patterned similar to adults except that the dorsum lacks the dark markings between the dorsolateral light stripes.

**ALABAMA DISTRIBUTION** Influence from this subspecies is detectable in coal skink populations in the northwestern-most portion of Alabama (Mount 1975). This area is essentially north of the Fall Line Hills and the adjacent portion of Sand Mountain in the northwestern quarter of Alabama. In number of scale rows at midbody, coal skinks from these areas may resemble *P. a. pluvialis*, but the labial striping is that of *P. a. anthracinus.*

**HABITS** Specimens from Alabama have been collected in hilly terrain in mixed pine-hardwood forests, forests on sandy soils, and rocky areas. Most are found under logs or rocks in close proximity to water. The Northern Coal Skink wriggles across the top of leaf litter when it moves and consume arthropods. Mating occurs in early spring, generally a month earlier than other members of the genus in Alabama (Trauth 1994). Mating involves males trailing females as they become receptive. A male then bites a receptive female on the neck or shoulder and wraps his tail around hers. Once inseminated, females then deposit eight to thirteen eggs during April and May (Trauth 1994). The nest is a depression under a fallen log, and the female attends the nest.

**CONSERVATION AND MANAGEMENT** The Northern Coal Skink is ranked Priority 2 (High Conservation Concern) by ADCNR (Mirarchi, Bailey, Haggerty, and Best 2004). Therefore, it is illegal to possess the Northern Coal Skinks in the state of Alabama without a scientific collecting permit. These secretive animals are difficult to detect in their environment. They do not appear to be abundant but are found widely across the state. Because they tend to be found in steep ravines near streams, streamside management zones designed for wildlife are likely to be beneficial for populations of *P. a. anthracinus.*

## Southern Coal Skink
*Plestiodon anthracinus pluvialis* (Cope, 1880)

DESCRIPTION This subspecies attains a maximum snout-vent length of around 65 mm and has scales that are smooth, shiny, overlapping, and cycloid. Scutellation of the head lacks postnasals and includes a single postmental (two postmentals in *P. fasciatus* and *P. laticeps*). The median subcaudals are wider than those in adjacent rows. Each side of the body has a dark lateral band, bordered below by a faded light stripe and above by a bold dorsolateral light stripe. Many individuals have an unbranched mid-dorsal light stripe that, when present, is less conspicuous than the dorsolateral stripes. In this subspecies, the head possesses posterior upper labials that have light centers and dark sutures, creating a spotted pattern rather than the light stripe passing through the posterior upper labials of *P. a. anthracinus*. Additionally, the scale rows at midbody usually are twenty-six or more (twenty-five or fewer in *P. a. anthracinus*). Juveniles are dark and lack any evidence of the pattern seen in *P. a. anthracinus*.

ALABAMA DISTRIBUTION *P. a. pluvialis* has been collected in central Alabama in the lower portions of the Ridge and Valley region (Shelby and Bibb Counties), the Talladega Upland (Chilton County), along the southern edge of Sand Mountain (St. Clair County), and in the Fall Line Hills (Tuscaloosa and Bibb Counties). In western Alabama, records are available from the Lower Coastal Plain (Mobile County),

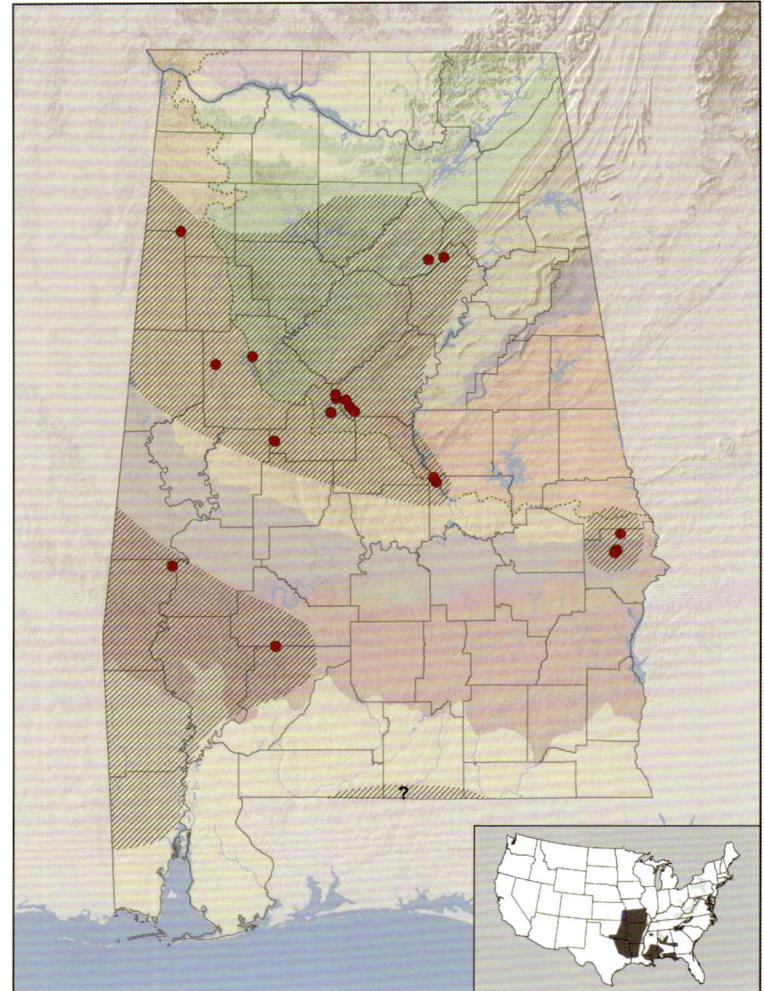

Distribution of Southern Coal Skink, *Plestiodon anthracinus pluvialis*. The presumed range of the subspecies in Alabama is indicated by hatching. Question mark indicates an unconfirmed sighting by a knowledgeable colleague. Inset map depicts approximate range in the United States, with dark shading indicating the range of *P. a. pluvialis* and light shading indicating the range of the other subspecies.

the Red Hills (Monroe County), and the transitional zone between the Black Belt and Red Hills (Choctaw County). The only records for eastern Alabama are for two localities in Russell County in the Fall Line Hills. It is not known whether the gaps indicated on the distributional map are real or only apparent. This subspecies has been documented from pitcher plant (*Sarracenia*) bogs in the Florida Panhandle close to the Alabama state line, and it may ultimately turn up in those habitats in extreme southern Alabama.

**HABITS** Specimens of *P. a. pluvialis* have been collected in hilly terrain in mixed pine-hardwood forests, forests on sandy soils, and rocky areas. Most are found under logs or rocks in close proximity to water. This subspecies resembles, at first glance, an oversized *Scincella lateralis*, but when alarmed it reacts quite differently. Whereas *Scincella* usually scurries underneath the ground litter, the coal skink usually runs or wriggles across the top. Mating occurs in early spring, generally a month earlier than other members of the genus in Alabama (Trauth 1994). Mating involves males trailing females as they become receptive. A male then bites a receptive female on the neck or shoulder and wraps his tail around hers. The pre-coital investigation and bite are of shorter duration in this subspecies than in *P. fasciatus* (Pyron and Camp 2007). Once inseminated, females then deposit four to thirteen eggs in April and May (Trauth 1994). The nest is a depression under a fallen log, and the female attends the nest.

**CONSERVATION AND MANAGEMENT** The Southern Coal Skink is ranked Priority 2 (High Conservation Concern) by ADCNR (Mirarchi, Bailey, Haggerty, and Best 2004), making it illegal to possess the subspecies in the state of Alabama without a scientific collecting permit. These secretive animals are difficult to detect in their environment. They do not appear to be abundant but are found widely across the state. Because they tend to be found in steep ravines near streams, streamside management zones designed for wildlife likely benefit populations of *P. a. pluvialis*. In the Coastal Plains, fire may be important in maintaining appropriate habitat for Southern Coal Skinks (Means 2004a).

# Northern Mole Skink

*Plestiodon egregius similis* (McConkey, 1957)

DESCRIPTION This is a small, slender skink attaining a maximum snout-vent length of about 55 mm, with limbs that are reduced in length and bulkiness. The scales are smooth, overlapping, and cycloid. The head is elongate and pointed, has six upper labials, lacks a primary temporal, and has an ear opening that is partially closed. There are eighteen to twenty-one scale rows at midbody. The dorsal ground color is gray, tan, or brown with two dorsolateral light stripes that begin on the head and may extend the length of the body or may end as far forward as the shoulders. The dorsolateral light stripes are evident from hatching and neither widen nor diverge. The tail is red, reddish orange, orange, or reddish brown. Adult males often have a yellow or orange suffusion on the lower sides and their lips, and average body length is slightly less than for females. Because of its unusual body shape, Northern Mole Skinks are not likely to be confused with any other skink.

ALABAMA DISTRIBUTION Northern Mole Skinks have a patchy distribution in the Piedmont, Ridge and Valley, and Fall Line Hills westward to the Black Warrior River; then the subspecies occurs sporadically southward through the Lower Coastal Plain from the southeastern corner of the state westward to at least Baldwin County. There are no records from the Black Belt.

HABITS This secretive skink spends most of its time underground and is rarely seen abroad. Historically, a frequent ecological associate of the Northern Mole Skink in the Coastal Plain was the Southeastern Pocket

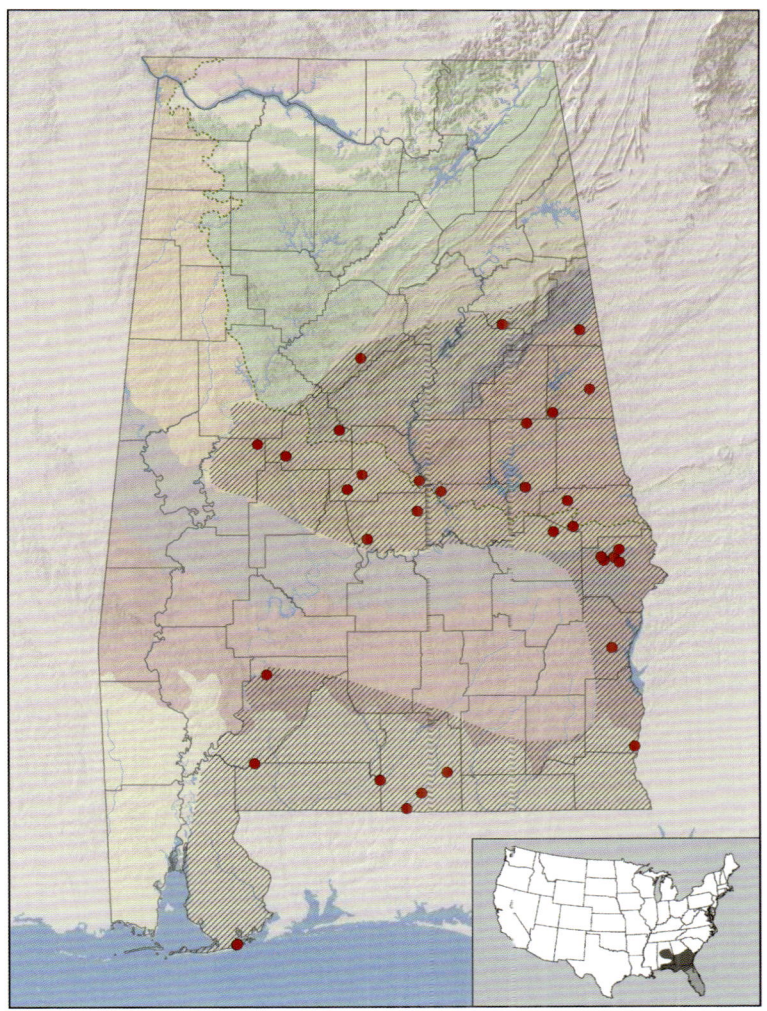

Distribution of Northern Mole Skink, *Plestiodon egregius similis*. The presumed range of the subspecies in Alabama is indicated by hatching. Inset map depicts approximate range in the United States, with dark shading indicating the range of *P. e. similis* and light shading indicating the range of all other subspecies.

Gopher, *Geomys pinetis*, an increasingly scarce burrowing mammal locally called salamander by many residents of southern Alabama. The pocket gopher digs horizontal tunnels beneath the surface and, at fairly regular intervals, pushes up the excess soil to form mounds on the surface. The burrows apparently provide ideal microhabitats for Northern Mole Skinks, and on warm, sunny days during late winter and early spring, the skinks may increase their body temperatures by positioning themselves within a mound. Raking through the mounds at such times is one of the best ways to locate these skinks. When a

Northern Mole Skink is spotted, it must be secured quickly or it will quickly wriggle out of sight. These skinks are occasionally found under logs, rocks, or other sheltering objects.

Unlike the other skinks of Alabama, Northern Mole Skinks reproduce during fall and winter (September through December), when males find receptive females by following scent trails deposited by the females. A male bites loose skin along the anterior half of a receptive female and immediately strokes her dorsum with his adjacent front limb. Eventually, he wraps his body in circular loops around the female's body until his cloaca is adjacent to hers. Once inseminated, the female retains sperm that are used to fertilize eggs that are ovulated the following spring (May). From two to nine elliptical, whitish eggs with thin leathery shells are laid in nests dug by the female to depths of about 4 inches (100 mm). Nests are large enough to contain the eggs and the female, who attends the eggs until they hatch. During this time the female turns the eggs and licks them. Hatchlings grow to adult size during their first year and likely breed the next. Females achieve slightly larger sizes than males. Northern Mole Skinks feed on crickets, spiders, roaches, and other small arthropods.

**CONSERVATION AND MANAGEMENT** This subspecies has rarely been seen in recent decades. During this time, pocket gophers have declined in the Lower Coastal Plain of Alabama, probably due to reduction in the vast expanses of open, fire-maintained habitat that these mammals prefer. Mole skinks are not provided special protection by state law. Management for this subspecies may require repatriation of pocket gophers across the southern tier of counties. Thinning of pine stands, reduction of shrubs, and expansion of the grass layer are all primary management goals for this subspecies. Fire, especially in the growing season, is likely to be a key management tool for Northern Mole Skinks. This subspecies may be more abundant in clear-cuts that have been burned, roller chopped, and seeded on mechanically prepared mounds than in mature forest (Greenberg et al. 1994).

**TAXONOMY** This species is the sister taxon to *Plestiodon reynoldsi* (Schmitz et al. 2004), a nearly limbless sand-swimming species endemic to central Florida. Five subspecies of *P. egregius* are recognized, one of which occurs in Alabama. Previous authors have placed this species in the genus *Eumeces*.

## Southeastern Five-lined Skink

*Plestiodon inexpectatus* (Taylor, 1932)

**DESCRIPTION** These skinks are medium in size, attaining a maximum snout-vent length of about 90 mm. The scales are smooth, shiny, overlapping, and cycloid. The head has two postmentals, and seven or eight upper labials plus two enlarged postlabials. The median subcaudals are about the same width as those in adjacent rows, differentiating this species from *P. fasciatus* and *P. laticeps* (median subcaudals are noticeably wider than those in adjacent rows). The body of juveniles and young adult females (except for the venter) is black or blue-black with five yellow or orange-yellow stripes (a mid-dorsal one, bifurcating anteriorly at back of head and extending onto the snout; two dorsolaterals, beginning over the eyes; and two laterals, beginning under the eyes). The mid-dorsal stripe is relatively narrow when compared with that of *P. fasciatus*. The tail of juveniles and young adult females is bright blue. Adults, especially adult males, become tan or brownish in color with broad, dark lateral stripes. In breeding males, the head becomes somewhat swollen and suffused with red.

**ALABAMA DISTRIBUTION** Apparently, this species is found statewide, except for its absence from much of the Tennessee Valley and Highland Rim. It is extremely patchy in its distribution elsewhere in Alabama.

**HABITS** The Southeastern Five-lined Skink is most abundant in dry habitats, such as ridge tops, and well-drained, sandy places. It shuns heavily shaded, mesic ravines and coves, damp stream edges, and

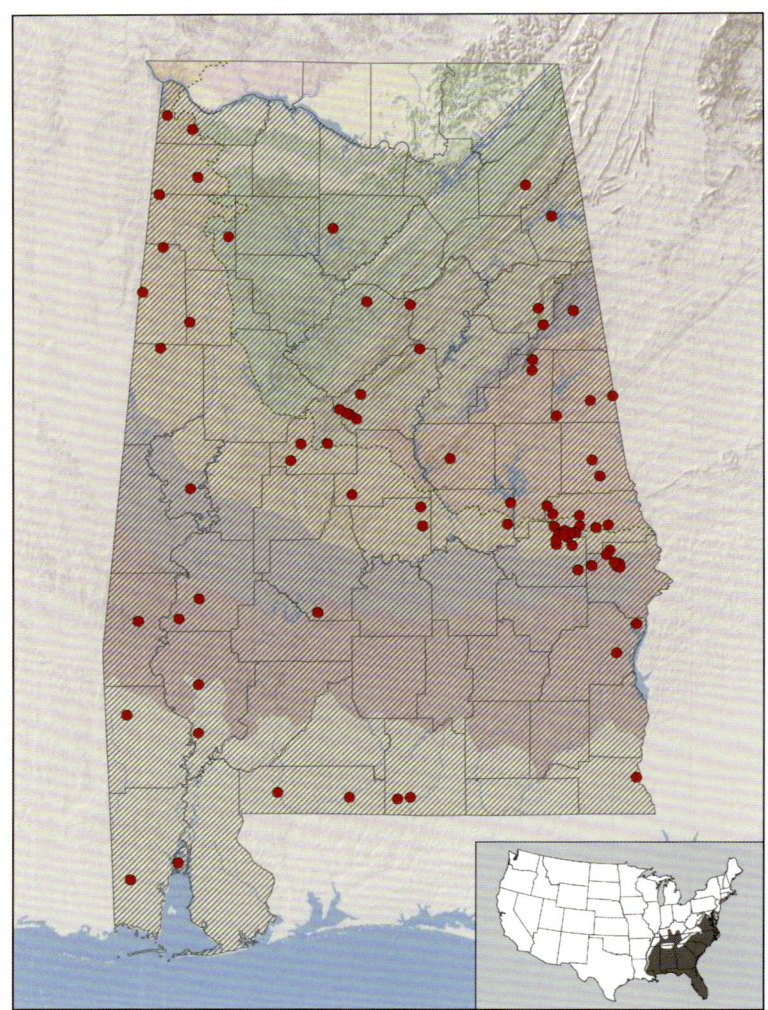

Distribution of Southeastern Five-lined Skink, *Plestiodon inexpectatus.* The presumed range of the species in Alabama is indicated by hatching. Inset map depicts approximate range in the United States.

other such situations where habitat conditions are optimal for *P. fasciatus.* The ranges of ecological tolerance for the two species overlap somewhat, however, and they are frequently found together. Rotting logs and stumps and rock piles are favored microhabitats of *P. inexpectatus,* and in such places, the female lays eggs in a protected cavity. Data on nesting of this species document six to eleven eggs, which are attended by the female. The Southeastern Five-lined Skink feeds on a variety of arthropods and possibly on other forms of invertebrate life as well. It is almost wholly terrestrial, and in climbing ability is somewhat intermediate between *P. fasciatus* and *P. laticeps*.

**CONSERVATION AND MANAGEMENT** Southeastern Five-lined Skinks were formerly common in the state but are now uncommon. Because of this decline, the species is ranked Priority 2 (High Conservation Concern) by ADCNR (Mirarchi, Bailey, Haggerty, and Best 2004). For this reason, it is unlawful to possess this species in the state of Alabama without a special collecting permit. However, this species can do well in urban settings in xeric areas. Management activities that retain snags and fallen logs, especially in open grassy areas of forests, are appropriate for this species. Frequent fire is likely to enhance habitat for Southeastern Five-lined Skinks.

**TAXONOMY** This species is most closely related to the ancestor of *P. fasciatus* + *P. septentrionalis* (Schmitz et al. 2004) and appears to have evolved in, and then expanded out of, Florida (Richmond 2006). No subspecies are recognized within this species, and no geographically contiguous genetic lineages have been identified (Richmond 2006). Previous authors have placed this species in the genus *Eumeces*.

## Broad-headed Skink

*Plestiodon laticeps* (Schneider, 1801)

**DESCRIPTION** These skinks are the largest members of the family in North America, attaining a maximum snout-vent length of about 130 mm. The head is large, especially so in adult males. The chin has two postmental scales, eight upper labials, and at most one enlarged post-labial. The scales are smooth, shiny, overlapping, and cycloid, and the median subcaudals are noticeably wider than scales in adjacent rows (as in *P. fasciatus* but not *P. inexpectatus*). The body of juveniles (except for venter) is black or blue-black with five yellowish to orange stripes extending the length of the body. The mid-dorsal light stripe bifurcates anteriorly at the back of the head and extends onto the snout. This stripe is relatively narrow when compared with that of *P. fasciatus*. There is a pair of dorsolateral stripes, one on each side, that begin over the eyes. Finally, a pair of lateral stripes is present, beginning under each eye. The tail of juveniles and young adult females is bright blue. The juvenile pattern persists in adult females, except that the light stripes fade with age, losing sharp contrast with the ground color and becoming obliterated in the case of the mid-dorsal stripe in some individuals. Adult females have a broad, dark lateral stripe. Adult males become virtually uniform brown or tan except for the head, which becomes red and considerably swollen.

**ALABAMA DISTRIBUTION** This species is found throughout the state.

Adult female Broad-headed Skink, *Plestiodon laticeps*, with eggs, Lowndes County, AL

**HABITS** In Alabama, the Broad-headed Skink frequents pine forests that have hollow trees or large rotting stumps or logs, living in the cavities that these provide. It climbs with much greater facility than Five-lined or Southeastern Five-lined Skinks and is elusive and often difficult to capture. When seized, adults bite viciously; a large male will occasionally break the skin slightly The bite is non-venomous, despite persistent folklore to the contrary.

Mating typically occurs in spring (May). During this time males develop bright orange heads and are territorial. The enlarged heads of males are used to bite rival males during territorial disputes, and it is not unusual for adult males to be scarred from previous fights. Males find receptive females by following scent trails deposited by those females. Males with the brightest orange heads tend to have higher levels of circulating testosterone, and this hormone increases the persistence with which males trail females. These features allow males to gain mates through mate guarding (Cooper and Vitt 1997). Once inseminated, females create a nest in a cavity under the bark of a dead tree or in a rotting stump. The eggs are white with thin, calcareous shells and number from six to twenty-two per clutch. The clutch is attended by the female until the juveniles hatch. Food of the Broad-headed Skink includes insects, spiders, snails, amphipods, and an occasional small vertebrate.

**CONSERVATION AND MANAGEMENT** This is a common species throughout Alabama and does well in urban settings. Because it is common, it receives no special protection under state law Management activities that retain snags and fallen logs, especially in open grassy areas of pine forests, are beneficial for this species.

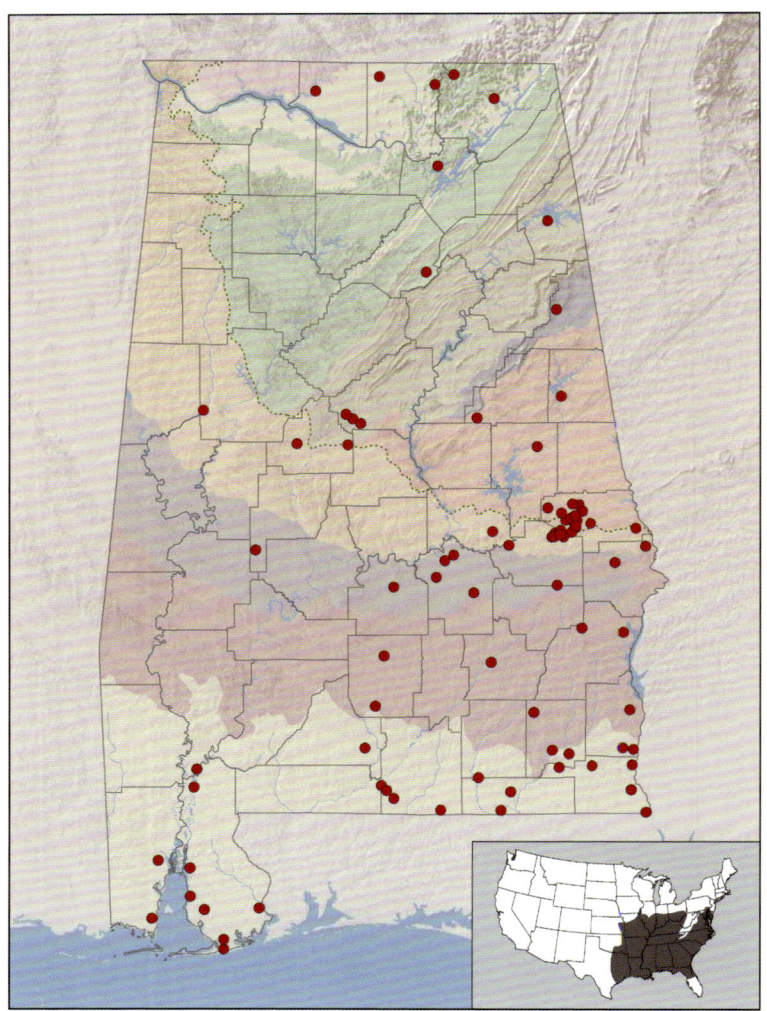

Distribution of Broad-headed Skink, *Plestiodon laticeps*. This species is assumed to occur throughout Alabama. Inset map depicts approximate range in the United States.

**TAXONOMY** This species is the basal lineage of a radiation of Five-lined Skinks that range from the eastern United States to the Rocky Mountains and south through northern Mexico (Schmitz et al. 2004; Richmond 2006). No subspecies currently are recognized, but mitochondrial data document two lineages, one east and one west of the Mississippi River (Richmond 2006). Specimens from Alabama belong to the eastern lineage. Previous authors have placed this species in the genus *Eumeces*.

# Five-lined Skink

*Plestiodon fasciatus* (Linnaeus, 1758)

**DESCRIPTION** This is a medium-sized skink attaining a maximum snout-vent length of about 85 mm. Its scales are smooth, shiny, overlapping, and cycloid. The head of this species possesses two postmental scales, seven upper labials, and two postlabials. A row of median subcaudals are noticeably wider than scales in adjacent rows to either side (similar in this respect to one of its two close relatives in Alabama, *P. laticeps*, but differing from the other, *P. inexpectatus*, in which median subcaudals are equal in width to those in adjacent rows). The body of juveniles and young adult females (except for the venter) is black or blue-black with five conspicuous yellowish stripes. The mid-dorsal stripe bifurcates anteriorly at the back of the head and extends onto the snout; this stripe is somewhat wider than that of *P. inexpectatus* or *P. laticeps*. A dorsolateral stripe occurs along each side, beginning over the eyes and continuing as a lateral stripe. The tails of juveniles are bright blue above. Juvenile colors and pattern become faded with increasing age, especially in adult males, which may become virtually uniform brownish. Adults often have a broad, dark, lateral stripe. The heads of adult males become somewhat swollen and reddish in color.

Juvenile Five-lined Skink, *Plestiodon fasciatus*, Shelby County, AL

**ALABAMA DISTRIBUTION** This species is found throughout the state.

**HABITS** The Five-lined Skink, our most common skink, is most often found in mesic forest habitats where it frequents rotting logs and stumps, rocky places, and trash piles. It is also common along the banks of streams in and around piles of woody debris deposited during periods of high water. Other favored habitats are abandoned farm houses and outbuildings. This species, along with *P. inexpectatus* and *P. laticeps*, is called scorpion throughout much of rural Alabama and erroneously is believed to be venomous. The tail of this species, as in all skinks in Alabama, breaks off easily and, when detached, wiggles vigorously, becoming a source of distraction to a potential predator. When the tail is lost, a new, albeit shorter, one is regenerated.

Movements of *P. fasciatus* average 5.1 m during most of the active season, but adult males increase movements during spring mate-seeking to an average of 17.5 m (Fitch and von Achen 1977). *P. fasciatus* nests in May and June in rotting logs or stumps, under rocks, and in cavities in sawdust piles (Trauth 1994). The female attends the four to fifteen whitish eggs during incubation and assists the young in hatching. The diet of the Five-lined Skink consists almost entirely of arthropods.

**CONSERVATION AND MANAGEMENT** This is a common species throughout Alabama that does well in urban settings. Because it is so common, it receives no special protection under state law. Management activities that retain snags and fallen logs, especially in forested areas, are appropriate for this species. These skinks may play a role in the transmission cycle of Lyme disease by being a host that dilutes the disease and makes it less likely to spread to humans (Giery and Ostfeld 2007).

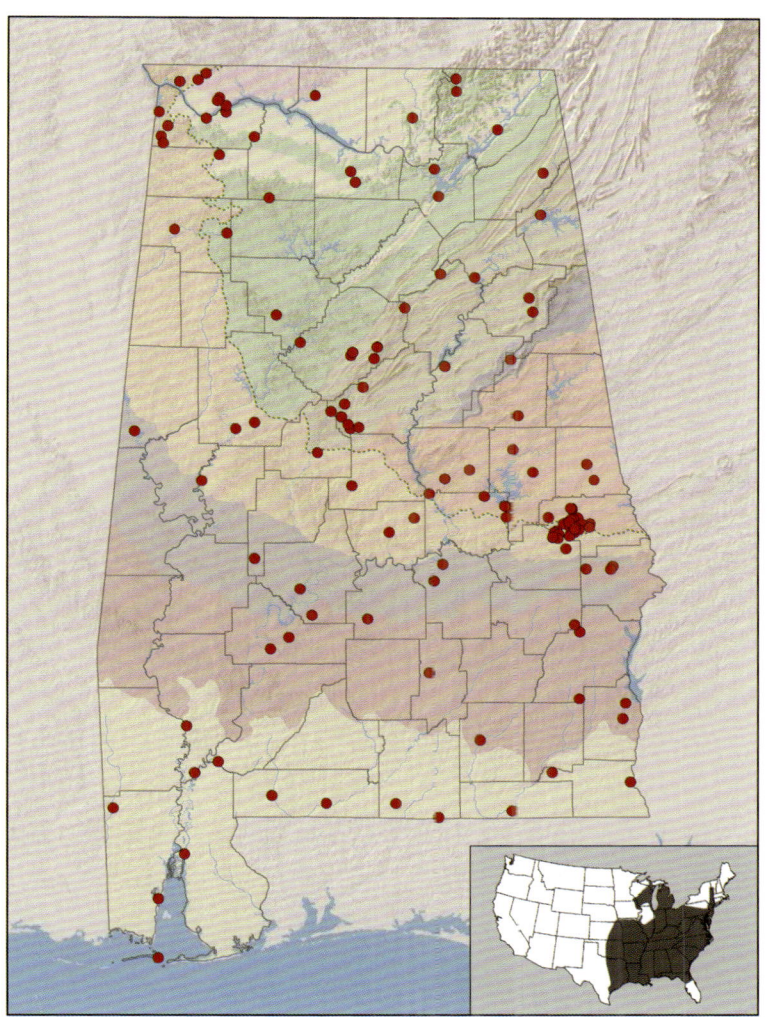

Distribution of Five-lined Skink, *Plestiodon fasciatus*. This species is assumed to occur throughout Alabama. Inset map depicts approximate range in the United States.

**TAXONOMY** This wide-ranging eastern species is the sister taxon to *P. septentrionalis*, the Prairie Skink (Schmitz et al. 2004). No subspecies are recognized but at least two genetic clades comprise the species; these are separated by the Mississippi River (Richmond 2006), and Alabama specimens belong to the eastern clade. Specimens from North Carolina may represent a separate lineage within the eastern clade and up to four lineages may characterize the western clade (Howes et al. 2006). Previous authors have placed this species in the genera *Eumeces* and *Lacerta*.

# Family Teiidae

A New World group with 10 genera and about 130 species, this family contains the racerunners, whiptails, ameivas, and tegus. Some South American forms are quite large and are a source of leather for the boot industry. All are wide-ranging, diurnal, carnivorous forms that taste the environment frequently with their tongues. They are helio-thermic, seeking open areas or sun spots in forests to maintain their body temperature at an elevated level that allows them to move rapidly. Only one genus, *Aspidoscelis*, is native to the United States. This family is most closely related to the family Gymnophthalmidae of the New World tropics (Conrad 2008).

## Whiptails
### Genus *Aspidoscelis* (Wagler, 1830)

We follow Reeder et al. (2002) in using this genus for the racerunners and whiptails of the desert southwest of the United States and extend-ing eastward to the southeast United States and southward to the dry forests of Costa Rica. About forty species are placed in the genus, only one of which, the Eastern Six-lined Racerunner (*Aspidoscelis sexlin-eata sexlineata*), occurs in Alabama. Members of this genus are found in arid areas, where they are known for their great speed as they dash from shrub to shrub. All have a pattern of light stripes on the dorsum separated by dark gray or brown markings that may be dissected into a checkerboard pattern in some species. The genus is famous for the many desert species that are parthenogenetic. These all-female species can be diploid or triploid and are created from the mismating of species that have typical, bisexual reproduction. In the parthenogenetic species, reproduction is clonal, with no need for males because eggs are capable of undergoing chromosome duplication followed by meiosis to yield vi-able diploid or triploid eggs. Each egg produces a genetically identical offspring of the female parent. Although the species found in Alabama reproduces via sexual reproduction, it participated in some of the mis-matings that yielded parthenogenetic species in the desert Southwest.

## Eastern Six-lined Racerunner
*Aspidoscelis sexlineata sexlineata* (Linnaeus, 1766)

**DESCRIPTION** This medium-sized subspecies has a long tail and attains a maximum snout-vent length of around 75 mm. The dorsal scales are tiny, imparting a velvety texture to the skin, while the ventral scales are large, quadrangular, and arranged in eight longitudinal rows. The dorsal ground color is olive, brownish, or grayish with six well-defined yellow longitudinal stripes on the body. Juveniles have a light blue tail, but this color is lost by the second year, when the tail becomes uniformly light brown or gray. In gross color pattern, this subspecies is vaguely reminiscent of the five-lined skinks of the genus *Plestiodon*, but that genus has no more than five light stripes and a shiny appearance (six lines and dull in *Aspidoscelis*).

**ALABAMA DISTRIBUTION** Eastern Six-lined Racerunners are found throughout the state. They are widespread in the Coastal Plain, becoming patchy in distribution in sandy areas along rivers and in regions of sandy soils in the rest of the state.

**HABITS** This familiar squamate, often called sand-streak or sand-runner, inhabits dry, relatively open places, such as fields, open dirt roads, wide banks of rivers or creeks, and barren waste places. It relies chiefly on speed for protection but, if pressed, will usually disappear into the burrow of a small rodent or mole or into one it has dug for itself. Dietary items include a variety of arthropods living in leaf litter.

Racerunners are active and alert, on the move constantly when abroad, and operate most efficiently at high temperatures. It is one

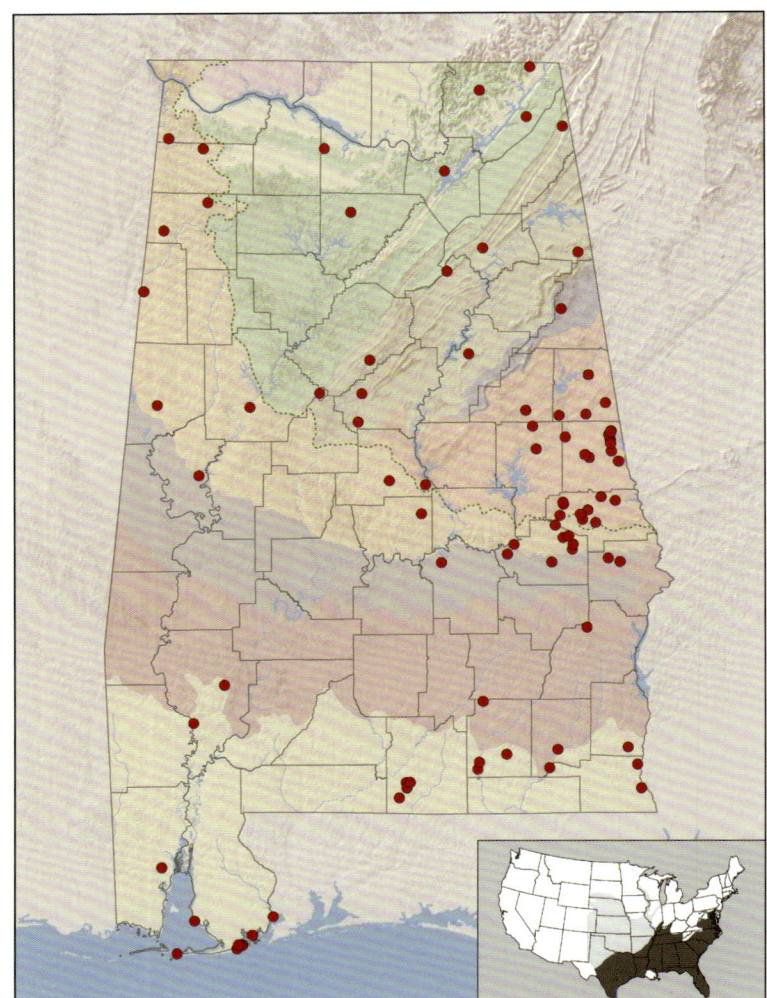

Distribution of Eastern Six-lined Racerunner, *Aspidoscelis sexlineata sexlineata*. This subspecies is assumed to occur throughout Alabama. Inset map depicts approximate range in the United States, with dark shading indicating the range of *A. s. sexlineata* and light shading indicating the range of all other subspecies.

of the few terrestrial reptiles in Alabama that voluntarily exposes itself to full sunlight at midday during hot summer weather. In fact, activity temperatures for this subspecies in the southeastern United States average about 37°C, and the critical thermal maximum is about 48°C (Witz 2001). These temperatures are higher than those for other squamates and suggest that Eastern Six-lined Racerunners operate at temperatures quite close to lethal temperatures. Adults are active from mid-April (males) or mid-May (females) through August, with juveniles emerging as early as mid-March and remaining active into November. They spend the winter period of inactivity in burrows dug within the home range of each individual. These burrows are deeper than the shallow burrows used at night during the active season. Males and females store fats during the season of activity that are used to survive during winter and to begin the reproductive cycle in late winter before emergence in May (Etheridge et al. 1986). Both sexes exhaust this stored energy during reproduction in early spring. During mating the male will bite the female at midbody and then wrap his body around hers in a circle that places his cloacal opening next to that of the female. Referred to as the doughnut position, this reproductive posture is found in all bisexual species of the genus. This act induces the female to ovulate her eggs at the same time that the male deposits fresh sperm into her reproductive tract. Females deposit eggs in a shallow nest cavity dug in friable soil, frequently on red clay slopes with southern or western exposures, near a burrow maintained as a refuge by that female (Trauth 1981). The clutch consists of one to five (typically three) soft-shelled, whitish eggs; two clutches may be deposited per season (Trauth 1981). After covering the eggs, the female takes no further interest in the nest. In droughty years, females frequently produce one fewer egg per clutch, probably because of shortage of food required to produce an egg (Trauth 1982).

**CONSERVATION AND MANAGEMENT** The Eastern Six-lined Racerunner is adapted to open pine savannas, where frequent fire is known to maintain high population densities of these lizards (Mushinsky 1985). Such habitats are found in Alabama, and the subspecies was common, especially in the Coastal Plain. However, these lizards are no longer common even in the appropriate habitat. The timing of this decline is correlated with an increase in the abundance of the Red Imported Fire Ant (*Solenopsis invicta*), an invasive pest species that can chew through egg shells of Eastern Six-lined Racerunners, killing the developing offspring (Mount et al. 1981). Despite this apparently negative interaction, the overall distribution of *A. sexlineata* in the state has not been altered. For these reasons, Eastern Six-lined Racerunners are listed as Priority 4 (Low Conservation Concern) by ADCNR, a 2014 modification of the designation in Mirarchi, Bailey, Haggerty, and Best (2004). Control of fire ants may be required to maintain this subspecies on small patches where ant predation might limit population size. Eastern Six-lined Racerunners can be more abundant in clear-cuts that have been burned, roller chopped, and seeded on mechanically prepared mounds than they are in mature forest (Greenberg et al. 1994). Management activities that thin dense pines stands and retain open habitat with frequent fire are helpful in increasing population density of this subspecies (Steen, McGee, et al. 2010).

**TAXONOMY** Two subspecies of this lizard are recognized, one of which occurs in Alabama. Previous authors have placed this species in the genera *Cnemidophorus* and *Lacerta*.

# Glass Lizards and Their Relatives

## Group Anguimorpha

We follow Wiens et al. (2010) in considering anguimorphans to contain six families comprised of about two hundred species. All members possess deeply forked, projectile tongues and a loose symphysis of the lower jaws, allowing each mandible to move independently. They are found in the New and Old Worlds and include forms with four, two, or no legs. Scientists seeking phylogenetic reconstructions have struggled to find a consistent placement for this radiation within squamates. Those that focus on morphology tend to place snakes within this radiation because the tongue and jaw characters portend the chemosensory system and extreme mobility of the mandibles observed in snakes (Lee 2005b). However, recent reconstructions based on morphology alone place anguimorphans as the sister taxon to Scincoidea + Lacertoidea (Conrad 2008); reconstructions based on molecular data alone place anguimorphans in a quadrichotomy with Iguania, Serpentes, and Lacertoidea + Amphisbaenia (Townsend et al. 2004) or as the sister taxon to iguanians + anguimorphans (Wiens et al. 2010); those that combine molecular and morphological data place anguimorphans as the sister taxon to snakes (Wiens et al. 2010). Regardless of whether snakes are included or excluded, anguimorphan squamates include mosasaurs, a charismatic fossil radiation of aquatic reptiles that were important predators of near-shore marine regions of Alabama. Mosasaur fossils are found on all continents and first appeared 90 million years ago, becoming extinct 65 million years ago, at the same time as the dinosaurs. The great diversity of mosasaurs present in Alabama's fossil deposits suggests that the state played an important and lengthy role in the evolution of anguimorphan squamates.

# Family Anguidae

This family contains limbed and limbless squamates with large, plate-like scales and a fold along each side of the body. Twelve genera and about 120 species belong to this family, with members being found in both hemispheres but being absent from Australia and Africa south of the Sahara. Members of this family are the sister family to Xeno-sauridae (Conrad 2008; Wiens et al. 2010), the knob-scaled lizards of Mexico and Central America. Some anguids lay eggs, while others give birth to live young. A lateral fold of skin separating the dorsal from the ventral halves of each animal is diagnostic of the family and is thought to participate in allowing expansion of the body cavity to allow breathing to occur when an individual is covered by loose soil or thick grassy vegetation. Two genera occur in the United States, one of which is represented in Alabama.

## Glass Lizards
### Genus *Ophisaurus* (Daudin, 1803)

We follow Nguyen et al. (2011) in restricting membership within this genus to the limbless anguids called glass lizards, glass snakes, or jointed snakes of eastern North America and Mexico. The genus is sister to *Anguis* + *Anniella* + *Dopasia* + *Pseudopus*, limbless anguids of northern Africa, Asia, Europe, and western North America. A popular herpetological myth concerns a "snake" that shatters into pieces when struck. Later, it is said, the pieces miraculously reunite, and the "snake" crawls away. There is little doubt that the limbless glass lizards are at the root of this fallacy. Their long tails break readily ("like glass"), and it is likely that a blow across the tail region would cause breakage at several places. Although the animal will ultimately regenerate a new tail, there is, of course, no chance that the detached pieces will come back together. Glass lizards are easily distinguished from snakes in that they possess eyelids and ear openings. Six species are recognized within the genus, three of which are found in Alabama.

## Key to the Species of *Ophisaurus* of Alabama

**1a** Fewer than ninety-six scales along lateral fold (to cloacal opening); supralabials usually in contact with eye.

      *Ophisaurus mimicus*—Mimic Glass Lizard . . . . page 104.

**1b** More than ninety-six scales along lateral fold (to cloacal opening); supralabials usually separated from eye by circumorbitals. Go to **2**.

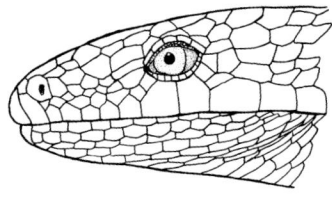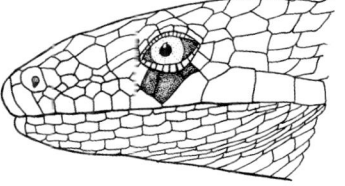

*From left to right:*

Lateral view of head of a Mimic Glass Lizard (*Ophisaurus mimicus*)

Lateral view of head of an Eastern Slender Glass Lizard (*Ophisaurus attenuatus*)

**2a** Anterior area immediately below lateral fold without dark markings that align to create stripes; dorsum having neither cross bars nor a mid-dorsal dark stripe; adults with "salt-and-pepper" markings of turquoise, black, and white.

      *Ophisaurus ventralis*—Eastern Glass Lizard . . . . page 107.

**2b** Anterior area immediately below lateral fold with one or two (most often the latter) longitudinal rows of dark spots, forming stripes; dorsum either with a mid-dorsal dark stripe or with light, black-bordered cross-bars.

      *Ophisaurus attenuatus longicaudus*—Eastern Slender Glass Lizard . . . .
        page 110.

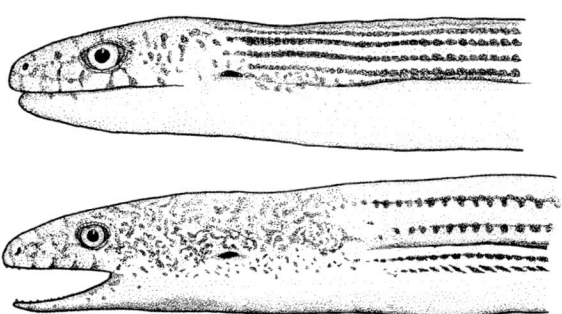

*From left to right:*

Lateral view of anterior body of an Eastern Glass Lizard (*Ophisaurus ventralis*); based on AUM 2213

Lateral view of anterior body of an Eastern Slender Glass Lizard (*Ophisaurus attenuatus*); based on AUM 3037

# Mimic Glass Lizard
*Ophisaurus mimicus* (Palmer, 1987)

DESCRIPTION  Like all members of the genus, Mimic Glass Lizards are legless and have an unusually long tail. The species attains a maximum snout-vent length of about 180 mm and a maximum total length of about 660 mm. The scales are large, rectangular, keeled, and shiny. They are arranged in longitudinal rows and there are fewer than ninety-seven scales along the lateral fold (second scale row above fold), a number that is lower than that for *O. a. longicaudus* or *O. ventralis*. One or two supralabials enter the orbit in most specimens, a feature that also will distinguish this species from the other two species in Alabama. Additionally, the head is noticeably narrow and elongate and becomes yellowish toward the tip of the snout. The dorsum is brown, becoming dark brown along the sides. Anteriorly, there are ten to seventeen heavily dissected light bands (one scale row wide) that cross the dorsum and extend onto the sides, where they create a greenish series of light spots against a dark background. Posteriorly, these lateral light markings become two to three light stripes, bordered by dark brown or black. Dark spots are present below the lateral fold and these align to form a weak dark stripe, as in *O. a. longicaudus*.

ALABAMA DISTRIBUTION  This is the rarest glass lizard in Alabama, being recorded from only three counties and five total localities. Mimic Glass Lizards are found in a narrow zone along the southernmost tier of counties where the Lower Coastal Plain has habitats suitable for pitcher plants.

**HABITS** Mimic Glass Lizards are associated with flatwoods or pitcher plant bogs. Thus, this species prefers wetter habitats than do Alabama's other two species of glass lizards. Many museum specimens have been collected on roads and, as with other members of the genus, many more males than females are present in collections, suggesting that males are more active than females (Palmer and Braswell 1995). Based on records from elsewhere within its geographic range (Palmer and Braswell 1995), this species may be active during any month of the year, likely because it is found only in habitats close enough to the coast to avoid hard freezes. Nothing is known of the diet of this species, but it is presumed to consist of insects and spiders. Similarly, nothing is known of reproduction in this species, but it is presumed to be oviparous, laying eggs under logs or other cover objects with the female remaining to attend the nest, as in other members of the genus.

**CONSERVATION AND MANAGEMENT** The Mimic Glass Lizard was ranked Priority 1 (Highest Conservation Concern) by ADCNR in 2012, making it illegal to possess Mimic Glass Lizards in the state of Alabama without a special permit. This species probably has always been rare throughout its range, but practices that have converted pitcher plant bogs into ponds or drained and planted them in pines undoubtedly have reduced and dissected the ancestral range of this species in Alabama. Two specimens were discovered in a survey of the herpetofauna of the Conecuh National Forests, and both were captured in a funnel trap in Crawford Bog, the largest and best-managed pitcher plant area in the region (Guyer et al. 2007). This is a species of open areas filled with a rich diversity of wetland herbaceous plants. Like other members of the genus, it is unlikely to recognize approaching fire and is likely to perish in such fires (Kaufman et al 2007). However, this

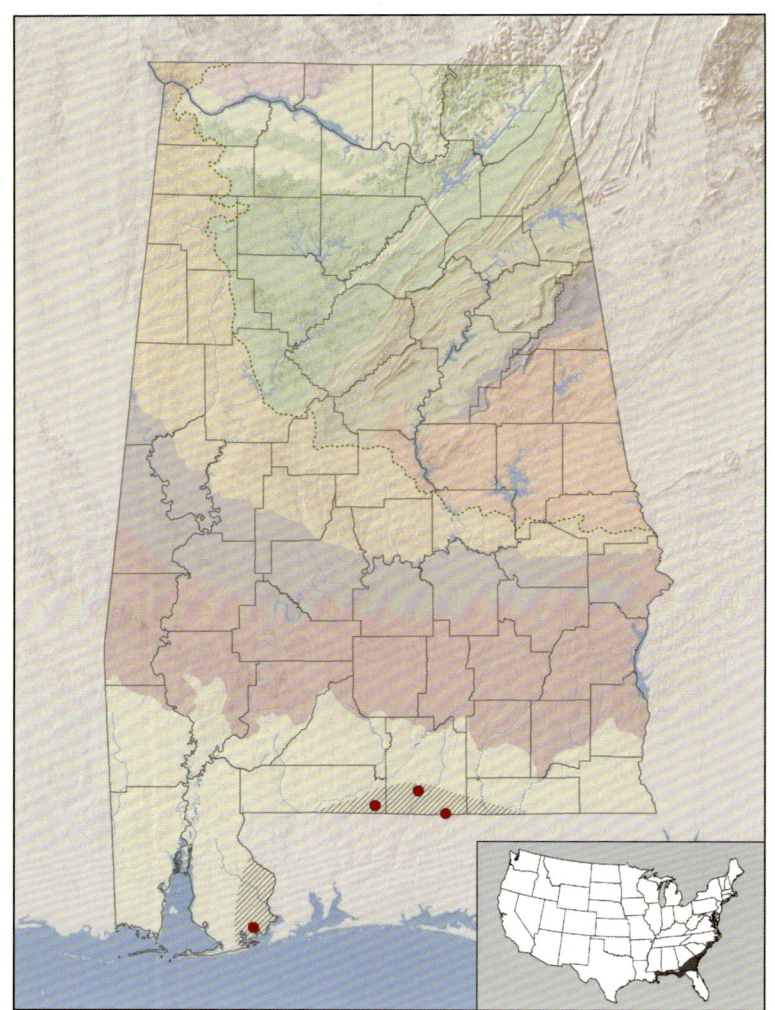

Distribution of Mimic Glass Lizard, *Ophisaurus mimicus*. The presumed range of the species in Alabama is indicated by hatching. Inset map depicts approximate range in the United States.

management tool is likely to enhance the habitat for subsequent generations of this species. So, fire is likely to have an overall beneficial effect on populations of Mimic Glass Lizards (Jensen 2004a).

**Taxonomy** We follow Palmer (1987) in separating this as a distinct species. No subspecies are recognized for this species.

Eastern Glass Lizard, *Ophisaurus ventralis*, Covington County, AL

# Eastern Glass Lizard

*Ophisaurus ventralis* (Linnaeus, 1766)

**DESCRIPTION** Eastern Glass Lizards are elongate, long-tailed, legless squamates that attain a maximum snout-vent length of around 290 mm and a total length of around 1,080 mm. The scales are square to rectangular, hard and glossy, forming straight longitudinal and transverse series. The supralabials are prevented from contacting the orbit by a ring of circumorbitals, a feature that will distinguish this species from *O. mimicus*. A deep groove extends along each side of the body separating dorsal from ventral scales. The dorsum is tan with two to three lateral dark stripes in young, becoming brownish or greenish with age, and lacking a mid-dorsal stripe in all individuals regardless of age. Most individuals in Alabama have several longitudinal rows of small, dark spots that are separated by white markings that are peripherally located on scales. This gives the animal a salt-and-pepper appearance that frequently has a greenish or turquoise cast. The sides above the lateral groove have a longitudinal dark band containing rows of light flecks, but these markings may be obscure or absent on dark individuals. In most individuals, there are no dark spots at the anterior end of the animal below the lateral fold, so there are no dark lateral stripes, as in *O a. longicaudus* and *O. mimicus* In rare individuals that possess dark markings, these make, at best, a single diffuse dark line and only on the anteriormost portion of the body.

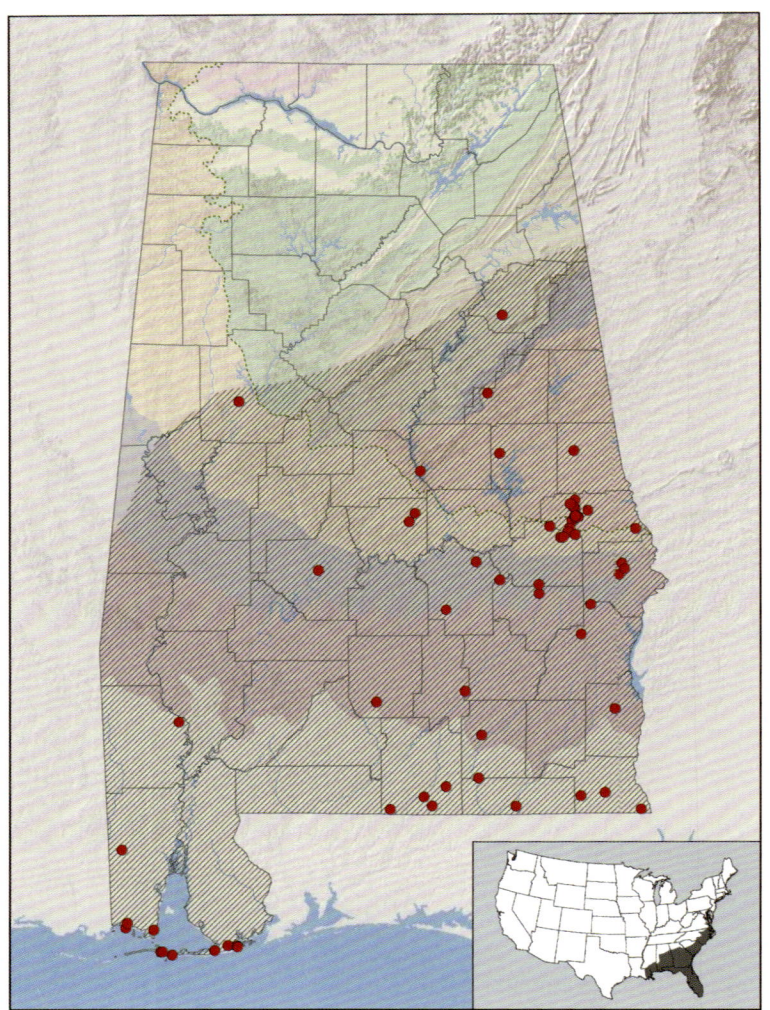

Distribution of Eastern Glass Lizard, *Ophisaurus ventralis*. The presumed range of the species in Alabama is indicated by hatching. Inset map depicts approximate range in the United States.

**ALABAMA DISTRIBUTION** This species is distributed throughout most of the Coastal Plain and has a spotty distribution in the Piedmont, Talladega Upland, and Ridge and Valley regions.

**HABITS** Although infrequently encountered, this is by far the most common glass lizard of the Alabama Coastal Plain, except possibly in the northwestern portion. Eastern Glass Lizards occur in a variety of habitats but seem to prefer damp or mesic situations. They often turn up in overgrown vacant lots and under piles of debris in residential areas. Early morning and late afternoon are favored times of surface activity. The numbers found dead on highways indicate that they move about extensively. The diet consists mostly of insects and spiders; several reports state that other lizards and small snakes are also eaten. Females lay eggs in protected places and attend the clutch until hatching. Observations from Alabama document seven to eight eggs per clutch and female attendance of it (Mount 1975); McConkey (1954) documents up to ten eggs per clutch.

**CONSERVATION AND MANAGEMENT** This species is still encountered throughout the state, and therefore, the species receives no special protection under state law. Eastern Glass Lizards occupy open grassland areas. Like other members of the genus, the species does not appear to recognize approaching fire and is frequently killed by fire (Means and Campbell 1981; Kaufman et al 2007). However, this management tool is likely to enhance the habitat for subsequent generations of the species. So, fire is likely to have an overall beneficial effect on populations of Eastern Glass Lizards.

**TAXONOMY** No subspecies are recognized for this species. Previous authors have placed this species in the genus *Anguis*.

# Eastern Slender Glass Lizard

*Ophisaurus attenuatus longicaudus* (McConkey, 1952)

DESCRIPTION Eastern Slender Glass Lizards are elongate, legless squamates that attain a maximum snout-vent length of around 330 mm and a total length of around 1,065 mm. Thus, the tail is approximately two-thirds of the body length. However, tails are frequently broken, especially in older individuals, and the replacement tail is not as long as the original. The scales are square to rectangular, hard and glossy, forming straight longitudinal and transverse series. A deep groove extends along each side of the body, separating dorsal from ventral scales. A ring of circumorbitals, a feature that distinguishes this subspecies from *O. mimicus*, prevents the supralabials from contacting the orbit. The dorsal ground color is tan to brownish. In juveniles and some adults, there is a median dark stripe extending the length of the body. In adults and some juveniles, irregular light, black-bordered crossbars are present; these become more pronounced with increasing age. White markings on the dorsum, when present, are centrally located on scales. The sides above the lateral groove are dark brown with two light stripes; these sometimes become obliterated in large adults. The white venter usually has two rows of dark spots forming longitudinal stripes immediately below the lateral grooves, a feature that distinguishes this subspecies from *O. ventralis*.

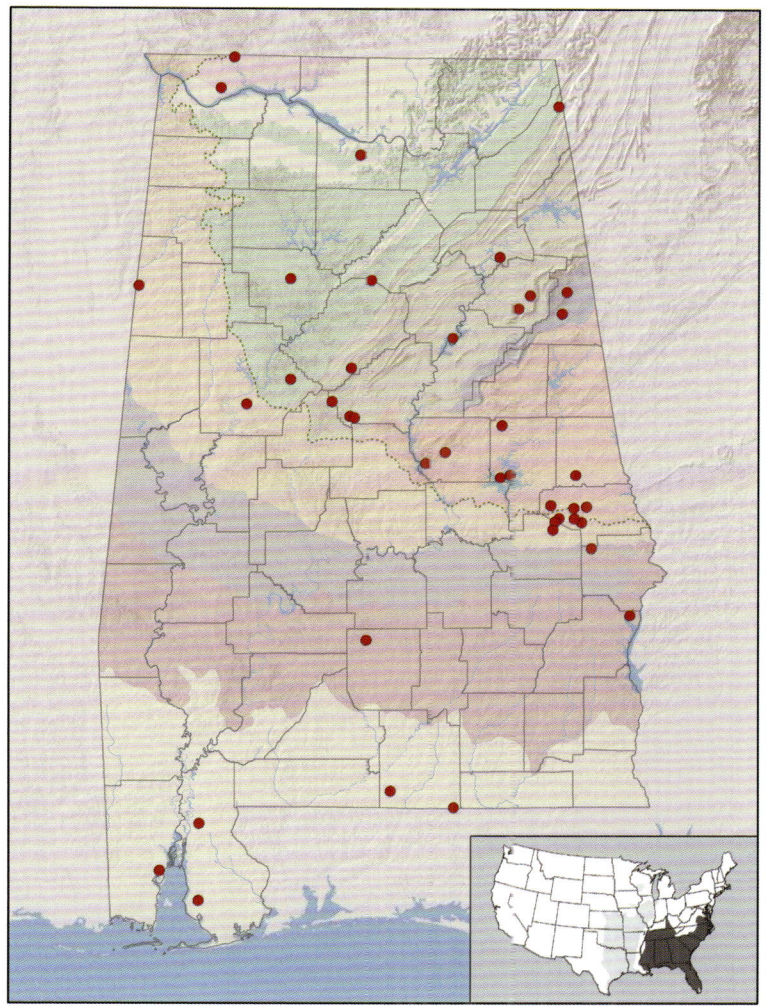

Distribution of Eastern Slender Glass Lizard, *Ophisaurus attenuatus longicaudus*. This subspecies is assumed to occur throughout Alabama. Inset map depicts approximate range in the United States, with dark shading indicating the range of *O. a. longicaudus* and light shading indicating the range of the other subspecies.

**ALABAMA DISTRIBUTION** This subspecies occurs throughout the state but is more common in the upland provinces.

**HABITS** The Eastern Slender Glass Lizard is most commonly encountered in brushy, cut-over woodlands, around abandoned farms, and along stream courses. It is given to burrowing and is often unearthed by plowing. In southern Alabama, where the subspecies is sympatric with the Eastern Glass Lizard, *O. ventralis*, it is not particularly common and is seldom found except in well-drained situations where the soil is loose and friable. The preferred body temperatures of Eastern Slender Glass Lizards are 31–32°C (Johnson and Voight 1978), values that are similar to other diurnal squamates. Mating season is thought to be in April and May (Trauth 1984). During this time, males and females are active, and this is the time of year that most individuals are encountered. Female Eastern Slender Glass Lizards lay eggs and attend the nest until the eggs hatch. Clutch size in *O. a. longicaudus* ranges from six to seventeen (Fitch 1970) and averages twelve (Trauth 1984). Insects, spiders, snails, and small vertebrates are consumed by these lizards.

CONSERVATION AND MANAGEMENT This subspecies has become rare throughout the state and was ranked Priority 2 (High Conservation Concern) by ADCNR in 2012. Like other members of the genus, Eastern Slender Glass Lizards do not appear to recognize approaching fire, and fire frequently kills individuals (Kaufman et al. 2007). However, this management tool is likely to enhance the habitat for subsequent generations of the subspecies. So, fire is likely to have an overall beneficial effect on populations of Eastern Slender Glass Lizards.

TAXONOMY Two subspecies of this lizard are recognized, one of which occurs in Alabama.

# Snakes

## Group Serpentes

Snakes are limbless squamates that have unique ears and eyes, jaw joints that have a mobile pair of squamosal bones, and an extremely loose symphysis of the lower jaws. These features give snakes an unusual ability to detect low-frequency sounds, like footsteps of approaching enemies, and an ability to swallow exceptionally large prey. However, they have relatively poor vision for a reptile, likely because the eyes are derived from a burrowing or aquatic ancestor.

Despite being diagnosed by such diverse characteristics, the evolutionary origin of snakes from within the group Squamata remains a challenge to systematic herpetology. Based on morphology alone, this radiation is either nested within anguimorphans (Lee 2005b) or is the sister taxon to amphisbaenians (Conrad 2008). Molecular data alone place snakes in a basal quadrachotomy with Lacertoidea + Amphisbaenia, Anguimorpha, and Iguania. Molecular and morphological data combined place snakes as the sister taxon to Anguimorpha (Wiens et al. 2010). Thus, additional information will be required to settle questions regarding the phylogenetic origin of snakes. However, current data indicate that this transition involved gradual elongation of the body by addition of trunk and caudal vertebrae. Whatever caused such elongation also caused correlated limb reduction and eventual loss, features that are evident in many independent squamate lineages (Wiens and Slingluff 2001). Traditionally, the process of becoming limbless in squamates is thought to have been associated with burrowing, and in snakes, some think the ancestral species developed unique eyes and ears as adaptations to burrowing (Gans 1975). However, support for this hypothesis is not evident from examination of limb loss across phylogenetic trees of squamates (Wiens and Slingluff 2001). Additionally, the consistent phylogenetic association of the snake radiation with mosasaurs, elongate fossil marine squamates, is consistent with an aquatic origin of snakes, an environment that might also select for the unique conditions of the eyes and ears of snakes (Lee 2005a).

*Paleophis* and *Pterosphenus*, fossil genera of marine aquatic snakes, are known from Alabama from up to 50 million years ago. These sea snakes were not related to any living sea snakes but instead belonged to the boid radiation of the modern snake fauna. This fossil evidence documents a lengthy association of snakes occupying diverse niches in Alabama.

Snakes may be active year-round in warm tropical regions, but in temperate regions, most species overwinter in protected places, usually underground. In Alabama, some species, such as the Eastern Diamondback Rattlesnake, Cottonmouth, and Eastern Gartersnake, will emerge occasionally during warm spells in winter to bask or feed, while others, the Eastern Coachwhip for example, are almost never seen abroad during the winter. Still others, such as Eastern Indigo and Southern Hog-nosed Snakes essentially are active year-round, emerging during sunny days during the winter to engage in courtship and mating.

In snakes, reproductive males locate receptive females chiefly by scent, following chemical trails deposited by females as they traverse the environment or as they emerge from overwintering sites. Males use their tongues to gather these pheromones from the environment and detect them in the vomeronasal organ, a pair of chemosensory depressions on the roof of the mouth that receive the two tines of the forked tongue. This mate-seeking phase of reproduction is followed by a courtship phase in which a single or multiple males crawl on top of and entwine around the female. During this phase, the male rubs his head forward and backward along the female's back while flicking his tongue at a high rate. The tongue flicking of the male may be stimulated by a secretion exuded by the female from between her dorsal scales. Courtship culminates in copulation, which is accomplished by insertion of one of a pair of hemipenes, which is everted from the male's vent and inserted into the cloaca of the female. Sperm storage is widely observed in snakes (Sever and Hamlett 2001), and this allows for multiple paternity of offspring produced by females during each reproductive event (Uller and Olsson 2008). Some species of snakes reproduce by laying eggs, others by giving birth to young. In Alabama, about half of all snake species produce eggs. In most of these species, the female deposits eggs in a nest and then provides no further care. But, maternal attendance of the nest is practiced in a few snakes, including mudsnakes among Alabama species. The other half of snakes in Alabama reproduce via viviparity in which females retain eggs in the uterine tract, with no shell being produced, and transfer of gases and nutrients occurs across a placental membrane. Although female parents may attend a litter of snakes for several days after birth, as has been observed in members of the family Viperidae, no further parental care is observed. So, the folk legend that females of some snakes swallow young to protect them has no basis in fact.

In most snake species, the males, on average, have longer tails than females, and, correspondingly, the average number of subcaudal scutes is greater in males than females. With experience, one can often determine the sex of a snake by noting relative width of the underside of the base of the tail, which is wider in males than in females. Other than this feature, no consistent external feature differentiates males from females.

Snakes are exclusively predaceous and are designed to swallow prey items that are remarkably larger in girth than the head of the predator. This allows snakes to gain energy for annual activity by consuming a few meals each year that are widely spaced in time. Growth continues throughout life, although the rate declines substantially after an individual reaches sexual maturity.

Worldwide, there are about 3,600 species of snakes. The vast majority are non-venomous. Forty-four species are native to Alabama, only six of which are dangerously venomous to humans. Among our harmless species, however, are several that will bite viciously when cornered or handled, including the Black Racer, Eastern Coachwhip, and water snakes of the genus *Nerodia*. Others, such as the Rough Greensnake, Mudsnake, Rainbow Snake, and Hog-nosed Snake, can seldom be induced to bite. Some of these, such as the Hog-nosed Snake, have enlarged fangs at the back of the jaws that are associated with venoms used to immobilize prey. Effects of the venoms on humans are largely unstudied but are known to produce intense pain. For these reasons, bites from any species should be avoided. Additionally, the rear end of most snake species is to be avoided as well because of foul-smelling secretions from the anal glands that may be smeared upon the captor.

## KEY TO THE FAMILIES OF SNAKES OF ALABAMA

**1a** No elongate teeth (fangs) at anterior end of maxilla.

Family **Colubridae—Colubrid Snakes** . . . . **page 117.**

**1b** Pair of elongate teeth at anterior end of maxilla, creating fangs. Go to **2.**

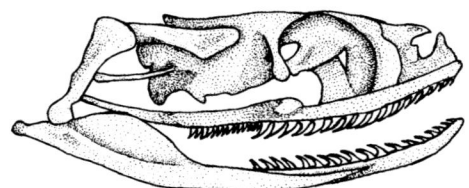

Lateral view of skull of a Watersnake (*Nerodia*)

**2a** Heat-sensitive pit present on side of face between external nares and orbit; maxilla mobile, rotating when mouth closed to direct long fangs posteriorly in mouth cavity.

Family **Viperidae—Pit Vipers** . . . . **page 311.**

**2b** No heat-sensitive pit on each side of face; maxilla not mobile and fangs are short, remaining in a vertical orientation when mouth closed.

Family **Elapidae—Coral Snakes** . . . . **page 350.**

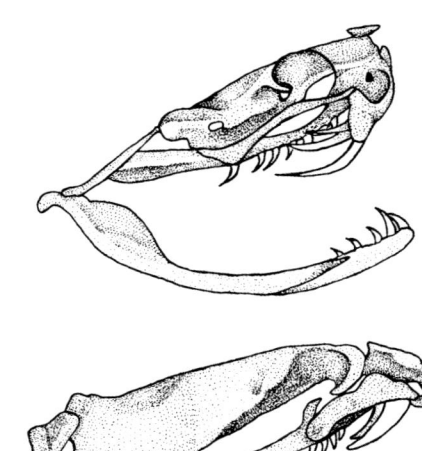

*From top to bottom:*

Lateral view of skull of a Rattlesnake (*Crotalus*)

Lateral view of skull of a Coralsnake (*Micrurus*)

# Family Colubridae

The concept of this family continues to be refined with recent taxonomic works. Mount (1975) included in this family all advanced snakes except those placed in the venomous families Elapidae and Viperidae, as did most authors of that era. However, recent data forced removal of some former lineages of colubrids when the families Viperidae and Elapidae were demonstrated to be nested within the former concept of Colubridae. Zaher et al. (2009) recommended restricting membership in the family Colubridae to those advanced snakes diagnosed by the presence of a particular conformation of the sulcus spermaticus on the hemipenes. However, this characteristic is not visible in most field settings and is lost in some members, making it difficult to implement this restricted concept of Colubridae. Therefore, we take the more conservative approach recommended by Pyron et al. (2011) of including in this family the most recent common ancestor of *Nerodia* and *Coluber* and all descendants of that ancestor. This family contains about 250 genera and about 1,780 species that are divided into 7 subfamilies, 3 of which (Colubrinae, Dipsadinae, and Natricinae) occur in Alabama. Zaher et al. (2009) provide diagnostic characteristics for each of these based on hemipenis morphology, but these characteristics exhibit frequent reversals, leaving no consistent characters that can be used to diagnose these subfamilies in the field. Therefore, to assist in identifying these subfamilies for snakes from Alabama, we provide sets of features that characterize each subfamily (appendix 7).

# Subfamily Colubrinae

We follow Pyron et al. (2011) in restricting this group to the most recent common ancestor of *Ahaetulla* and *Lampropeltis* and all descendants of that ancestor. The subfamily generally is diagnosed by the presence of hemipenes in males in which the sulcus spermaticus is a single structure derived from the right branch of a divided sulcus in the ancestor to derived snakes (Zaher et al. 2009). Members of this subfamily lack any modification of teeth to assist in delivering venom, generally lay eggs, and may constrict prey or swallow them without first killing them. The subfamily contains 119 genera, 7 of which occur in Alabama, and about 120 species, 13 of which occur in Alabama.

**1a** No loreal; dorsum tan with light ring on neck bordered anteriorly and posteriorly by dark brown.

Genus *Tantilla*—**Black-headed, Crowned, and Flat-headed Snakes . . . . page 120.**

**1b** Loreal present; dorsum lacking light ring on neck. Go to **2**.

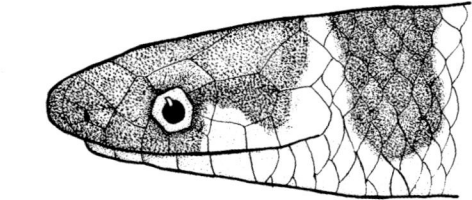

Dorsolateral view of head of a Southeastern Crowned Snake (*Tantilla coronata*)

**2a** Dorsal scale rows greater than or equal to twenty-five. Go to **3**.

**2b** Dorsal scale rows less than twenty-five. Go to **4**.

**3a** Rostral scale enlarged, protruding from between internasals; four prefrontals; ventral scutes not turning upward abruptly near their lateral ends to form an angle; dorsum uniform black or light brown with dark brown blotches; dorsal scales keeled.

Genus *Pituophis*—**Bullsnakes, Pinesnakes, and Gophersnakes . . . . page 123.**

**3b** Rostral scale not enlarged and protruding from between internasals; two prefrontals; ventral scutes turning upward abruptly near their lateral ends to form an angle; dorsal scales weakly keeled.

Genus *Pantherophis*—**North American Ratsnakes . . . . page 135.**

*From left to right:*

Frontal view of head of a Pinesnake (*Pituophis*)

Cross section and ventrolateral view of a Kingsnake (*Lampropeltis*)

Cross section and ventrolateral view of a Ratsnake (*Pantherophis*)

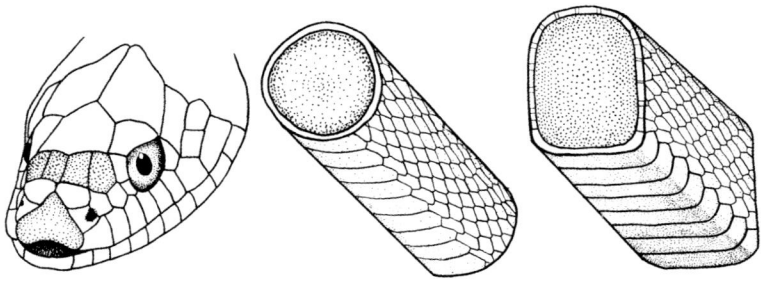

**4a** Dorsal scale rows seventeen. Go to **5**.

**4b** Dorsal scale rows nineteen to twenty-one. Go to **7**.

**5a** Scales keeled; dorsal color uniform bright green (bluish in preservative).

   Genus *Opheodrys*—**Greensnakes** .... page 144.

**5b** Scales smooth; dorsum of adults uniform black or black anteriorly and tan posteriorly or tan with brown blotches in juveniles. Go to **6**.

*From left to right:*

Dorsal view of keeled scales

Dorsal view of smooth scales

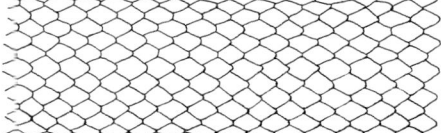

**6a** One preocular; upper labials of juveniles with bold dark band below eye.

   Genus *Drymarchon*—**Indigo Snakes** .... page 147.

**6b** Two preoculars; upper labials of juveniles without a bold dark band below eye.

   Genus *Coluber*—**Racers, Coachwhips and Whipsnakes** .... page 153.

*From left to right:*

Lateral view of head of an Indigo Snake (*Drymarchon*)

Lateral view of head of a Racer (*Coluber*)

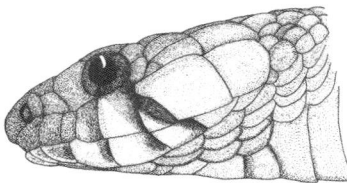
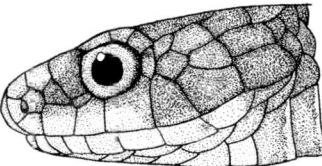

**7a** Rostral scale enlarged and projecting well beyond lower jaw; color pattern of bold red, black, and yellow bands that do not encircle body.

   Genus *Cemophora*—**Scarletsnakes** .... page 163.

**7b** Rostral scale neither enlarged nor projecting well beyond lower jaw; color variable, but if red, black, and yellow markings are present then black rings encircle body.

   Genus *Lampropeltis*—**Kingsnakes.** .... page 166.

*From left to right:*

Color pattern of bands of a Scarletsnake (*Cemophora coccinea*)

Color pattern of rings of a Scarlet Kingsnake (*Lampropeltis elapsoides*)

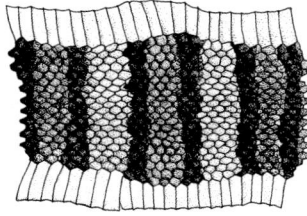
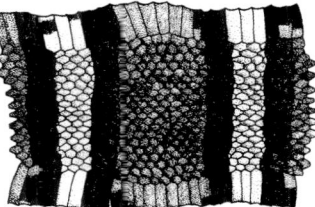

## Black-Headed, Crowned, and Flat-Headed Snakes
### Genus *Tantilla* (Baird and Girard, 1853)

This genus contains small, thin snakes that typically have a light brown body with a dark brown head. Most live under logs or rocks, where they consume arthropods. The sixty-four species in the genus are found from the United States to Argentina. The phylogenetic position of this genus within Colubrinae is unsettled, with some analyses placing it basal to a lineage of small desert and tropical species of snakes that eat invertebrates (*Chilomeniscus, Chionactis, Conopsis, Pseudoficimia, Simpholis, Sonora,* and *Stenorrhina*; Pyron et al. 2011), and others placing it with racers of the genus *Coluber* (Pyron et al. 2013). Thus, the origin of the single species of *Tantilla* found in Alabama is likely uncertain.

Southeastern Crowned Snake, *Tantilla coronata*, Tatnall County, GA

## Southeastern Crowned Snake
### Genus *Tantilla coronata* (Baird and Girard, 1853)

DESCRIPTION This is a small, delicate snake attaining a maximum total length of about 215 mm. The head is slightly or not at all distinct from the neck, and there are six or seven upper labials. The anal scute is divided. Dorsal scalation consists of smooth scales that are arranged in fifteen rows at midbody. In color, the top of the head is dark brown to black followed by a light parietal band that is followed by a black collar that contrasts with the uniform tan to pinkish brown pigmentation of the rest of the body.

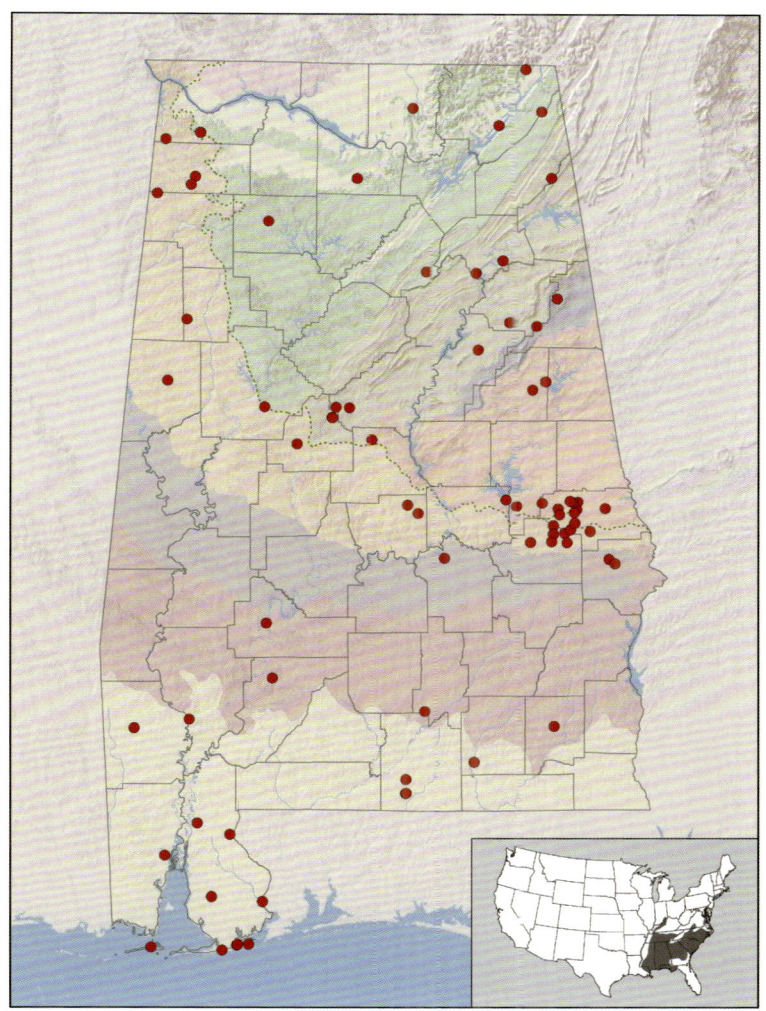

Distribution of Southeastern Crowned Snake, *Tantilla coronata*. This species is assumed to occur throughout Alabama. Inset map depicts approximate range in the United States.

**ALABAMA DISTRIBUTION** This species is found throughout the state but is spotty in its distribution within the state.

**HABITS** Southeastern Crowned Snakes are small, harmless serpents that inhabit dry hillsides, ridge tops, and other xeric habitats. These snakes are fond of hiding under rocks, logs, piles of debris, and in rotting stumps. They frequently are associated with microhabitats occupied by the Rough Earthsnake. Activity in such habitats peaks in June and July (Semlitsch et al. 1981). Males are particularly active from July through October, perhaps because this is when they search for mates. Females produce a single clutch of eggs each year with one to four eggs per clutch. The eggs are cylindrical and are deposited in June through early July (Aldridge and Semlitsch 1992). Southeastern Crowned Snakes are diet specialists, consuming only centipedes (Todd et al. 2008).

CONSERVATION AND MANAGEMENT This species can be locally abundant throughout the state and for that reason receives no special protection by state law. Habitat management that increases or maintains open forests on rocky slopes should be sufficient to retain stable populations of this species. Southeastern Crowned Snakes are more abundant in thinned forests than unthinned ones and are less abundant in clear-cuts in which litter has been removed than clear-cut areas with litter retained (Todd and Andrews 2008).

TAXONOMY This species is considered to have no subspecific variation. Previous authors have placed this species in the genera *Homalocranion* and *Homalocranium*.

# Bullsnakes, Pinesnakes, and Gophersnakes
## Genus *Pituophis* (Holbrook, 1842)

This genus contains the Pinesnakes, Gophersnakes, and Bullsnakes of North America, Mexico, and northern Central America. The members of this genus are large constrictors that live in open savanna and forest areas. They are not venomous but can be aggressive, often flattening the neck laterally, vocalizing, and twitching their tail when approached by humans. The genus is the sister taxon to *Pantherophis*, a genus of similar size and distribution. Six species are placed in *Pituophis*, and Alabama has one of them.

## Eastern Pinesnake
*Pituophis melanoleucus* (Daudin, 1803)

TAXONOMY  The Pinesnake, Bullsnake, and Gophersnake complex was once thought to consist of one polytypic species (Mount 1975). However, distinct mitochondrial lineages within this complex suggest that it consists of at least three species, *P. melanoleucus* for eastern forms referred to as pinesnakes, *P. catenifer* for western forms referred to as Gophersnakes or Bullsnakes, and *P. ruthveni* for the distinctive Louisiana Pinesnake (Rodríguez-Robles and de Jesús-Escobar 2000). Three subspecies are recognized within *P. melanoleucus*, all of which occur in Alabama. Two of the three, *P. m. melanoleucus* and *P. m. lodingi* are monophyletic, whereas the third, *P. m. mugitus* is a basal paraphyletic assemblage of individuals leading to the other two taxa (Rodríguez-Robles and de Jesús-Robles 2000). Previous authors have placed this species in the genera *Churchilla* and *Coluber*.

## Key to the Subspecies of *Pituophis melanoleucus* of Alabama

**1a** Dorsal color dark brown to black; dorsal blotches, if present, obscure; venter uniform black or dark brown.

   *Pituophis melanoleucus lodingi*—Black Pinesnake . . . . page 125.

**1b** Dorsal color not dark brown or black; dorsal blotches evident, at least posteriorly; venter uniform white or white with irregular brown spots. Go to 2.

Color pattern of a Black Pinesnake (*Pituophis melanoleucus lodingi*); based on AUM 10156

**2a** Dorsal blotches not contrasting sharply with ground color; venter light, but not glistening white, frequently with numerous irregular brown spots.

   *Pituophis melanoleucus mugitus*—Florida Pinesnake . . . . page 128.

**2b** Dorsal blotches contrasting sharply with light ground color, at least posteriorly; venter glistening white with a row of distinct black spots down each side.

   *Pituophis melanoleucus melanoleucus*—Northern Pinesnake . . . .
   page 132.

*From left to right:*

Color pattern of a Northern Pinesnake (*Pituophis melanoleucus melanoleucus*)

Color patterns of a Florida Pinesnake (*Pituophis melanoleucus mugitus*)

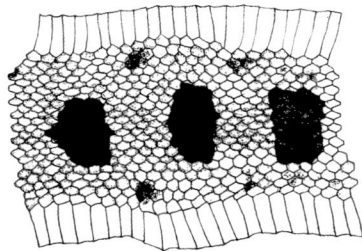 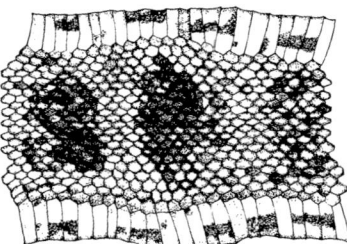

## Black Pinesnake
*Pituophis melanoleucus lodingi* (Blanchard, 1924)

DESCRIPTION  Black Pinesnakes are large in size, attaining a maximum total length of about 1,875 mm, with a fairly short tail. Most are notably bulky in shape. The head is small and is only slightly wider than the neck. The anal scute is undivided. The dorsal scales are keeled except for some along the lowermost row. The rostral scale is enlarged and rounded, curving backward between the internasals and ending in a point. Additionally, there are four prefrontals. In dorsal coloration, adults typically are nearly uniform dark brown to black. A trace of pattern is visible on some individuals, with blotches becoming evident at and near the tail. A few white scales may be present on some individuals. The venter is dark brown to black, occasionally with a few light markings, mainly near the tail. Young are similar in color to adults except that they are lighter, frequently showing faint blotches, and have more light scales. Subcaudals average fifty-two in males and thirty in females.

ALABAMA DISTRIBUTION  This subspecies is found in southwestern Alabama in the Lower Coastal Plain and Red Hills regions west of the Alabama River. Specimens are known only from Baldwin, Mobile, Clarke, and Washington Counties, but it is probable that the range includes southern Choctaw County as well. Individuals captured at the Solon Dixon Forestry Education Center on the border of Escambia and Covington Counties range in color pattern from near-perfect *P. m. lodingi*

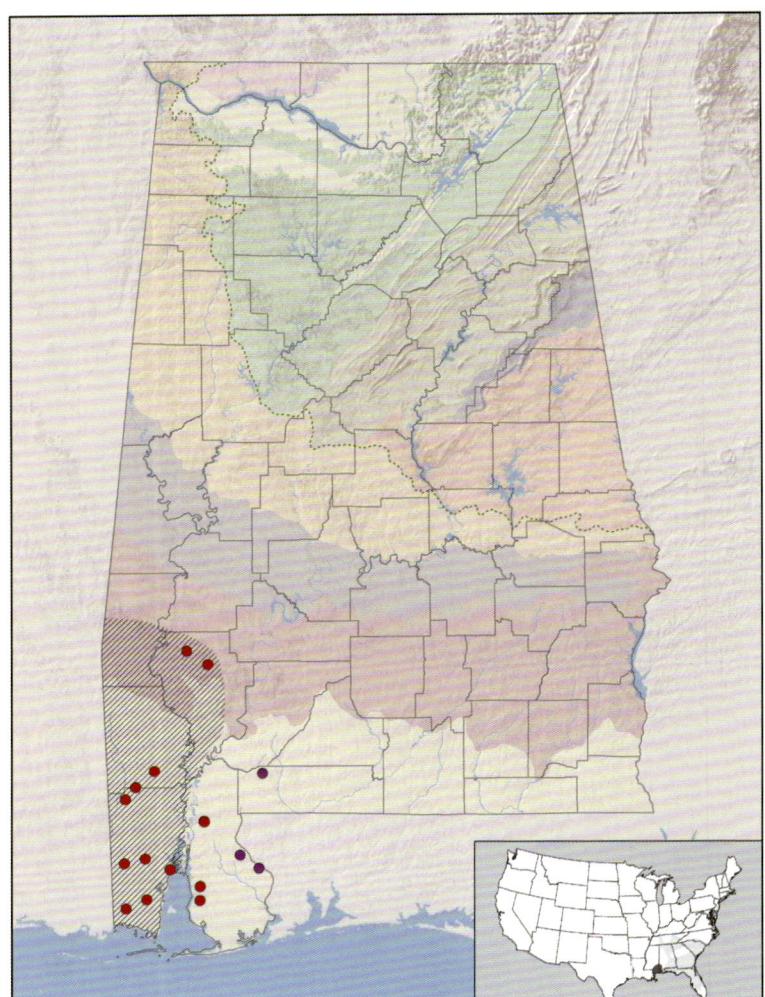

Distribution of Black Pinesnake, *Pituophis melanoleucus lodingi*. The presumed range of the subspecies in Alabama is indicated by hatching. Solid purple dots indicate intergrades between *P. m. lodingi* and *P. m. mugitus*. Inset map depicts approximate range in the United States, with dark shading indicating the range of *P. m. lodingi* and light shading indicating the range of all other subspecies.

to near-perfect *P. m. mugitus*, suggesting that the influence of Black Pinesnakes extends at least that far eastward. Specimens from the remainder of Covington County, while being extremely variable in color, are all referable to *P. m. mugitus*. The Black Belt in Alabama apparently is uninhabited by Pinesnakes. If so, the ranges of *P. m. lodingi* and *P. m. melanoleucus* are widely separated, and it is doubtful that these forms intergrade.

**HABITS** Black Pinesnakes occupy open pine forests in uplands that have reduced areas of roads, agricultural fields, and urban developments

(Baxley et al. 2011). Most individuals occur in longleaf pine–turkey oak sandhill associations or dry pine flatwoods. The subspecies is known to overwinter in stump holes (Rudolph et al. 2007), and these features are used extensively during the active season. In fact, 65 percent of observations of radio-tracked Black Pinesnakes were underground (Duran 1998). The home range of these snakes can be as large as 395 ha and home ranges are placed in areas with unusually high densities of Cotton Rats (*Sigmodon hispidus*) (Baxley and Qualls 2009). Remarkably, Gopher Tortoise burrows are not used by this subspecies, despite their availability in the habitat. Mating is thought to occur in April, and eggs are thought to be laid in late May to mid-July (Mount 1986). Food is presumed to be primarily rodents but includes birds and bird eggs.

CONSERVATION AND MANAGEMENT The Black Pinesnake was listed as threatened under the ESA (2015) and is ranked Priority 1 (Highest Conservation Concern) by ADCNR (Mirarchi, Bailey, Haggerty, and Best 2004). Therefore, it is unlawful to possess the subspecies in the state of Alabama without a scientific collecting permit. This action is associated with the observation that these snakes are rare and becoming rarer (Mount 1986). This subspecies is likely to have been affected by decades of short-term rotation forestry practices for the pulp wood industry, reduction in fire frequency, stump removal on logged longleaf tracts, and urban sprawl within Mobile County, where Black Pinesnake sightings have historically been most frequent. To maintain this subspecies in the state, extensive areas of fire-maintained longleaf pine savanna will be required (Baxley et al. 2011). Nelson and Bailey (2004) recommend that tracts of at least 1,000 acres (400 ha) in relatively roadless areas are needed. Mature pines are required to provide fine fuels to carry fire, as are warm-season grasses. Depending on the size of remaining populations, captive rearing of this subspecies for repatriation on managed lands may be appropriate. Additionally, wildlife underpasses along key roadways may be necessary to reduce road mortality (Nelson and Bailey 2004). The Frank Boyd Wildlife Management Area and the Fred T. Stimpson Wildlife Sanctuary are key public or leased properties for conservation of this subspecies within the state. But public education, especially for owners of large forested tracts in Washington County, and creation of incentives for maintenance of such tracts with frequent fire will be crucial for retaining Black Pinesnakes in Alabama (Nelson and Bailey 2004).

# Florida Pinesnake
*Pituophis melanoleucus mugitus* (Barbour, 1921)

**DESCRIPTION** The Florida Pinesnake is the largest member of the genus in Alabama, attaining a maximum total length of about 2,285 mm with a fairly short tail. Most are bulky in shape with a head that is small and only slightly wider than the neck. The anal scute is undivided. The dorsal scales are keeled except for some along the lowermost row. The rostral scale is enlarged and rounded, curving backward between the internasals and ending in a point. Additionally, there are four prefrontals. The dorsal ground color of Florida Pinesnakes is gray anteriorly to rusty brown posteriorly, and the dorsal blotches are virtually non-existent anteriorly, becoming more distinct, but still faded, posteriorly (compared to the distinct blotching in *P. m. melanoleucus*). The venter is white, but not glistening white, as in *P. m. melanoleucus*, and there are brownish spots appearing at irregular intervals on the lateral edges of the ventral scutes (regular black spots in *P. m. melanoleucus*). Subcaudals average fifty-two in males and thirty in females.

**ALABAMA DISTRIBUTION** This subspecies has a patchy distribution in the southeastern quadrant of the state. It is present in the Lower Coastal Plain but appears to be absent from the Black Belt, a feature that restricts this subspecies along the southern boundary of the state. However, the Black Belt disintegrates near the eastern border of the state, and Florida Pinesnakes occur in counties along the

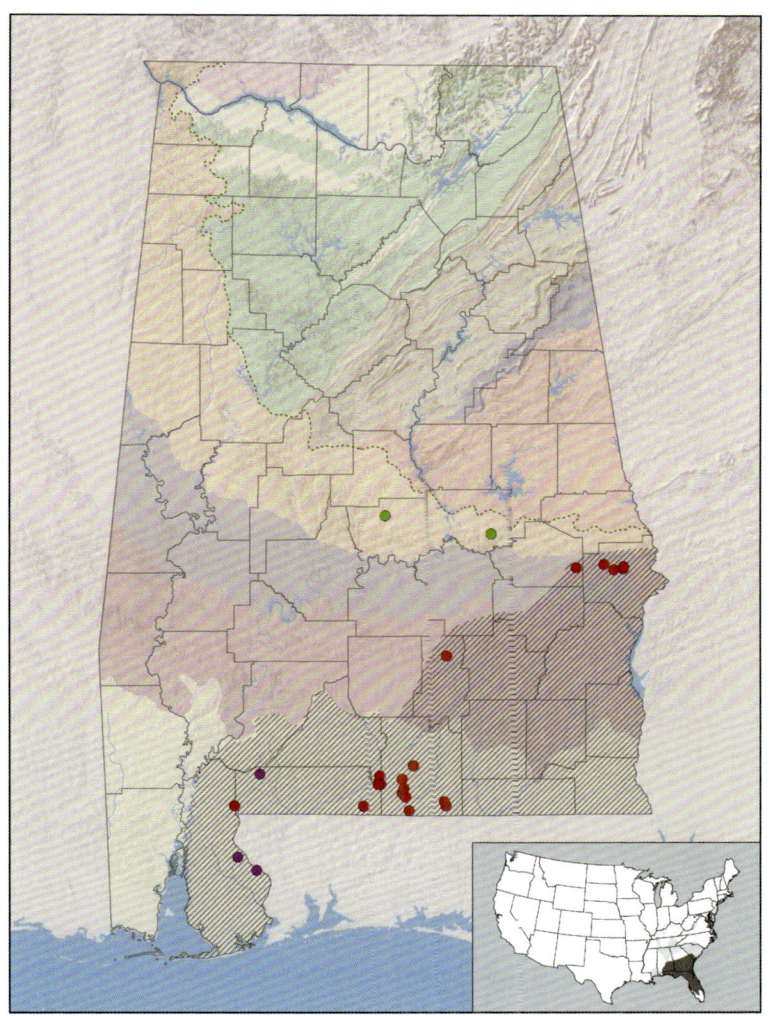

Distribution of Florida Pinesnake, *Pituophis melanoleucus mugitus*. The presumed range of the subspecies in Alabama is indicated by hatching. Solid purple dots indicate intergrades between *P. m. lodingi* and *P. m. mugitus*. Solid green dots indicate intergrades between *P. m. melanoleucus* and *P. m. mugitus*. Inset map depicts approximate range in the United States, with dark shading indicating the range of *P. m. mugitus* and light shading indicating the range of all other subspecies.

Chattahoochee River as far north as Russell County. Intergrades between *P. m. mugitus* and *P. m. lodingi* are found in Escambia and Baldwin Counties, and intergradient specimens between *P. m. mugitus* and *P. m. melanoleucus* are known from Autauga and Elmore Counties.

**HABITS** Most of the specimens from below the Fall Line have come from longleaf pine–turkey oak sandhill associations or from dry pine flatwoods areas. In fact, 85 percent of observations of radio-tracked snakes in Florida were in longleaf pine–turkey oak forests or old pastures (Franz 1984). Miller (2008) documented an association of these snakes with mixed pine-hardwood forests at a landscape scale but no other habitat associations at finer scales. Florida Pinesnakes are often found in association with the Southeastern Pocket Gopher (*Geomys pinetis*), a fossorial mammal upon which it preys and in whose burrows it takes shelter. These snakes are most active in May through July, have home ranges that average about 60 ha for both sexes, and are below ground in Pocket Gopher or Gopher Tortoise burrows during 75–85 percent of observations (Franz 1984; Miller 2008). However, males are more active than females during spring activity (Miller 2008). Clutches of four to eight eggs are produced, and these are laid in underground chambers in July and August, frequently within Gopher Tortoise burrows. The eggs are cream to white and are oblong with a leathery shell. They hatch during September and October.

The primary dietary items of these snakes are small mammals. However, they also consume ground-nesting birds and bird eggs.

**CONSERVATION AND MANAGEMENT** In 2015, the Florida Pinesnake was petitioned for protection as threatened under the ESA and the subspecies is ranked Priority 2 (High Conservation Concern) by ADCNR (Mirarchi, Bailey, Haggerty, and Best 2004). Therefore, it is unlawful to possess the subspecies in the state of Alabama without a scientific collecting permit. This action is associated with the observation that these snakes are rare and becoming rarer (Means 2004b). State statutes outlawing gassing of Gopher Tortoise burrows to collect rattlesnakes, if enforced, should decrease mortality of Florida Pinesnakes. This subspecies is also likely to be adversely affected by decades of short-term rotation forestry practices for the pulpwood industry, reduction in fire frequency, stump removal on logged longleaf tracts, and the decline of the Southeastern Pocket Gopher. To maintain Florida Pinesnakes in the state, extensive areas of fire-maintained longleaf pine savanna will be required. The Conecuh National Forest, Perdido River Longleaf Hills Tract, Geneva State Forest, Solon Dixon Forestry Education Center, and Fort Rucker Military Reservation appear to provide adequate opportunities to maintain viable populations of this subspecies on public lands. Mature pines will be needed to provide fine fuels to carry fire, as will warm-season grasses. Implementation of frequent (one- to three-year rotation), low-intensity, growing season fires across large tracts (1,000 acres) will be required to maintain appropriate habitat for Florida Pinesnakes (Means 2004b). Depending on the size of the remaining population, captive rearing of this subspecies for repatriation on managed lands may be appropriate. Repatriation of Southeastern Pocket Gophers to areas that now lack this species may also be needed to restore a key food resource and shelter for Florida Pinesnakes.

## Northern Pinesnake

*Pituophis melanoleucus melanoleucus* (Daudin, 1803)

**DESCRIPTION** These snakes are large in size, attaining a maximum to- tal length of about 2,110 mm with a fairly short tail. Most are notably bulky in shape. The head is small and is only slightly wider than the neck. The anal scute is undivided. The dorsal scales are keeled except for some along the lowermost row. The rostral scale is enlarged and rounded, curving backward between the internasals and ending in a point. Additionally, there are four prefrontals. In dorsal coloration, a gray, cream, or yellowish ground color offsets the twenty-five to thirty-one dark blotches or saddles (excluding tail markings) extend- ing down the mid-dorsum. The blotches are somewhat poorly defined and blackish anteriorly but become sharply defined and dark reddish brown toward the tail (they are more faded in *P. m. mugitus*). Along each side are a series of dark blotches that are also more distinct pos- teriorly. The venter of this subspecies is glistening white (dull white in *P. m. mugitus*) with a row of distinct, black spots down each side, each spot involving the lateral portions of one to three ventral scutes anteri- orly and three to five ventral scutes posteriorly (irregular brown spots in *P. m. mugitus*). The tail is marked with reddish saddles or rings. Subcaudals average fifty-two in males and thirty in females.

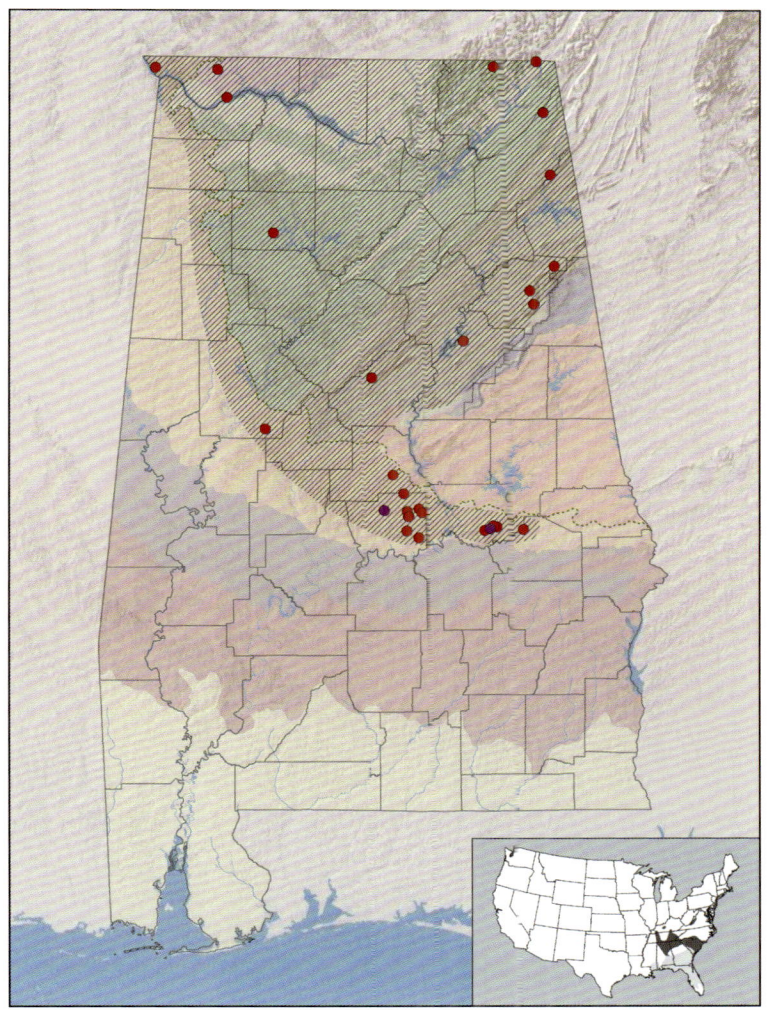

Distribution of Northern Pinesnake, *Pituophis melanoleucus melanoleucus*. The presumed range of the subspecies in Alabama is indicated by hatching. Solid purple dots indicate intergrades between *P. m. melanoleucus* and *P. m. mugitus*. Inset map depicts approximate range in the United States, with dark shading indicating the range of *P. m. melanoleucus* and light shading indicating the range of all other subspecies.

**ALABAMA DISTRIBUTION** This subspecies has a spotty distribution in the mountains, ridges, and plateaus above the Fall Line, exclusive of the Piedmont, and in sandy areas of the Fall Line Hills region eastward to the Coosa River. The subspecies intergrades with *P. m. mugitus* in scattered areas of suitable habitat within the Fall Line Hills of the Coosa River.

**HABITS** Ecological requirements of the Northern Pinesnake are poorly understood in Alabama. Nowhere, apparently, are these serpents commonly observed, perhaps because of their fossorial tendencies and propensity to hide when approached by humans. Northern Pinesnakes are encountered most frequently in open areas of sandy soils, such as the vicinity of Ider, DeKalb County, where several individuals have been collected. This area likely was dominated by *Pinus virginiana* forests and/or oak savannas, forest types that may be crucial for this subspecies.

Northern Pinesnake males move over longer distances than females in spring, probably in search of mates. Movements of both sexes become reduced during autumn. Home range size averages about 60 ha with a core area of about 8 ha, and snakes avoid late successional forests while seeking early successional fields (Gerald et al. 2006). Pinesnakes are oviparous, and in *P. m. melanoleucus*, clutches of seven to twenty-four eggs have been reported. The eggs are large, ranging in length from 50 to 64 mm. Nests have been found in cavities or burrows in sandy soil several inches below the surface, usually in open areas. Pinesnakes are known to feed on small mammals, birds, and bird eggs. Young individuals may eat lizards. Temperament varies considerably from snake to snake. Many will hiss loudly and strike at a molester while twitching the tail but will usually keep the mouth closed. Others are even-tempered and docile.

CONSERVATION AND MANAGEMENT The Northern Pinesnake is ranked Priority 2 (High Conservation Concern) by ADCNR (Mirarchi, Bailey, Haggerty, and Best 2004), and therefore, it is unlawful to possess this subspecies in the state of Alabama without a scientific collecting permit. Maintenance of open pine forests and grasslands in areas of deep, loose soils are keys to retaining this subspecies in the state. In current fragmented landscapes, individuals may be forced to make long movements through unfavorable patches to find widely spaced areas of appropriate open habitat (Gerald et al. 2006). This difficulty likely increases mortality from predators and road kill. Little River Canyon Nature Preserve, the Shoal Creek and Oakmulgee Districts of the Talladega National Forest, Bankhead National Forest, and the Mountain Longleaf National Wildlife Refuge are public lands where conservation efforts should be focused (Godwin 2004b).

# North American Ratsnakes
## Genus *Pantherophis* (Linnaeus, 1766)

We follow Utiger et al. (2002) in separating the Old World genus *Elaphe* from the New World genus *Pantherophis*. This New World genus contains the Ratsnakes, Cornsnakes, and Fox Snakes, a lineage of large constrictors of North America, Mexico, and Central America. This lineage is the sister radiation to the genus *Pituophis* (Pyron et al. 2013), a second radiation of large constrictors of North America. *Pantherophis* contains nine species, two of which are found in Alabama, with a third (*Pantherophis vulpina*) being described from Pleistocene deposits in Colbert County (Holman 1995).

### Key to the Species of *Pantherophis* of Alabama

**1a** Dorsal blotches red or orange; neck bands uniting on head in a spear-point.

    *Pantherophis guttatus*—**Red Cornsnake** . . . . **page 136.**

**1b** Dorsal blotches, if present, dark gray to brownish or nearly black; no neck bands uniting on head to form a spear-point.

    *Pantherophis obsoletus*—**Gray Ratsnake** . . . . **page 139.**

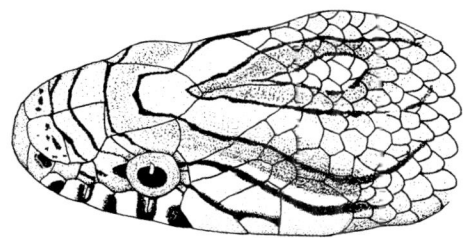

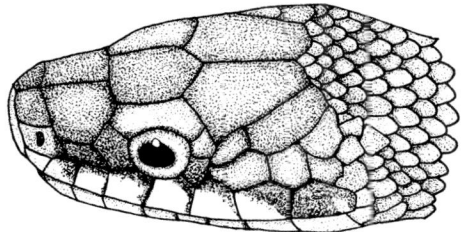

*From top to bottom:*

Dorsolateral view of head of a Red Cornsnake (*Pantherophis guttatus*); based on AUM 16257

Dorsolateral view of head of a Gray Ratsnake (*Pantherophis obsoletus*); based on AUM 5070

Red Cornsnake,
*Pantherophis
guttatus*, Marshall
County, AL

# Red Cornsnake
*Pantherophis guttatus* (Linnaeus, 1766)

DESCRIPTION This is a medium-sized to rather large snake attaining a maximum length of about 1,830 mm. The head is distinct from the neck, and the anal scute is divided. The dorsal scales are smooth to weakly keeled and usually are arranged in twenty-seven rows at mid-body. The ventrals turn upward sharply near the lateral ends. This is a handsome snake, patterned dorsally with black-bordered red or dark orange median blotches on a reddish-orange or brownish-orange ground color. The mid-dorsal blotches alternate with a series of smaller lateral blotches that tend to become elongate and run together longitudinally. A third series of blotches involves the ends of the ventrals and the adjacent two to three dorsal scale rows. Atop the head, two convergent stripes form a spear-point shape that is directed anteriorly. The venter is white, conspicuously marked with quadrangular black blotches that form a distinct checkerboard pattern. The under surface of the tail has two rows of black spots that tend either to coalesce, forming convergent longitudinal stripes, or to expand to involve the entire surface. Geographic variation in color is pronounced in Alabama Red Cornsnakes. Individuals from the Lower Coastal Plain are marked with vivid orange and red markings on the juveniles, persisting in undiminished intensity into adulthood. Those from the northernmost provinces are darker and less vividly marked. Variation in the intervening area appears to be clinal.

Red Cornsnake,
*Pantherophis guttatus*, Santa Rosa
County, FL

**ALABAMA DISTRIBUTION**  This species is found throughout the state.

**HABITS**  This snake occurs in greatest abundance around abandoned farms and other places where small rodents are likely to thrive. It is also found in pine and hardwood forests and rocky hillsides. Mice are the chief food of adults; juveniles feed on lizards and small frogs. Red Cornsnakes are rather strongly nocturnal and are most often seen crossing roads at night. Data on reproduction are scarce, and most reports deal with captives. Mating occurs in spring months. Females then deposit ten to thirty elongate eggs in rotting stumps or mats of vegetation.

**CONSERVATION AND MANAGEMENT**  Perhaps because of its association with small farms, where rodents were plentiful, this species formerly was abundant statewide. However, it has declined in abundance in close association with both the reduction of small-scale farming and the invasion of Red Imported Fire Ants into Alabama. Its close relative, *P. obsoletus*, remains abundant, perhaps because it is more arboreal, inhabiting trees where fire ant predation on snake eggs would be minimized. Red Cornsnakes continue to be detected in field surveys across the state but never in abundance. From this, we infer that the species is not in danger of extirpation, although they may have vanished from some areas without our notice. Studies of the key factors regulating abundance are needed to understand how best to manage

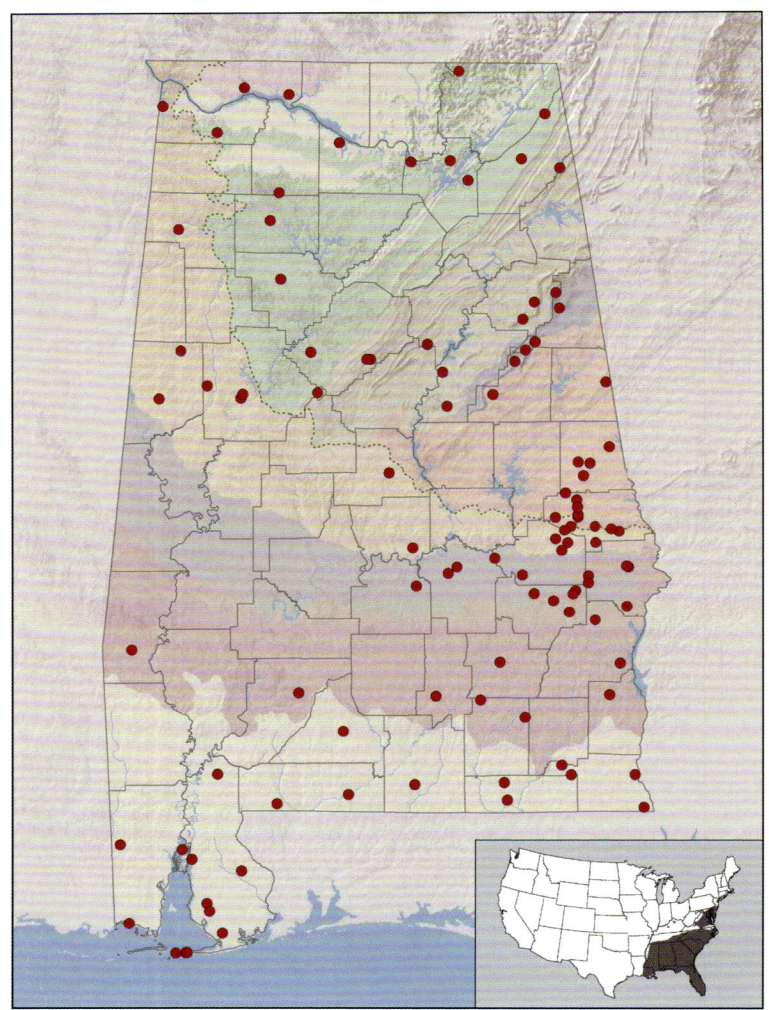

Distribution of
Red Cornsnake,
*Pantherophis gut-
tatus.* This species
is assumed to
occur throughout
Alabama. Inset map
depicts approxi-
mate range in the
United States.

remaining populations. In addition to fire ants, emerging pathogens, such as *Cryptosporidium,* may reduce populations (Kimbell et al. 1999).

**TAXONOMY** No subspecies are recognized. Previous authors have placed this species in the genera *Callopeltis, Coluber, Coryphodon. Elaphe, Ela-phis, Georgia, Pituophis,* and *Scotophis.*

Gray Ratsnake, *Pantherophis obsoletus*, Limestone County, AL

## Gray Ratsnake

*Pantherophis obsoletus* (Duméril, Bibron, and Duméril, 1854)

**DESCRIPTION** Gray Ratsnakes are large, moderately stout snakes attaining a maximum length of about 2,565 mm with a relatively short tail. The head is fairly distinct from the neck, and the anal scute is divided. The dorsal scales are keeled, except for the lateral-most four or five rows, and usually are arranged in twenty-five rows at midbody. The ventral scutes turn sharply upward near their lateral ends. The dorsal ground color of juveniles is gray to brown or yellowish brown with a series of median dark gray or dark brown blotches. The anterior-most blotches usually have longitudinal extensions from their corners that may unite with extensions from neighboring blotches. Along the sides are a series of dark blotches that alternate with the mid-dorsal blotches; anteriorly, these lateral blotches tend to join together to form a longitudinal stripe on each side. Finally, a third series of lateral blotches is present, involving the lowermost one to three scale rows on each side, and continue on to the lateral ends of the ventrals. In adults this pattern becomes faded with age and, in the northeast corner of the state (Calhoun, Cherokee, Cleburne, DeKalb, Jackson, and Randolph Counties), darken to a near-uniform black or dark brown color. The venter of individuals for most of the state is white with diffuse-edged quadrangular dark gray

Gray Ratsnake, *Pantherophis obsoletus*, Bibb County, AL

to brownish blotches and dark stippling, the latter becoming more intense posteriorly. In specimens from the northeast corner of the state, the venter becomes uniform dark. The undersurfaces of the tail have two convergent rows of dark spots that coalesce longitudinally to form stripes or expand to involve the entire surface.

Variation in pigmentation among Alabama populations is marked. In southeastern Alabama, most individuals are much lighter in color than are those from elsewhere in the state. The ground color is silvery gray, and in most individuals, the blotches are only slightly darker than the ground color. Northward, the ground color tends to become darker and less uniform with some individuals showing a distinctly brownish cast. In the western half of the state, at least one-half of all individuals are decidedly brownish in overall appearance. In the westernmost tier of counties below Marion County, brownish orange contributes to the ground color of some and the blotches are often somewhat obscure. Subcaudals average eighty-five in males and seventy-eight in females but with substantial overlap between the sexes.

**ALABAMA DISTRIBUTION** This species is found throughout the state.

**HABITS** *P. obsoletus* is familiar to nearly all of our rural residents. In most parts of the state, it is called chicken snake because of its tendency to raid chicken coops. The species is named white oak runner or oak snake by many people in southeastern Alabama, because of its

Gray Ratsnake, *Pantherophis obsoletus*, Baldwin County, AL

light gray appearance, and black ratsnake in northern Alabama, because of its dark gray appearance. It occurs in most kinds of terrestrial habitats but attains greatest densities in areas where forests and farmland are intermixed and small rodents and nesting birds are relatively abundant. The Gray Ratsnake is one of a few large snakes that frequently inhabit well-established residential areas in and around the edges of cities and towns in Alabama. It is an accomplished climber, and its presence in a tree is often revealed by a flock of screaming blue jays or scolding wrens, chickadees, and tufted titmice.

The whitish, oblong eggs are laid in a variety of places, including arboreal cavities of hardwood trees. These nest locations may reduce predation on nests by fire ants, a feature of reproductive ecology not shared by the far-less-abundant and ground-nesting *P. guttatus*. Clutch size is usually between twelve and eighteen and, because females can store sperm, may be sired by several males (Blouin-Demers et al. 2005). Nest temperatures need to be warm, and limited availability of such nest cavities may cause communal nesting in northern populations (Blouin-Demers et al. 2004). The food of the young consists mainly of lizards and small frogs. Large individuals prefer warm-blooded animals and eggs. Predation success is increased in areas that mimic the open forest edges preferred by this species, and aspects of both sit-and-wait and wide foraging are exhibited (Mullin et al. 1998). Despite this snake's occasional depredations on chicken houses, Gray Ratsnakes

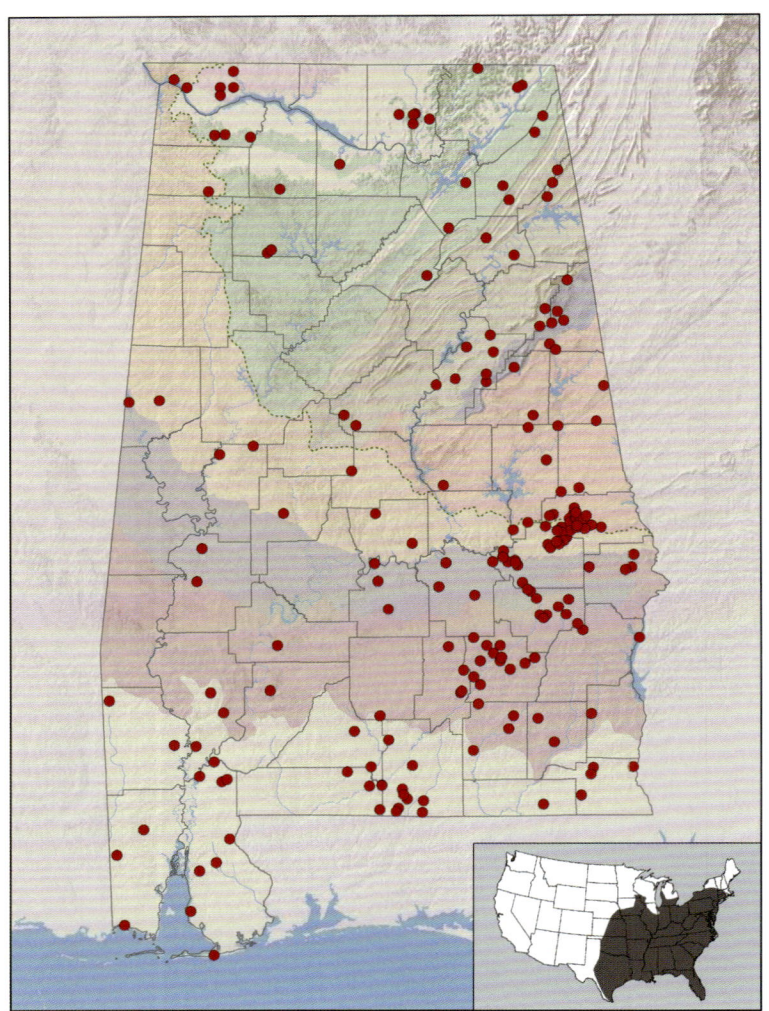

Distribution of Gray Ratsnake, *Pantherophis obsoletus*. This species is assumed to occur throughout Alabama. Inset map depicts approximate range in the United States.

usually are considered beneficial because of their value in consuming rats and mice. Gray Ratsnakes vary considerably in temperament. Many individuals, when threatened, defend themselves vigorously, striking repeatedly at the offender. Others allow themselves to be picked up by the body and handled without becoming overly alarmed.

**CONSERVATION AND MANAGEMENT** The Gray Ratsnake is common and maintains seemingly viable populations in areas altered heavily by humans. Therefore, habitat loss and fragmentation are not major problems for this species. For these reasons, the species receives no special protection by state law, and no altered management is needed to protect it. Of particular note is the observation that this arboreal species has not suffered the population decline of *P. guttatus*, its more terrestrial sister species.

TAXONOMY The taxonomic status of Ratsnakes in Alabama is confusing. Mount (1975) recognized eight subspecies throughout the range, one of which is now recognized as a distinct species, *P. bairdi* of Texas. Burbrink et al. (2000) assessed the remaining seven subspecies and found the differences between them consisted of pattern classes of individuals, which did not conform to the signal of four distinct lineages diagnosed by the mitochondrial genome. Because of this, Burbrink (2001) recommended rejection of the traditional subspecies in favor of elevating the four mitochondrial lineages to species status. Color pattern and scale counts of these taxa differ in modal categories but broadly overlap. Nevertheless, according to Burbrink's (2001) classification, all Alabama specimens belong to a single species, *P. spiloides*, that is separated from an eastern sister species, *P. alleghaniensis*, at the Chattahoochee River. However, these two species show evidence of interbreeding in the northern parts of the geographic range (northeastern United States), and there is no known loss of fitness associated with this hybridization (Gibbs et al. 2006). Because of this evidence of gene flow, we retain the traditional taxonomy of considering these to be a single species, *P. obsoleta*, but accept Burbrink's (2001) analyses that the taxon we consider to be *P. obsoleta* lacks subspecific variation in Alabama. Previous authors have placed these snakes in the genera *Coluber, Elaphe, Georgia*, and *Scotophis*.

## Greensnakes
### Genus *Opheodrys* (Fitzinger, 1843)

This genus contains two species of exceptionally thin, bright green snakes that are diurnal, arboreal predators of insects and their larvae. These snakes are found from northern Mexico to southern Canada and are related to the genus *Oxybelis*, a Neotropical radiation of vine snakes (Pyron et al. 2013). Greensnakes are represented in Alabama by the Northern Rough Greensnake (*Opheodrys aestivus aestivus*).

Northern Rough Greensnake, *Opheodrys aestivus aestivus*, Limestone County, AL

## Northern Rough Greensnake
*Opheodrys aestivus aestivus* (Linnaeus, 1766)

**DESCRIPTION** This extremely slender snake attains a maximum total length of around 1,160 mm, with the tail being exceptionally long. The head is slender and pointed with a distinctly thin neck. The dorsal scales are keeled and arranged in seventeen rows at midbody. The anal scute is divided. In dorsal coloration, these snakes are uniform light green, becoming bluish in preservative. The belly is uniform yellow or white shading to yellowish. No other snake species in Alabama is as boldly green as this subspecies.

**ALABAMA DISTRIBUTION** Northern Rough Greensnakes are found throughout the state and are fairly common in every region.

**HABITS** This docile snake is among the most familiar of all our species. It is unmistakable, and its life is frequently spared even by those who are inclined to kill most snakes on sight. The Northern Rough

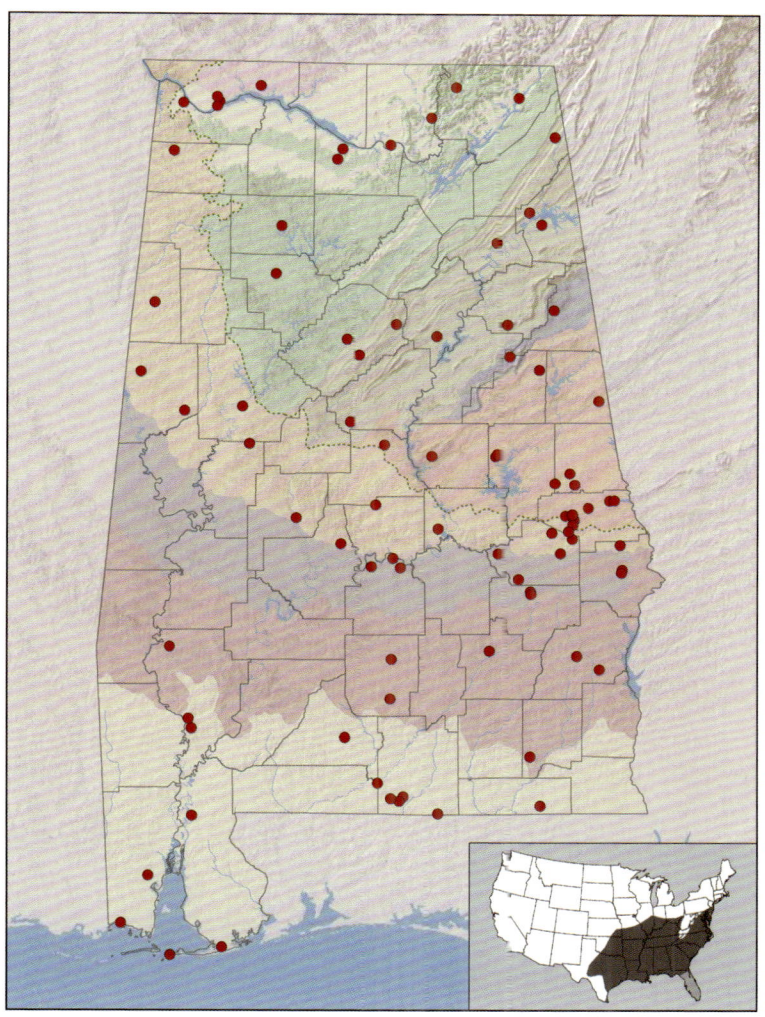

Distribution of Northern Rough Greensnake, *Opheodrys aestivus aestivus*. This subspecies is assumed to occur throughout Alabama. Inset map depicts approximate range in the United States, with dark shading indicating the range of *O. a. aestivus* and light shading indicating the range of all other subspecies.

Greensnake is often found among shrubbery and overhanging vegetation around lakes and streams, where it is active by day (Plummer 1981). Although an agile climber, it moves rather slowly and relies chiefly on concealment for protection. It feeds mainly on soft-bodied arthropods, particularly spiders, caterpillars, grasshoppers, crickets, and odonates (Plummer 1981). These snakes tend not to bask, instead remaining concealed in vegetation (Plummer 1993). Their movements are restricted to areas with a diameter of about 70 m, shifting only about 50 m from year to year (Plummer 1997). The Northern Rough

Greensnake lays two to twelve elongate eggs in holes in living trees or under rocks and other surface objects (Mount 1975; Plummer 1990), but these sites may be so limited that females nest communally (Palmer 1978) or reuse nest sites in successive years (Plummer 1990). Males are more commonly encountered in early spring, perhaps associated with mate seeking, and females are more commonly encountered in late spring, perhaps associated with nest seeking (Plummer 1985). Northern Rough Greensnakes are noted for their docile nature and do not bite.

CONSERVATION AND MANAGEMENT This subspecies is fairly common throughout the state and likely has had its habitat enhanced by creation of farm ponds with forested edges. Management activities that retain forested edges of creeks and ponds and that restrict use of insecticides should enhance the habitat for Northern Rough Greensnakes. However, population density in this subspecies is sensitive to drought (Plummer 1997), and rapid death may result for individuals infected with the protozoan parasite *Cryptosporidium* (Brower and Cranfield 2001). So, this subspecies may be sensitive to some aspects of global climate change and the introduction of novel pathogens.

TAXONOMY Two subspecies are recognized within this species, one of which is found in Alabama. Previous authors have placed this species in the genera *Coluber, Cyclophis, Herpetodryas,* and *Leptophis.*

# Indigo Snakes
## Genus *Drymarchon* (Fitzinger, 1843)

This genus contains large, active, diurnal predators referred to as indigo snakes or cribos. They have extremely powerful jaws, which are used to subdue their prey by crushing the brain case. The four species in the genus are found from northern South America through Middle America to the southeastern United States. Indigo snakes share an evolutionary history with the tropical genera *Chironius* and *Spilotes*, the widespread New World genus *Trimorphodon*, and the southwestern desert genus *Phyllorhynchus* (Pyron et al. 2013). One species is present in Alabama.

Eastern Indigo Snake, *Drymarchon couperi*, Coffee County, GA

## Eastern Indigo Snake
*Drymarchon couperi* (Holbrook, 1842)

DESCRIPTION Eastern Indigo Snakes are large, fairly stout snakes attaining a maximum total length of 2,630 mm. Their head is slightly, if at all, distinct from the neck, and the anal scute is undivided. Scales on the dorsum are large, smooth, and shiny and arranged in seventeen rows at midbody. In color, indigo snakes are a uniform lustrous blue-black, except for some reddish- or cream-colored suffusion on the chin, throat, and cheeks. The redness differs significantly among clutches, suggesting that this component of color is under genetic control (Deitloff et al. 2013). Hatchlings are uniform dark gray with a series of thin (one scale wide) light bands that cross the body along its length. The color becomes uniform black within a month.

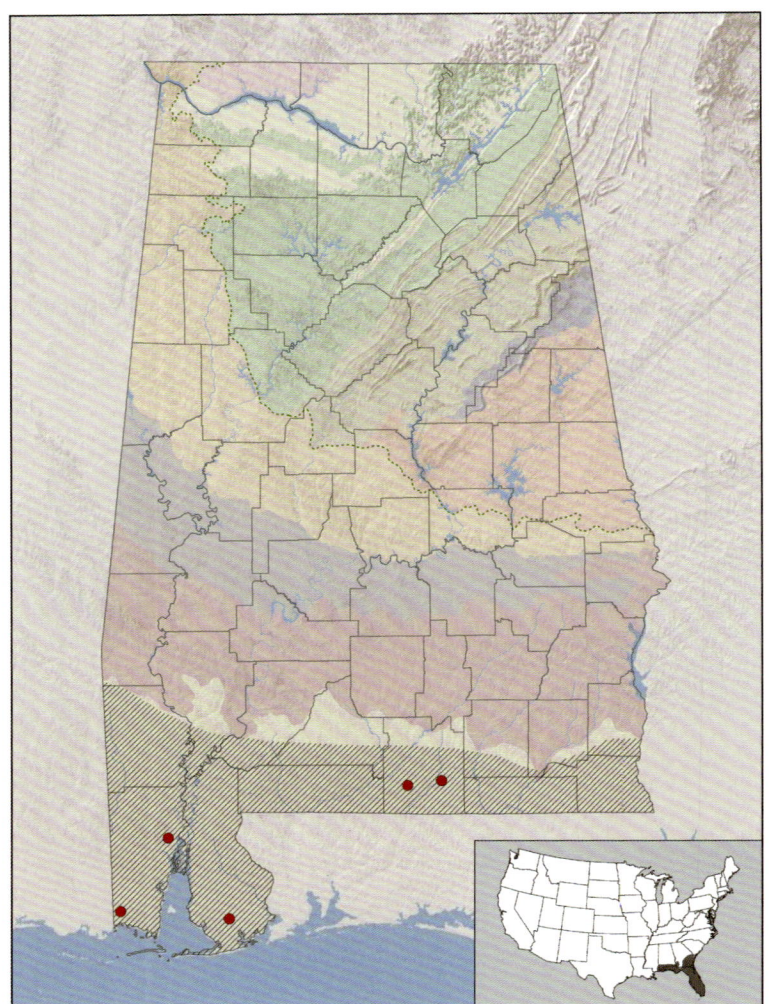

Distribution of Eastern Indigo Snake, *Drymarchon couperi*. The presumed range of the species in Alabama is indicated by hatching. Inset map depicts approximate range in the United States.

**ALABAMA DISTRIBUTION** This species is documented from Alabama by literature records of individuals from Baldwin, Covington, and Mobile Counties (Mount 1975). In the late 1970s and early 1980s, captive bred snakes were released in Autauga, Baldwin, Covington, Escambia, Mobile, and Washington Counties (Hart 2002). Since the end of that repatriation study in 1986, putative sightings of this species have been recorded for Coffee, Covington, Escambia, Mobile, and Washington Counties (Hart 2002), but no known populations persist in these counties.

**HABITS** Eastern Indigo Snakes are large, conspicuous, and relatively slow inhabitants of longleaf pine ecosystems and associated drainages. The species is unusual in its ability to remain active during winter months and in its extensive movements during the season of activity. In northern Florida and southern Georgia, indigo snakes are found locally in a few desolate areas where Gopher Tortoise burrows occur in close proximity to stream or swamp-edge habitat. In these areas, the snakes of both sexes use Gopher Tortoise burrows during winter. Throughout the active and dormant seasons, females tend to use abandoned burrows more than do males (Hyslop, Cooper, and Meyers 2009). Adult snakes tend to use upland habitat during spring and summer, whereas juveniles prefer wetland habitats (C. Smith 1987). Females tend to use open habitat more than males do (Hyslop, Cooper, and Meyers 2009). Adult males have huge home ranges that are on the order of 100 to 200 ha (Dodd and Barichivich 2007; Hyslop 2007). The extensive movement patterns associated with maintaining such large home ranges apparently increases mortality of the largest snakes relative to shorter ones (Hyslop, Meyers, et al. 2009).

Eastern Indigo Snakes feed on a wide variety of vertebrate animal life. Snakes, including venomous ones, frogs, small mammals, birds, and even young turtles are all known dietary items. However, 85 percent of dietary items are frogs, snakes, Gopher Tortoises, and rodents. Snake prey are killed by biting the head and crushing the skull between the powerful jaws of the indigo snake, an event that immobilizes the prey. Some snakes have been photographed dragging the prey item by the head before consuming it. Mating of Eastern Indigo Snakes, at latitudes comparable to those of south Alabama, is thought to occur from October through February (Speake 1986), when males wander widely in search of Gopher Tortoise burrows occupied by female snakes. Oviposition typically occurs in May and June but extends from early March to late July (Speake et al. 1987). Gopher Tortoise burrows are thought to be the primary nesting sites of females (Diemer and Speake 1983) where from five to twelve eggs are deposited. The eggs are large, elongate, and protected by a shell that is noticeably granular on the exterior. Hatchlings typically emerge in August and September but may hatch as early as May and as late as October (C. Smith 1987). Although sex ratio may vary widely within a clutch, the overall ratio at hatching is 1:1. Growth is rapid during the first two years, with most individuals reaching lengths of about 1 m during that time period. Adult males grow to larger sizes than females and may take seven years to reach maximum size, two years longer than females (Stevenson et al. 2009). Most surveys of snakes are performed during sunny days in January and February, when nearly all individuals may emerge to either bask or seek mates. Populations sampled at this time are heavily male biased (2.1:1; Stevenson et al. 2009), but this likely results from the greater detectability of males.

CONSERVATION AND MANAGEMENT The Eastern Indigo Snake was listed as Threatened by the US Fish and Wildlife Service (1978), and therefore, this species is protected under the ESA. It is ranked Priority 1 (Highest Conservation Concern) by ADCNR (Mirarchi, Bailey, Haggerty, and Best 2004), making it unlawful to possess Eastern Indigo Snakes in the state of Alabama. Habitat loss and fragmentation undoubtedly are the primary factors resulting in the vastly reduced populations of this species. But, these snakes also suffered from indiscriminant killing by humans and over-collection for the pet trade. In Alabama, gassing of Gopher Tortoise burrows by humans collecting snakes for the rattlesnake roundup in Opp may have contributed to the extirpation of the snake from the state. This reduction in populations also characterizes the panhandle of Florida, where the Eastern Indigo Snake has been documented only from scattered recent detections.

Population viability analyses based on Florida sites indicate that large conservation areas with restricted fragmentation and limited human access are crucial for long-term conservation of the species. These models suggest that current reserves are not large and contiguous enough to maintain the species (Breininger et al. 2004). Therefore, corridors between reserve areas may be necessary in conservation planning for this species. Additionally, survival of adult (greater than or equal to three years old) females appears to be a key variable for achieving population viability (Breininger et al. 2004). Because road density and traffic flow on roads likely increases mortality of adult females, reserve areas for Eastern Indigo Snakes will need to minimize these threats. In Alabama, gassing of tortoise burrows by snake hunters was outlawed by a 2009 state law. This action removed a potential source of mortality and opened the possibility of repatriation of the species to the state.

Habitat management that improves the remaining longleaf pine forests of the Lower Coastal Plain is vital to returning Eastern Indigo Snakes to Alabama (Godwin 2004a). Thinning of dense stands and use of fire to reduce shrubs and enhance native grasses are the most important management techniques for improving longleaf habitat. Retention of large pine trees is a requirement of such management because these trees provide fine fuels in the form of seasonal deposition of pine needles. These fuels are required for fire management, the most cost-effective tool for re-creating and maintaining the open pine savannas that once covered much of south Alabama. Such practices will allow increased density of Gopher Tortoises, which should benefit snakes as well.

In the 1980s, a large-scale repatriation study was implemented by Dr. Dan Speake as a project of the Cooperative Fish and Wildlife Service Unit at Auburn University. In this project, Eastern Indigo Snakes captured from Georgia and Florida, as well as specimens confiscated by USFWS, were maintained and bred at Auburn. Eggs were incubated to hatching and then released at nine locations in Alabama. A total of 318 individual snakes were released, with an average of 35 per site. The primary goal of the project was creation of viable populations of Eastern Indigo Snakes. Unfortunately, this goal was not achieved. However, the secondary goal of having individual snakes survive long enough to reproduce was reached, at least at one site, and recaptures of released individuals were made at three sites. A second effort at repatriation of Eastern Indigo Snakes to Alabama commenced in 2010 under a joint project of USFWS, USFS, ADCNR, Auburn University, and the Orianne Society. Based on simple assumptions about likely patterns of survival, it was estimated that the release of a minimum of three hundred individuals over time was needed to generate a self-sustaining repatriated population at any one site. This project commenced in the Conecuh National Forest of Covington County in 2009, where at least three hundred captive-reared Eastern Indigo Snakes will be released by 2020. We hope that future summaries of the state herpetofauna will document the snake's successful return to state lists. The Perdido River Longleaf Hills Tract and Fort Rucker Military Installation are at present the only other public properties large enough to serve as conservation areas for Eastern Indigo Snakes in the state of Alabama.

TAXONOMY We follow Collins (1991) in elevating this taxon to species status. No subspecies are recognized within this species. However, the mitochondrial genome documents eastern and western lineages separated from each other by the central ridge of Florida (Krysko, Nuñez, et al. 2016). Because these lineages appeared to differ in the shape of an infralabial, Krysko, Granatosky, et al. (2016) elevate the western lineage to species status (*D. kolpobascilius*) with the eastern lineage retained in *D. couperi*. However, the snakes repatriated to the Conecuh National Forest, while being derived from the eastern lineage, have infralabials consistent with the shape attributed to the western lineage. Additionally, the huge home range sizes of male snakes, a feature that should allow them to cross barriers not crossed by females, limits the utility of taxonomic decisions founded on the mitochondrial genome. For these reasons, we retain the traditional view that Eastern Indigo Snakes represent a single species. Previous authors have placed this species in the genera *Coluber*, *Georgia*, and *Spilotes*.

# North American Racers, Coachwhips, and Whipsnakes
## Genus *Coluber* (Linnaeus, 1758)

This genus includes relatively large, slender, rapidly moving diurnal predators, often referred to as racers. Most have large eyes that are used to survey the environment, often with the head elevated above the ground. A second genus of racers, the genus *Masticophis,* has recently been demonstrated to be paraphyletic because *Coluber* is nested within it. For that reason, we expand *Coluber* to include all members of the former *Masticophis,* as recommended by Utiger et al. (2005). This concept of the genus *Coluber* has representatives in North America, Middle America, and Eurasia, including twenty species, eleven of which are in the New World, including two species found in Alabama.

### KEY TO THE SPECIES OF *COLUBER* OF ALABAMA

1a Tail long and whiplike; dorsum of adults black anteriorly and tan posteriorly.

    *Coluber flagellum flagellum*—Eastern Coachwhip . . . . **this page.**

1b Tail not long and whiplike; dorsum of adults plain black.

    *Coluber constrictor* ssp. . . . . **page 157.**

Adult Eastern Coachwhip, *Coluber flagellum flagellum*, Okaloosa County, FL

## Eastern Coachwhip
*Coluber flagellum flagellum* (Shaw, 1802)

**DESCRIPTION** Eastern Coachwhips are long, slender snakes attaining a maximum total length of about 2,590 mm. The head is fairly distinct from the neck, but narrow, and with obviously enlarged eyes. The tail

is exceptionally long and whiplike in appearance. The anal scute is divided. The dorsal scales are smooth and arranged in seventeen rows at midbody. In color, the head, neck, and anterior one-fourth to one-half of the body are dark brown to jet black, then grading into tan on the posterior one-half to two-thirds of the body. Individuals from west of Mobile Bay have darker heads and necks, and the dark markings are more extensive along the body than those from the rest of the state. The venter is dark brown anteriorly, shading to cream posteriorly. Juveniles have a tan dorsum with dark cross bands that are more pronounced anteriorly than posteriorly. The venter of juveniles is cream with a double row of spots in the neck region.

**ALABAMA DISTRIBUTION** This subspecies is found throughout the state except for some areas north of the Tennessee River.

**HABITS** In Alabama, the Eastern Coachwhip is usually found in dry, relatively open places. The most favorable habitats are uplands, where open woods intersperse with weedy fields. The Coachwhip is lithe and graceful, and among North American snakes, its speed and maneuverability are unmatched. It is an adept climber and often ascends into bushes and small trees when pursued. Its common name is based on the appearance of the tail, which is long and resembles the distal portion of a braided whip. Most Coachwhips put up a vigorous defense when cornered, often aiming their strikes at the adversary's face. It will not, however, chase down a person on horseback and flail both

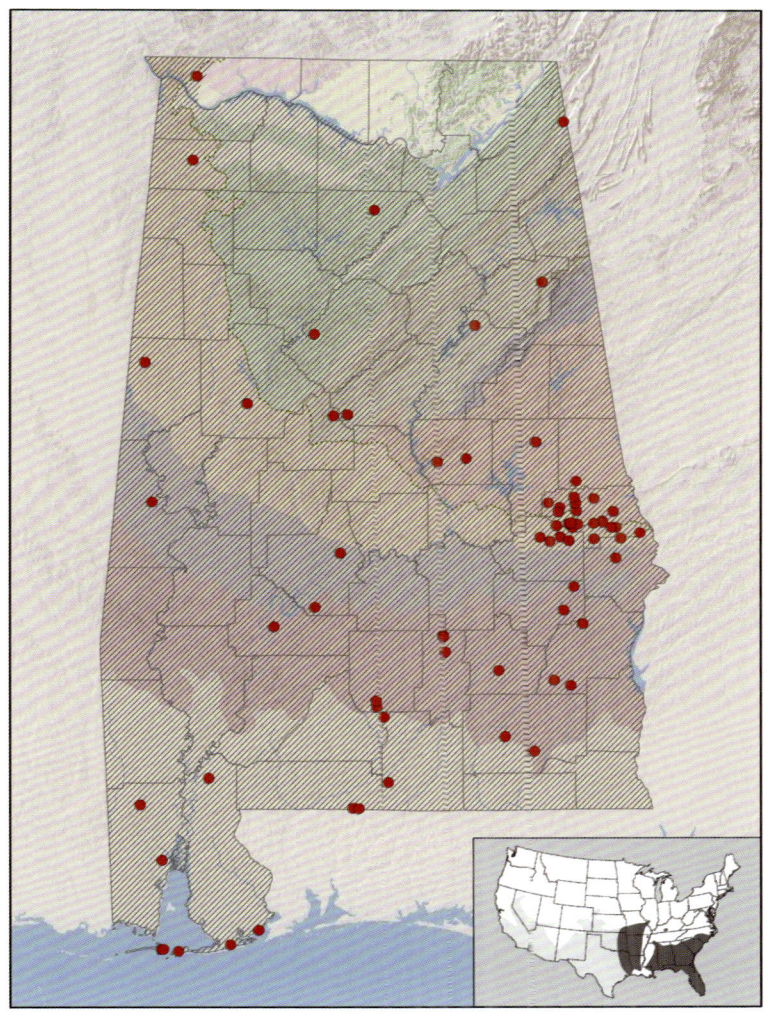

Distribution of Eastern Coachwhip, *Coluber flagellum flagellum*. The presumed range of the subspecies in Alabama is indicated by hatching. Inset map depicts approximate range in the United States, with dark shading indicating the range of *C. f. flagellum* and light shading indicating the range of all other subspecies.

horse and rider to death, as recounted in folklore. Some individual snakes take the opposite tact when captured, becoming limp and apparently lifeless, salivating profusely.

The eggs are oblong with a granular surface. Clutch size ranges from two to twelve, averaging seven eggs per clutch; males appear to have viable sperm throughout the active season (April through November), but testes are regressed during spring months (Goldberg 2002). Food consists of insects, lizards, small mammals, birds, and other snakes. These predatory snakes capture their prey during daylight hours by active movement. The subspecies has large eyes that detect prey movement, especially when these snakes elevate their heads above the ground. They grasp and swallow their prey whole without immobilizing them with venom or killing them by constriction. Hatchling Coachwhips, when offered a variety of chemicals, respond most strongly to odors from lizards and snakes, suggesting an innate attraction to these potential prey (Cooper et al. 1990). During the hottest parts of the day, individuals may switch to a sit-and-wait foraging strategy by hiding in the shade of dense vegetation and dashing out to capture lizards; this strategy appears to increase capture success and decrease handling time of the prey (Jones and Whitford 1989).

Active Eastern Coachwhips are known to move long distances (several kilometers), and they are known to cross major rivers and streams (Steen et al. 2007). For western subspecies, body temperatures achieved by these snakes are higher (33°C; Secor and Nagy 1994) than most other snakes. Active individuals average four hours of activity per day and move within home ranges that average 53 ha (Secor 1995). The eastern subspecies likely exhibits similar temperatures and activity patterns. Oak savanna vegetation is selected more frequently than early-successional pine plantations or forested seeps (Johnson et al. 2007). Gopher Tortoise and mammal burrows are key refuge sites, especially when an individual is about to shed its skin (Dodd and Barichivich 2007).

CONSERVATION AND MANAGEMENT Eastern Coachwhips are large and are persecuted because of the belief by many that they are aggressive and venomous. Alabama populations may have declined in recent years, and state regulations make it unlawful to possess the subspecies in the state of Alabama without a special permit. This subspecies requires open pine savanna vegetation that is enhanced by stand thinning and frequent fire. Growing season fires tend to generate the grass-dominated understory that perpetuates the food resources required by Eastern Coachwhips and the fuel needed for frequent burns.

TAXONOMY This species is the sister taxon to *C. constrictor* (Pyron et al. 2011), and three subspecies are currently recognized. One of these occurs in Alabama. Previous authors have placed this species in the genera *Basconion*, *Herpetodryas*, *Masticophis*, *Psammophis*, and *Zamensis*.

# North American Racer

*Coluber constrictor* (Linnaeus, 1758)

**TAXONOMY** This wide-ranging species is the sister taxon to *C. flagellum* (Pyron et al. 2011). In the United States, *C. constrictor* ranges from coast to coast and from the northern to the southern border. Until recently, this species was thought to contain ten subspecies. However, evidence from the mitochondrial genome identifies six clades that are strongly associated with major barriers, such as mountain ranges and rivers (Burbrink et al. 2008). Three clades, the Eastern, Florida Panhandle, and Gulf Coast clades, are known from Alabama. We infer that the Eastern clade is synonymous with *C. c. constrictor* and apply this name to specimens from populations possessing a short basal spine to the hemipenes of males, a likely synapomorphy for the subspecies. Three genomic clades are found across the range formerly associated with *C. c. priapus*, the subspecies characterized by an enlarged basal spine on the hemipenes of males. Unfortunately, these three genomic clades are paraphyletic, indicating that the enlarged basal spine likely is a shared primitive characteristic. Because the Peninsular Florida clade of Burbrink et al. (2008) includes the type locality for *C. c. priapus*, we associate *C. c. priapus* with that genomic clade and no longer consider the subspecies to be present within Alabama. Most Alabama specimens with enlarged basal hemipenial spines belong to the Panhandle Florida clade of Burbrink et al. (2008), a lineage that includes the type locality of *C. c. helvigularis*. We apply this name to all Alabama specimens below the Fall Line, combining the Panhandle Florida and Gulf Coast clades of Burbrink (2008) These two clades are separated along the border between Alabama and Mississippi except for one Gulf Coast specimen from extreme western Mobile County. Because this boundary does not appear to separate the lineages in nature we infer that the two clades represent historical entities currently experiencing rampant gene flow. We describe the two subspecies in a single account. Previous authors have placed this species in the genera *Bascanion, Bascanium, Coryphodon,* and *Zamensis*.

## Key to the Subspecies of *Coluber constrictor* of Alabama

**1a** Enlarged spine at base of male's hemipenis less than three times the length of the adjacent spine in the same row.

> *Coluber constrictor constrictor*—**Northern Black Racer** . . . . page 159.

**1b** Enlarged spine at base of male's hemipenis at least three times the length of the adjacent spine in the same row.

> *Coluber constrictor helvigularis*—**Southern Black Racer** . . . . page 159.

*From left to right:*

Hemipenis of a Northern Black Racer (*Coluber constrictor constrictor*)

Hemipenis of a Southern Black Racer (*Coluber constrictor helvigularis*)

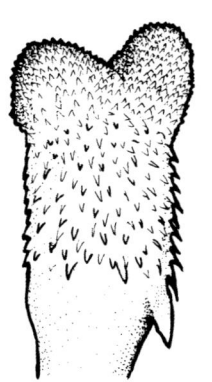

## Northern Black Racer
*Coluber constrictor constrictor* (Linnaeus, 1758)
and
## Southern Black Racer
*Coluber constrictor helvigularis* (Auffenberg, 1955)

**DESCRIPTION** This is a long, fairly slender snake attaining a maximum length of about 1,955 mm. The head is distinct from the neck and contains large eyes. The dorsal scales are smooth and these are arranged in seventeen rows at midbody. The anal scute is divided. A uniform dull black ground color characterizes the dorsum. Ventrally, the color shades from black to dark gray except for the chin, which is white or mottled with white and gray. Along the lateral edge of each ventral scute is a small semicircular or crescent-like dark spot that becomes indistinct or absent posteriorly. Juveniles up to one year of age differ in color from adults. These individuals are distinctly patterned with forty-eight to seventy-one saddle-shaped, dark gray, brown, or reddish blotches, alternating with single or paired lateral dark blotches. Regional variation occurs in the extent of white that appears on the upper labials, chin, and anterior ventrals. Populations in which these regions are mostly white occur in southwestern Alabama, northwestern

Alabama, and in the Talladega Upland and upper Piedmont of eastern Alabama. The extent of white is relatively restricted in most other areas of the state, especially for specimens from central eastern Alabama (Lee, Macon, Bullock, Tallapoosa, and Russell Counties).

**ALABAMA DISTRIBUTION** The Northern Black Racer is found north of the Fall Line. Male specimens with short basal spines on preparations of the hemipenes are known from Calhoun, Chilton, Clay, Cleburne, Coffee, Colbert, Jefferson, Lauderdale, Lawrence, Lee, Macon, Marion, Tallapoosa, and Walker Counties. The Southern Black Racer is found below the Fall Line. Male specimens with long spines on the hemipenes, consistent with this taxon, are known from Baldwin, Butler, Colbert, Conecuh, Covington, Crenshaw, Dallas, Escambia, Fayette, Franklin, Geneva, Henry, Houston, Mobile, and Washington Counties. Although evidence from the mitochondrial genome is lacking, we assume most of these are the Southern Black Racer.

**HABITS** The habits of Alabama's two subspecies of black racers are apparently similar. These are active, diurnal snakes and are familiar to most Alabama citizens. They occur in nearly all of our terrestrial habitats but are most common in open woods, forest edges, and along brushy margins of streams, swamps, and lakes. Folklore among local people suggests that a racer will chase a person. However, no such behavior has been documented. If chased by a human, however, these subspecies will put up a vigorous defense, twitching its tail in agitation

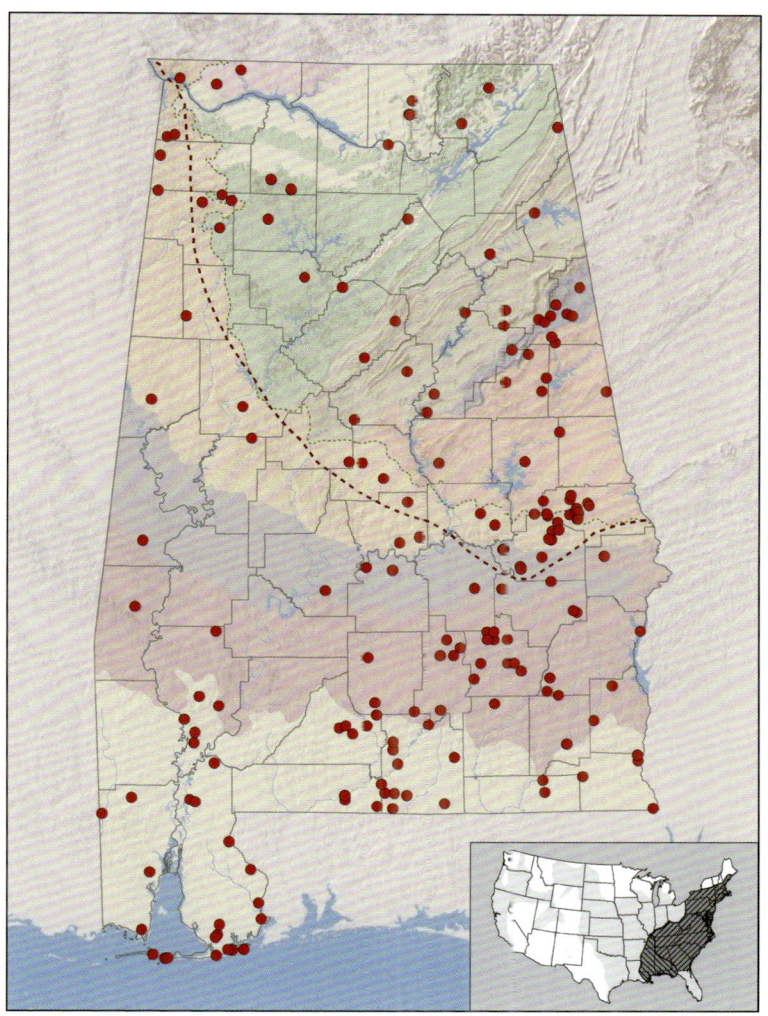

Distribution of the Northern and Southern Racer, *Coluber constrictor constrictor/C. c. helvigularis*. Dotted line indicates presumed boundaries between *the two subspecies*. Inset map depicts approximate range of *C. constrictor* in the United States, with subspecies occupying Alabama indicated by dark shading and all other subspecies indicated by light shading. Right-slant hatching indicates the range of *C. c. constrictor*; horizontal hatching indicates the range of *C. c. helvigularis*.

and striking viciously when cornered. The granular-surfaced, oblong eggs are laid in loose soil, in a rotting tree or stump, or under some sheltering object. The usual number per clutch is between ten and eighteen. Black racers eat snakes, small mammals, birds and bird eggs, frogs, toads, and lizards. The lack of dietary specialization in these subspecies has probably contributed to its apparent success. However, these snakes may compete for food with the Eastern Coachwhip, potentially resulting in smaller body size for populations of black racers that are sympatric with the larger competitor (Steen et al. 2013).

**CONSERVATION AND MANAGEMENT** Black racers are among the most common snakes in Alabama. These subspecies maintain populations in urban areas as well as agricultural areas. Therefore, they receive no special protection from state law. Additionally, no special management efforts appear necessary to maintain these subspecies in Alabama's herpetofauna.

# Scarletsnakes
## Genus *Cemophora* (Cope, 1869)

This genus contains brightly colored snakes that mimic Coralsnakes in color pattern but are easily distinguished from that taxon by the arrangement of colors (red bordered by black in *Cemophora*; red bordered by yellow in *Micrurus*) and the uniform light belly (black color continues onto belly of *Micrurus*). Additionally, this genus has an elongate snout in which the upper jaw protrudes far beyond the lower jaw because of a greatly enlarged rostral scale. This anatomical feature is shared with the genus *Rhinocheilus* of the desert Southwest, a genus that contains some members that are coral mimics and that, like *Cemophora*, consume squamate eggs. However, these features appear to be convergent since *Cemophora* is sister to a clade that includes the Kingsnakes (Pyron and Burbrink 2009a). *Cemophora* is composed of one polytypic species that is found throughout much of the Southeast, including Alabama. The Northern Scarletsnake (*Cemophora coccinea copei*) is the only subspecies within the state.

Northern Scarletsnake, *Cemophora coccinea copei*, Clarke County, AL

## Northern Scarletsnake
*Cemophora coccinea copei* (Jan, 1863)

DESCRIPTION This subspecies is a small to medium-sized snake, attaining a maximum total length of about 800 mm but averaging less than half that. The head is small and is not distinct from the neck. The dorsal scales are smooth and usually are arranged in nineteen rows at

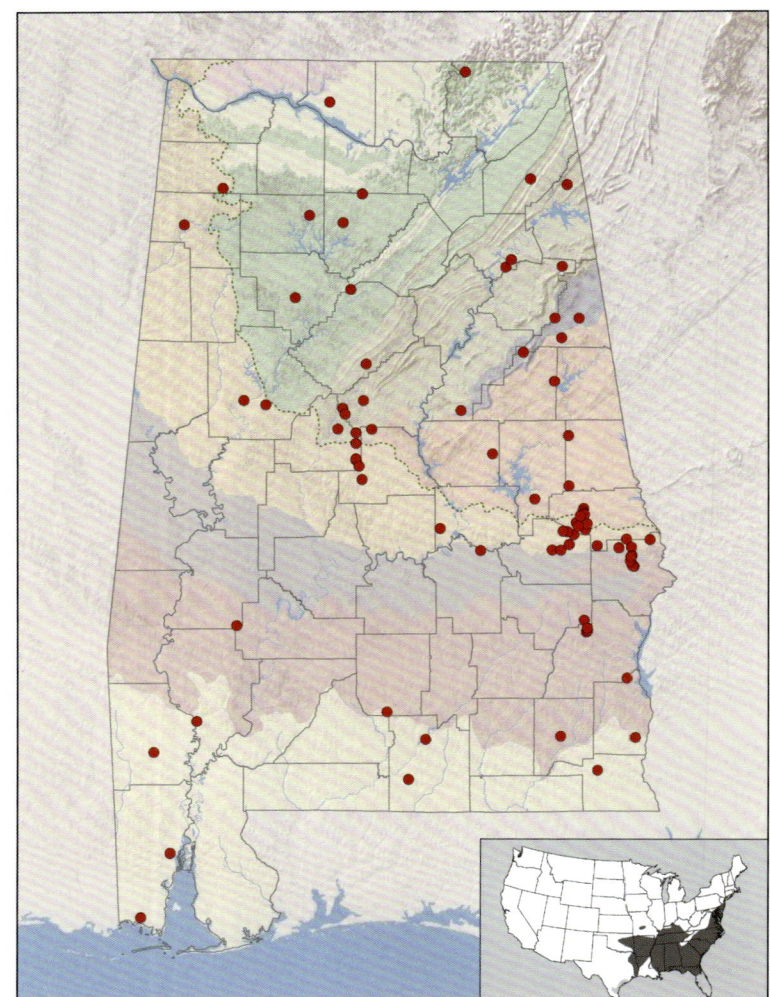

Distribution of Northern Scarletsnake, *Cemophora coccinea*. This subspecies is assumed to occur throughout Alabama. Inset map depicts approximate range in the United States, with dark shading indicating the range of *C. c. copei* and light shading indicating the range of all other subspecies.

midbody. The anal scute is single. An enlarged rostral scale creates a snout that is pointed and projects well beyond the lower jaw. In dorsal coloration, these snakes possess red saddles bordered by black on a ground color of white or light yellow. The snout is red and the venter is white without additional markings. The Northern Scarletsnake differs from other subspecies in having the following combination of characteristics: fewer than 185 ventrals, laterally closed body saddles, fewer than 6 upper labials, and the first black body blotch usually touching parietals or joining with a black head band.

**ALABAMA DISTRIBUTION** This subspecies is found throughout the state.

**HABITS** Relatively little is known of the habits of this secretive snake. It is seldom encountered during the day, although an occasional individual is found under a rock or under the bark of a rotting log. Northern Scarletsnakes occur in a variety of forested and partially forested habitat types, but in general seem to be most abundant in well-drained areas with relatively loose soil. Most specimens are collected at night on paved roads. The Northern Scarletsnake is oviparous, laying three to eight elongate, whitish eggs. Eggs of other reptiles, especially lizards, appear to be the preferred food of these snakes. Other items have been reported in the diet, including small lizards, amphibians, insects, and small mammals. The Northern Scarletsnake is innocuous in demeanor, being reluctant to bite.

**CONSERVATION AND MANAGEMENT** This subspecies is widespread but spotty in its distribution. It receives no special protection by state law because it appears to be common. Open, fire-maintained habitats appear to be preferred, probably because these features increase populations of fence lizards and skinks that likely produce the eggs that comprise the main dietary items of this snake.

**TAXONOMY** This species has three subspecies, one of which occurs in Alabama. Previous authors have placed this species in the genera *Coluber*, *Coronella*, *Rhinostoma*, and *Simotes*.

# Kingsnakes
## Genus *Lampropeltis* (Fitzinger, 1843)

This genus contains the Kingsnakes of the New World. These snakes are medium to large-sized colubrids, most of which are boldly colored, and all of which are constrictors. Kingsnakes are distributed across the entire United States and in Mexico and Central America. The closest living relatives are the genus *Cemophora*, a North American genus, perhaps suggesting a North American origin for the radiation (Pyron and Burbrink 2009a). The genus contains sixteen species, five of which occur in Alabama.

### Key to the Species of *Lampropeltis* of Alabama

**1a** Upper and lower labials boldly marked with light and dark bars.
> *Lampropeltis getula* . . . . . page 167.

**1b** Upper and lower labials variously marked, but never with bold light and dark bars. Go to **2**.

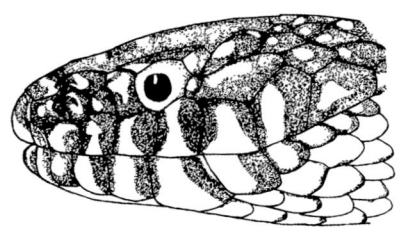

Lateral view of head of an Eastern Kingsnake (*Lampropeltis getula*); based on AUM 32928

**2a** Tip of snout bright red; dorsal pattern of bold red, black, and yellow or white rings.
> *Lampropeltis elapsoides*—Scarlet Kingsnake . . . . page 180.

**2b** Tip of snout gray; dorsal pattern of red to gray rings or saddles separated by white, yellow, or light gray ground color. Go to **3**.

**3a** Mid-dorsum with red or brown saddles bordered by black on a light gray ground color; neck lacking parallel brown linear markings on each side.
> *Lampropeltis triangulum*—Milksnake . . . . page 183.

**3b** Mid-dorsal brown to brick red saddles, these not bordered by black and often faded in adults, on a gray to tan ground color; neck with parallel brown linear markings on each side.
> *Lampropeltis calligaster*—Mole Kingsnake . . . . page 191.

# Eastern Kingsnake
*Lampropeltis getula* (Linnaeus, 1766)

**TAXONOMY** We follow the traditional concept of this species by considering it to include seven subspecies that span most of North America, with three subspecies, *L. g. getula*, *L. g. holbrooki*, and *L. g. nigra*, occurring in Alabama. Based on color patterns for specimens from Alabama, Mount (1975) recognized these three subspecies and demonstrated that all three intergrade: *L. g. getula* and *L. g. holbrooki* along a narrow northeast-to-southwest zone of contact, and *L. g. getula* and *L. g. nigra* as well as *L. g. holbrooki* and *L. g. nigra* along an east-to-west zone of contact. Based on an examination of the mitochondrial genome, Pyron and Burbrink (2009b) concluded that specimens referable to *L. g. holbrooki* and *L. g. nigra* from Alabama are more closely related to each other than either is to *L. g. getula*. Additionally, these authors showed that Alabama *L. g. holbrooki* specimens are more distantly related to specimens of that putative subspecies in other parts of its former range than they are to specimens of either *L. g. nigra* or *L. g. getula* from within the state of Alabama. Pyron and Burbrink (2009b) used their data to restrict *L. getula* to those specimens formerly placed in *L. g. getula*. They also placed specimens of *L. g. nigra* and *L. g. holbrooki* distributed east of the Mississippi River into the species *L. nigra*, while placing similarly colored snakes west of the Mississippi River (formerly part of *L. g. holbrooki*) into a separate species, *L. holbrooki*. These authors list color characteristics that they consider diagnostic of *L. getula* and *L. nigra* but acknowledge that specimens from Alabama document that their concept of *L. getula* and *L. nigra* hybridize in the state.

The conclusions drawn by Pyron and Burbrink (2009b) were clouded by Krysko et al. (2017), who examined two regions of the mitochondrial genome and one nuclear locus. Relationships recovered by Krysko et al. (2017) were incongruent with those of Pyron and Burbrink (2009a). Nevertheless, Krysko et al. (2017) retained *L. nigra* as a separate species, including in it specimens from northern Alabama, but placed speckled Kingsnakes from Alabama and Mississippi in their concept of *L. holbrooki*, despite recovering paraphyly for such a species.

Based on accumulated data, we are not convinced that species delimited by current genomic evidence improve our understanding of Eastern Kingsnake evolution. Gene trees currently estimated for the group are dominated by mitochondrial genes that are known to recover apparently monophyletic populations, between which gene flow is known to occur (e.g., Burbrink and Guiher 2015; Strickland et al. 2014). Species erected based on these genomic data violate our concept of what species should be. Therefore, we consider *L. getula* to represent a single species within Alabama and that three distinctive color morphs justify recognition of three subspecies, each of which experiences zones of intergradation with the other two subspecies. If intergradient forms were demonstrated to have lower fitness, then we would consider this as evidence indicating that the taxonomy of Pyron and Burbrink (2009b) should be followed. Of particular interest are specimens that Mount (1975) considered to be intergradient between *L. g. getula* and *L. g. nigra*. These specimens have light rings that are consistent with the pattern of *L. getula* presented in Pyron and Burbrink (2009b) but that are remarkably thin and faded. Pyron and Burbrink (2009b) considered these specimens to represent darkening in northern populations rather than hybridization. All AUM specimens from Cleburne County show the thin and faded banding pattern, but an individual from that county appearing in Pyron and Burbrink (2009b) has a mitochondrial genome that is consistent with their concept of *L. nigra*. Unfortunately, there is no museum specimen associated with the mitochondrial sample, and therefore, the color pattern of this snake cannot be ascertained. However, there are only two possibilities for its color—a black, ringless pattern (consistent with our concept of *L. g. nigra*) or a pattern of thin, faded, *L. g. getula*–like rings, consistent with Mount's (1975) conclusion that these subspecies freely intergrade. The presence of a *L. g. getula*–like pattern in a specimen possessing a mitochondrion from the *L. g. nigra* lineage is inconsistent with the description of Pyron and Burbrink (2009b), suggesting that faded rings result from darkening of northern populations. Previous authors have placed members of the snakes that we now consider to be *L. getula* in the genera *Coluber, Coronella, Ophibolus,* and *Triaeniopholis*.

## Key to the subspecies of *Lampropeltis getula* of Alabama

**1a** Body black with twenty-five to fifty bold, narrow (one to three scale rows) white or yellow cross bands (faded in northern populations); venter with bold, square dark brown markings (tending toward uniform dark brown in some individuals).

*Lampropeltis getula getula*—Eastern Kingsnake . . . . page 170.

**1b** Body black with yellow spots at center of some or all dorsal scales; if light cross bands present, these are very thin and fifty to one hundred in number; venter mottled with small light and dark markings. Go to 2.

**2a** Fewer than 50 percent of dark dorsal scales with light central spots.

*Lampropeltis getula nigra*—Black Kingsnake . . . page 174.

**2b** Nearly all dark dorsal scales with light central spots.

*Lampropeltis getula holbrooki*—Speckled Kingsnake . . . . page 177.

Eastern Kingsnake,
*Lampropeltis getula
getula*, Geneva
County, AL

## Eastern Kingsnake

*Lampropeltis getula getula* (Linnaeus, 1766)

**DESCRIPTION**  The Eastern Kingsnake is a large, relatively robust snake attaining a maximum total length of about 2,085 mm with a relatively short tail. The head is not, or but slightly, distinct from the neck, and the anal scute is undivided. The dorsal scales are smooth, shiny, and arranged in twenty-one or twenty-three rows at midbody. The basic body color is dark brown or black with twenty-three to thirty-six narrow, yellow or cream-colored transverse bands (these are occasionally broken) that typically divide on the sides and connect to a series of light spots. The belly is checkered with large yellow and dark brown squares, becoming nearly uniform dark brown in some individuals. Variation in the number and width of light bands among Alabama *L. g. getula* is considerable. Geographic variation in the latter is evident with band width tending to be greater in populations from the southeastern portion of the range (Barbour, Henry, Hale, Houston, Pike, Coffee, and Geneva Counties) than in those from northern portions. The narrowest bands tend to occur in populations inhabiting Lee, Macon, and Russell Counties.

**ALABAMA DISTRIBUTION**  Influence of this subspecies is found from the southeast corner of the state northward to Cherokee County. It also

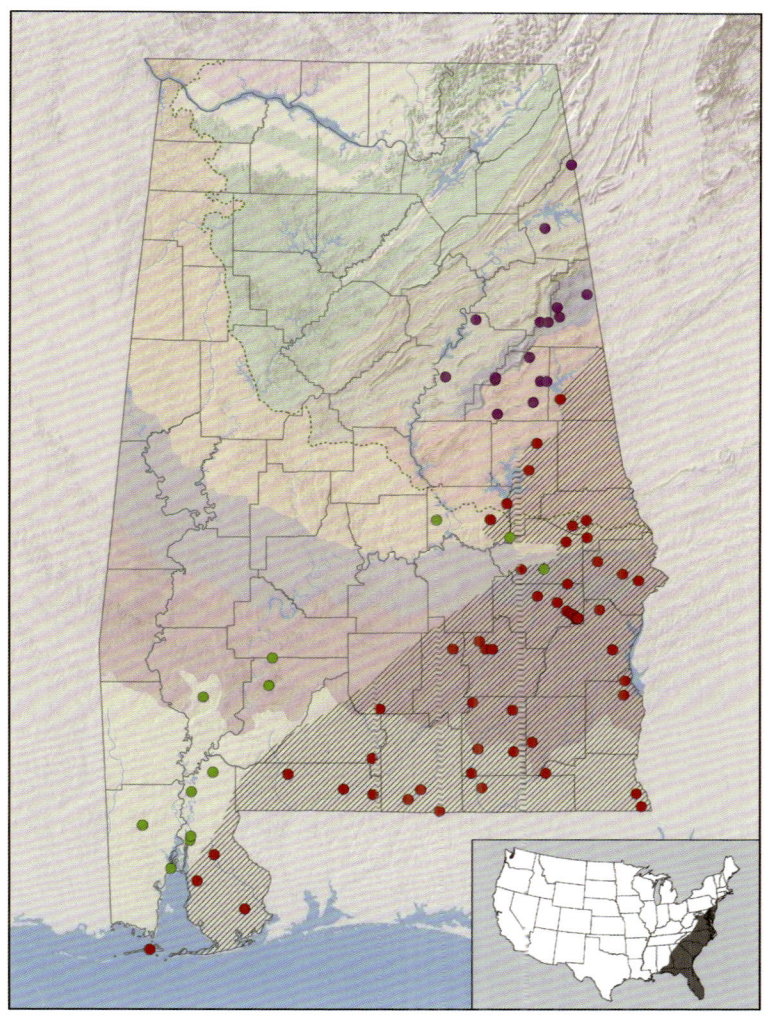

Distribution of Eastern Kingsnake, *Lampropeltis getula getula*. The presumed range of the subspecies in Alabama is indicated by hatching. Solid purple dots indicate intergrades between *L. g. getula* and *L. g. nigra*. Solid green dots indicate intergrades between *L. g. getula* and *L. g. holbrooki*. Inset map depicts approximate range in the United States.

extends from the southeastern corner westward along the southern tier of counties to Baldwin County and onto Dauphin Island in Mobile County. The drainage basins of the Chattahoochee, Choctawhatchee, Pea, Yellow, Conecuh, Perdido, and Styx Rivers comprise the bulk of the range of this subspecies in Alabama. However, influence of the Eastern Kingsnake is present in the Piedmont, Talladega Upland, and Ridge and Valley formations east of the Coosa River. The influence of intergradation with *L. g. holbrooki* is seen along a narrow zone of contact along the eastern side of the Coosa and Alabama River valleys.

Intergrades with *L. g. nigra* are found in Calhoun, Cherokee, Clay, Cleburne, and Talladega Counties.

Habits Within its range in Alabama, the Eastern Kingsnake occurs in nearly all of our terrestrial habitat types. Optimal habitats include abandoned farms, rural garbage dumps, edges of floodplains, and brushy margins of streams and swamps. The subspecies is diurnal and almost exclusively ground dwelling but spends most of its time in underground refugia (Linehan et al. 2010). Among available habitats, Eastern Kingsnakes are strongly associated with natural pine and hardwood forests, within which these snakes tend to be found in areas with the highest concentration of logs on the ground and woody vegetation (Steen, Linehan, and Smith 2010; Steen, McGee, et al. 2010). In these habitats, the subspecies frequents rodent burrows and stump holes as refuges. A conspicuous snake, the Eastern Kingsnake moves rather slowly but occupies annual home ranges of about 50 ha for both males and females (Linehan et al. 2010). Freshly captured individuals expel copious amounts of musk and may try to bite. If treated gently, however, they usually become calm and inoffensive.

Kingsnakes are constrictors but may swallow small prey alive. Kingsnakes eat snakes, including venomous ones, along with lizards, mice, and, on occasion, birds and amphibians. These snakes are immune to the effects of viperid bites and eat these snakes frequently enough that many residents of rural Alabama recognize Kingsnakes and refrain from killing them. Males are more frequently captured than are females and surface activity peaks during March through June (Linehan et al. 2010), suggesting mate searching in spring. Males that encounter other males during this time of year may engage in a combat dance in which the two will intertwine and attempt to use a loop of the body to pin to the ground the head of the rival. Females produce three to twenty-four eggs per clutch and deposit them in moist soil under logs or other surface objects in June or July. The eggs are oblong in shape, leathery in texture, adhere to each other, and hatch in August or September.

Conservation and Management The Eastern Kingsnake formerly was one of the most frequently encountered large snakes in the southern part of the state. Unfortunately, it has suffered a precipitous decline throughout its range, including at places protected from extreme habitat loss and fragmentation (Winne et al. 2007). This decline happened during the 1980s and 1990s and occurred in association with an increased frequency and severity of drought. Mount (1980) encountered this subspecies frequently in the Conecuh National Forest but noted declining populations by 1980. By 2005, it was so scarce there (if present at all) that it was not detected in two years of intensive sampling (Guyer et al. 2007). The loss of this predator at this site may be related to the change in status of copperheads, a snake that was rare during the survey conducted by Mount (1980) but was the most commonly trapped species in the survey by Guyer et al. (2007). Such a precipitous decline suggests that a disease or the spread of fire ants, perhaps exacerbated by increased drought, caused Eastern Kingsnake populations to crash. Because it is ranked Priority 2 (High Conservation Concern) by ADCNR (Mirarchi, Bailey, Haggerty, and Best 2004), it is unlawful to possess the subspecies in the state of Alabama without a special permit.

Repatriation efforts are likely to be necessary to return Eastern Kingsnakes to their position of prominence in Coastal Plain habitats. Before this can be accomplished, the cause of the population decline must be understood. Research on the effects of disease, parasites, and fire ants must be explored to determine if any of these is the key factor limiting population size. The Perdido River Longleaf Hills Tract, Conecuh National Forest, Geneva State Forest, Fort Rucker Military Installation, and the Chattahoochee State Forest are important public properties where conservation efforts can be concentrated. On these properties, use of growing season fire on a one- to three-year rotation will be important for maintaining quality habitat for Eastern Kingsnakes (Means 2004c).

Black Kingsnake, *Lampropeltis getula nigra*, Morgan County, AL

## Black Kingsnake

*Lampropeltis getula nigra* (Yarrow, 1882)

DESCRIPTION This is a large, relatively robust snake attaining a maximum total length of about 1,675 mm with a relatively short tail. The head is not, or but slightly, distinct from the neck, and the anal scute is undivided. The dorsal scales are smooth, shiny, and arranged in twenty-one or twenty-three rows at midbody. Black Kingsnakes lack light spots at the center of all or most dorsal scales, although light spots are found along the lateral scales. Numerous (fifty to one hundred), thin (less than one scale row), light bands may be present across the dorsum. The venter is marked with numerous small, irregular light and dark markings; these do not form the bold, square markings found in *L. g. getula*. Subcaudals average fifty in males and forty in females.

ALABAMA DISTRIBUTION Black Kingsnakes are found above the Fall Line as well as in the Tennessee River valley in the extreme northeast corner of the state. Specimens with color patterns suggesting intergradation with *L. g. holbrooki* are known from Colbert, Jackson, Limestone, and Tuscaloosa Counties. Specimens with color patterns suggesting intergradation with *L. g. getula* are known from Calhoun, Cherokee, Clay, Cleburne, and Talladega Counties.

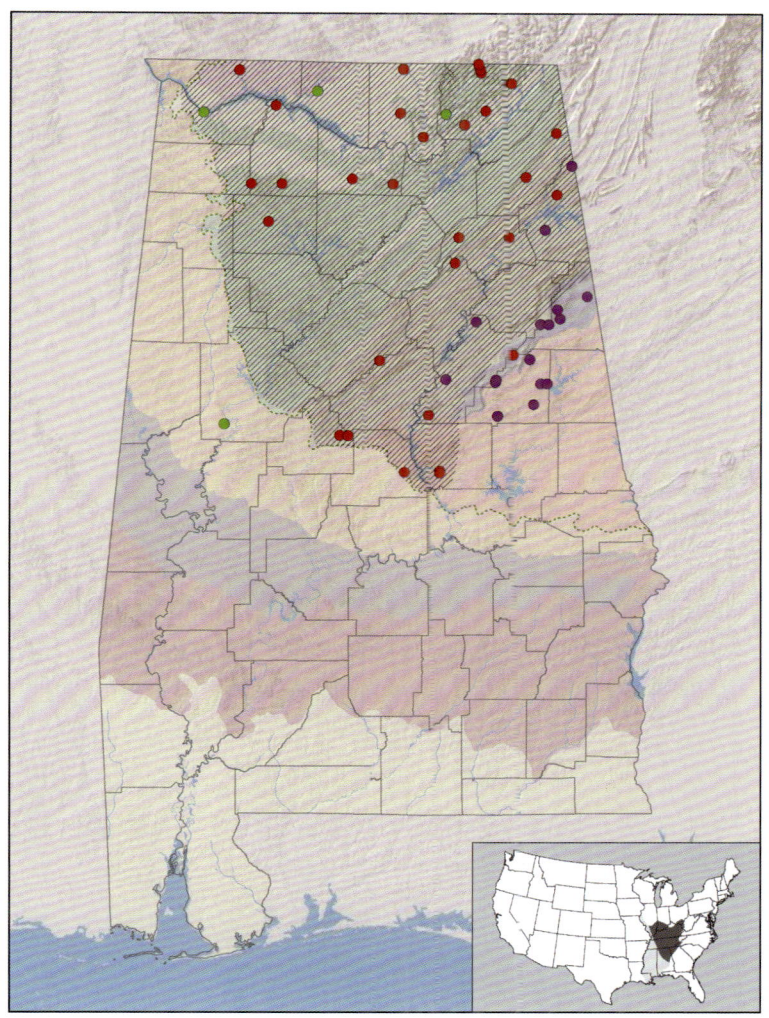

Distribution of Black Kingsnake, *Lampropeltis getula nigra*. The presumed range of the subspecies in Alabama is indicated by hatching. Solid purple dots indicate intergrades between *L. g. nigra* and *L. g. getula*. Solid green dots indicate intergrades between *L. g. nigra* and *L. g. holbrooki*. Inset map depicts approximate range in the United States, with dark shading indicating the range of *L. g. nigra* and light shading indicating the range of the other subspecies.

**HABITS** Black Kingsnakes occupy forested, grassland, and pasture areas, especially those near streams. swamps, and ponds. These snakes are diurnal in activity pattern and are most commonly observed during morning hours of summer months. Ctherwise, these slow-moving animals are encountered under surface objects. This subspecies experiences cold weather associated with high elevat.ons that likely force it into periods of torpor during winter.

Freshly captured individuals expel musk from the cloacal opening and may try to bite. However, they quickly calm and become inoffensive. Like all Kingsnakes, this subspecies constricts its prey, which includes primarily other snakes, including venomous ones, along with lizards, mice, and birds. Many residents of rural Alabama recognize Black Kingsnakes and refrain from killing them in order to limit populations of venomous snakes. Male Black Kingsnakes search for females in spring (March to June), when mating takes place. Females produce three to twenty-four eggs per clutch and deposit them in moist soil under logs or other surface objects in June or July. The eggs are oblong in shape, leathery in texture, adhere to each other, and hatch in August or September.

CONSERVATION AND MANAGEMENT The Black Kingsnake remains common in Alabama. However, body condition of this morph has declined in recent years in Tennessee populations (Faust and Blomquist 2011), which may portend a decline in abundance. Management techniques that open hardwood forests to create open grasslands are appropriate for enhancing the habitat for Black Kingsnakes. Power line rights of way may be key resources for this subspecies. But, if these habitats are mowed to maintain grassland vegetation under the power line then this management tool should be performed during winter to avoid excessive mortality to Black Kingsnakes that might be present during the active season.

## Speckled Kingsnake
*Lampropeltis getula holbrooki* (Yarrow, 1882)

DESCRIPTION This is a large, relatively robust snake attaining a maximum total length of about 1,675 mm with a relatively short tail. The head is not, or but slightly, distinct from the neck, and the anal scute is undivided. The dorsal scales are smooth, shiny, and arranged in twenty-one or twenty-three rows at midbody. Yellow or cream-colored dots occur at the center of nearly all dorsal scales. For specimens from the Alabama River, these light markings are particularly large in size. For specimens from the Tombigbee and Warrior Rivers, these light markings are quite small. Numerous (fifty to one hundred), thin (< one scale row), light bands may be present across the dorsum. The venter is marked with numerous small, irregular light and dark markings; these do not form the bold, square markings found in *L. g. getula*. Subcaudals average fifty in males and forty in females.

ALABAMA DISTRIBUTION Speckled Kingsnakes are found on the west side of Mobile Bay and occupy the Tombigbee, Alabama, Cahaba, Coosa, and Tennessee drainages. The subspecies hybridizes with *L. g. getula* along the Alabama, Tensaw, and Mobile Rivers. Specimens with color patterns suggesting intergradation with *L. g. nigra* are known from Colbert, Jackson, Limestone, and Tuscaloosa Counties.

HABITS Speckled Kingsnakes occupy forested, grassland, and pasture areas, especially those near streams, swamps, and ponds. These snakes are diurnal in activity pattern and are most commonly

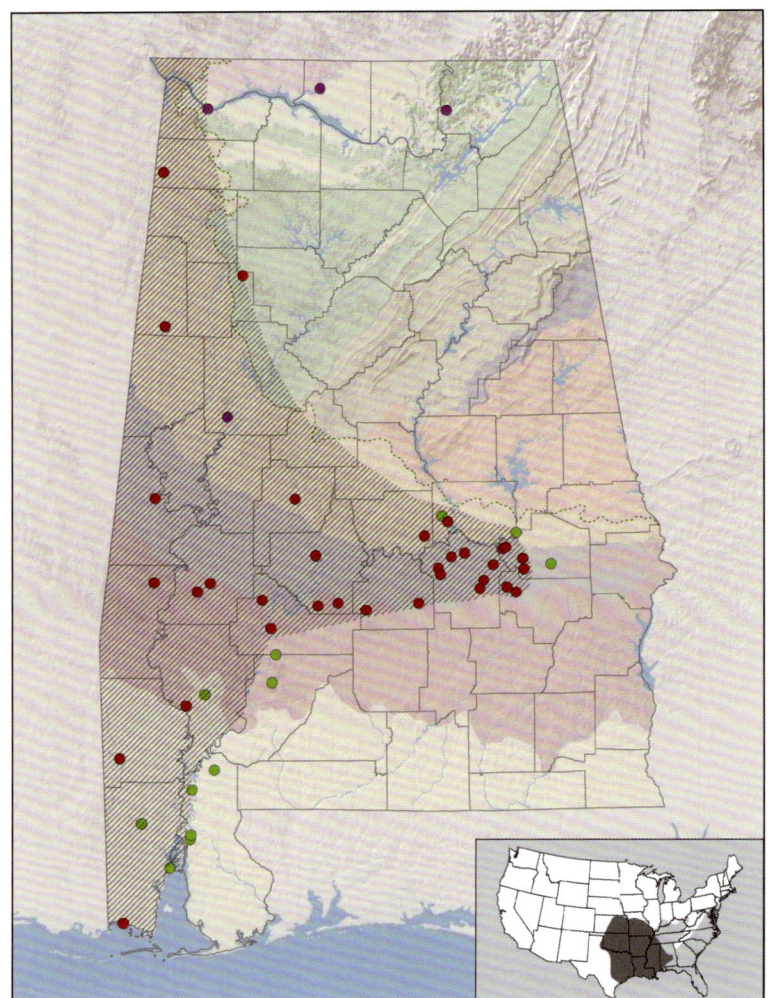

Distribution of Eastern Speckled Kingsnake, *Lampropeltis getula holbrooki*. The presumed range of the subspecies in Alabama is indicated by hatching. Solid purple dots indicate intergrades between *L. g. holbrooki* and *L. g. nigra*. Solid green dots indicate intergrades between *L. g. holbrooki* and *L. g. getula*. Inset map depicts approximate range in the United States, with dark shading indicating the range of *L. g. holbrooki* and light shading indicating the range of the other subspecies.

observed during morning hours of summer months. Otherwise, these slow-moving animals are encountered under surface objects. This subspecies, called guinea snake by many who know it, is well camouflaged in grassy situations and, until recently, was one of the more common serpents of the Black Belt. The subspecies is likely to be active year-round.

Freshly captured individuals expel musk from the cloacal opening and may try to bite. However, they quickly calm and become inoffensive. Like all Kingsnakes, this subspecies constricts its prey. Other snakes, including venomous ones, are a primary prey item along with lizards, mice, and birds. Many residents of rural Alabama recognize Speckled Kingsnakes and refrain from killing them in order to limit populations of venomous snakes. Male Speckled Kingsnakes search for females in spring (March to June), when mating takes place. Females produce three to twenty-four eggs per clutch and deposit them in moist soil under logs or other surface objects in June or July. The eggs are oblong in shape, leathery in texture, adhere to each other and hatch in August or September.

CONSERVATION AND MANAGEMENT Because it is ranked Priority 2 (High Conservation Concern) by ADCNR (Mirarchi, Bailey, Haggerty, and Best 2004), it is unlawful to possess this subspecies in the state of Alabama without a special permit. This conservation status was established because these snakes have declined in abundance in the past decade. Management techniques that open hardwood forests to create open grasslands are appropriate for enhancing the habitat for Speckled Kingsnakes. Restoration of substantial grasslands within Alabama's Black Belt is particularly important for retaining this subspecies within the state.

## Scarlet Kingsnake
*Lampropeltis elapsoides* (Holbrook, 1838)

**DESCRIPTION** Scarlet Kingsnakes are fairly small, strikingly marked serpents attaining a maximum total length of about 600 mm. The head is small, somewhat pointed, and only slightly distinct from the neck. The anal scute is undivided. The dorsal scales are smooth, shiny, and arranged in seventeen or nineteen rows at midbody. In color pattern, the head is red from the snout to about half way along the parietal scales and then follows a repeated pattern of black, yellow, black, and red rings along the length of the body. The relative width of each color varies markedly among individuals. The ringed pattern often becomes irregular on the venter, and the dorsal rings may occasionally become asymmetrical, forming ocelli instead of rings.

**ALABAMA DISTRIBUTION** This species is most abundant in the Lower Coastal Plain and Fall Line Hills regions of Alabama but is found sporadically above the Fall Line in northern parts of the state.

**HABITS** This small Kingsnake is extremely secretive and seldom ventures abroad during the day. Although an occasional individual is picked up on the road at night, most specimens are taken from rotting pine stumps, especially in spring months. Breeding occurs in spring (May to June) and females then lay two to nine eggs in summer (June to August). The eggs are elliptical in shape, adhere to each other, and have leathery shells. These eggs are placed in rotting stumps or under logs, where hatching occurs in late summer. This species appears to be

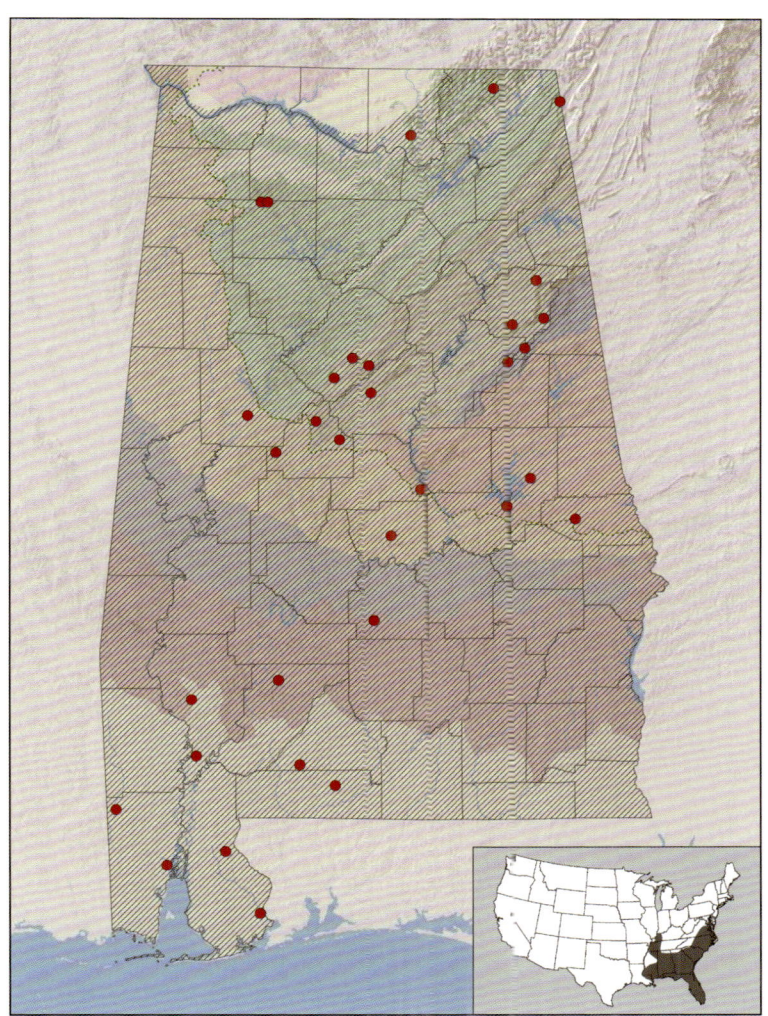

Distribution of Scarlet Kingsnake, *Lampropeltis elapsoides*. The presumed range of the species in Alabama is indicated by hatching. Inset map depicts approximate range in the United States.

capable of producing a second clutch of eggs within a season of activity (Tryon and Murphy 1982). Food items typically are ground skinks and other small lizards. Other dietary items are reported to include small snakes and mice. Scarlet Kingsnakes, along with the Northern Scarletsnake, are commonly mistaken for the venomous Eastern Coralsnake. Neither of the two mimics has the black nose of the Coralsnake, however, and on neither does the red and yellow (or white) dorsal markings come into contact with each other, as is the case on the Coralsnake. The two mimics can be distinguished from each other by the belly, which is uniform white in the Northern Scarletsnake and ringed with black, red, and yellow in the Scarlet Kingsnake.

**CONSERVATION AND MANAGEMENT** Scarlet Kingsnake populations are difficult to assess because the species is so secretive. However, specimens continue to be discovered in the state, suggesting that it is maintaining viable populations, despite collecting for the pet trade, predation by Red Imported Fire Ants, and extensive habitat loss and fragmentation, problems that are thought to reduce snake populations. Management activities that retain snags and fallen logs, which serve as hiding places for *L. elapsoides* and the skinks that it frequently consumes, should benefit this species.

**TAXONOMY** Until recently, this taxon was treated as a subspecies of *L. triangulum* (Mount 1975). However, it is phylogenetically a sister to the Mexican and western tricolored Kingsnakes rather than sharing a taxonomic affinity to *L. triangulum* (Pyron and Burbrink 2009a). For that reason, we consider it to be a species with no subspecific variation. Previous authors have placed this species in the genera *Coluber, Coronella,* and *Ophibolus.*

## Lampropeltis triangulum (Lacépède, 1788)

**TAXONOMY** The content of this species has changed since the publication of the classic monograph by Williams (1978). At that time, a single wide-ranging species was recognized, extending from Canada to northern South America. Recent evidence, however, documents that this old concept of the species must be rejected because it includes a variety of taxa that are not necessarily closely related (Bryson et al. 2007; Pyron and Burbrink 2009a). Because the type specimen is of the nominate subspecies and because this subspecies occurs in Alabama, this species name is likely to be a stable component of Alabama's herpetofauna regardless of how other former members of this species complex are allocated. We accept the elevation of the Scarlet Kingsnake, now *L. elapsoides,* from its rank as a subspecies in Williams's (1978) concept of *L. triangulum.* Scarlet Kingsnakes are found in sympatry with *L. triangulum* in the state of Alabama with no evidence of hybridization (Mount 1975). That leaves, however, two relatively distinct subspecies of *L. triangulum* in the state, *L. t. triangulum* and *L. t. syspila,* which do show evidence of intergradation. Until additional data prove otherwise, we accept these as subspecies of *L. triangulum*. Previous authors have placed members of this species in the genera *Ablabes, Coluber,* and *Coronella.*

## KEY TO THE SUBSPECIES OF *LAMPROPELTIS TRIANGULUM* OF ALABAMA

1a Large red mid-dorsal blotches extend laterally to the level of the first scale row; nape or back of head usually with a light band or collar.

*Lampropeltis triangulum syspila*—Red Milksnake . . . . page 184.

1b Mid-dorsal blotches brown, usually extending to within three to four scale rows of the ventral scutes; a mecial light spot or a Y- or V-shaped light area on nape or back of head.

*Lampropeltis triangulum triangulum*—Eastern Milksnake . . . .
page 188.

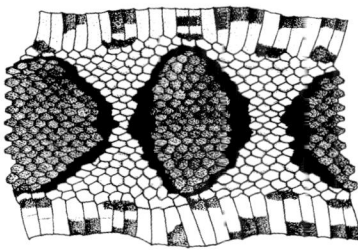

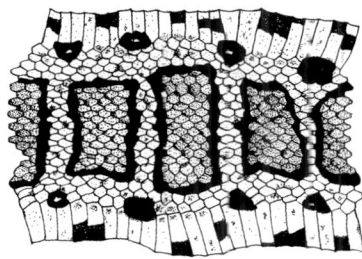

*From top to bottom:*

Color pattern of a Red Milksnake (*Lampropeltis triangulum syspila*); based on AUM 23112

Color pattern of an Eastern Milksnake (*Lampropeltis triangulum triangulum*); based on AUM 33392

Red Milksnake, *Lampropeltis triangulum syspila*, Lawrence County, AL

## Red Milksnake

*Lampropeltis triangulum syspila* (Cope, 1888)

**DESCRIPTION** This is an attractive, medium-sized snake that attains a maximum total length of around 990 mm. In general, this subspecies is similar to *L. t. triangulum* in scalation but differs rather markedly in color characteristics. Body bands, or saddles, typically are redder, wider, and fewer in number (sixteen to thirty-one with an average of twenty-three) in *L. t. syspila* and extend ventrally along each side to about the first scale row (as opposed to the third or fourth in *L. t. triangulum*). In general aspect, when viewed dorsally, these snakes appear to be banded or ringed as opposed to being blotched, as in *L. t. triangulum*. The light ground color between dorsal saddles is white, cream, yellowish, or light gray rather than the gray color of *L. t. triangulum*.

The head is variously patterned but with red predominating dorsally. The head ends with a thin black (one to two scale rows), thin white (three to four scale rows), thin black (one to two scale rows), and wide red (eight to nine scale rows) series of bands. The pattern is repeated for the entire length of the body with the red bands becoming more similar in width to the others on the tail. In older individuals, the red color may become brownish or grayish brown. The ventrolateral surface often has a series of small blotches, similar in constitution to the dorsal ones and alternating with them. The venter is light gray with large rectangular marks arranged irregularly or in checkerboard fashion.

*Above:*
Red Milksnake, *Lampropeltis triangulum syspila*, Winston County, AL

*Left:*
Intergrade Red Milksnake, *Lampropeltis triangulum syspila x triangulum*, Morgan County, AL

**ALABAMA DISTRIBUTION** This subspecies has been encountered most frequently in the Bankhead National Forest and is restricted to the Appalachian Plateaus north of the Sequatchie Valley. Individuals that appear to be intergradient with *L. t. triangulum* are found in Morgan County.

**HABITS** Optimal habitat is open woodland and woodland edge with an abundance of flat rocks and other types of cover under which the snakes spend much of their time. All Alabama specimens have been

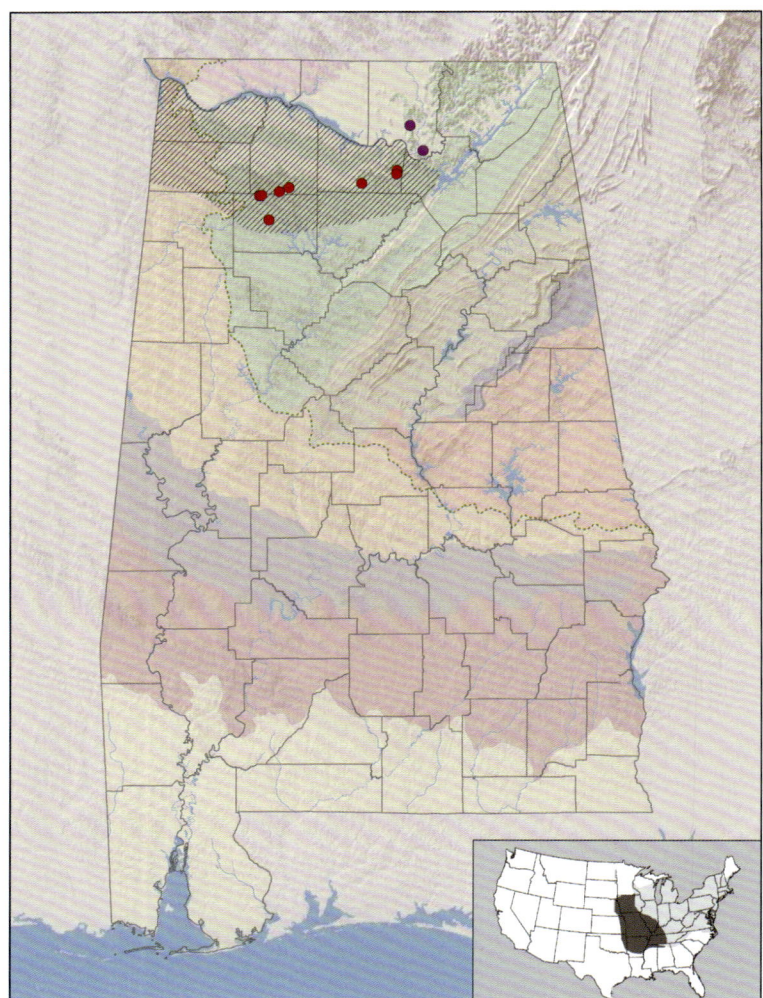

Distribution of Red Milksnake, *Lampropeltis triangulum syspila*. The presumed range of the subspecies in Alabama is indicated by hatching. Solid purple dots indicate intergrades between *L. t. triangulum* and *L. t. syspila*. Inset map depicts approximate range in the United States with dark shading indicating the range of *L. t. syspila* and light shading indicating the range of all other subspecies.

collected at relatively high elevations. Turning rocks and logs, and breaking open rotting logs in relatively open, dry woodlands is how most individuals are found. Areas near rock outcrops seem to be favored, perhaps because these areas are used as hibernacula. Essentially all Alabama specimens have been collected during April and May. Several of the snakes collected in Alabama have lost a part of their tails, perhaps due to frostbite. The number of eggs laid by the Red Milksnake averages around seven or eight, and nests are likely placed under rocks or rotting logs. Food items include lizards, snakes, and small mammals, with the reptilian component being most important.

**CONSERVATION AND MANAGEMENT** This subspecies receives no special protection under state law but is tracked as imperiled by the Alabama Natural Heritage system. Populations in Alabama are centered on public lands, much of which are managed as a wilderness area in the Bankhead National Forest. Therefore, these snakes may be secure within the state. Any activities, such as mining, that disturb rock outcrops are likely to disrupt hibernacula that may be used habitually. Such sites are likely to be limiting to this and other snake species. So, care should be taken to protect these geological formations.

Eastern Milksnake,
*Lampropeltis trian-gulum triangulum*,
Madison County, AL

## Eastern Milksnake

*Lampropeltis triangulum triangulum* (Lacépède, 1788)

**DESCRIPTION** Members of this subspecies are medium-sized snakes attaining a maximum total length of about 1,140 mm. The head is at most slightly distinct from the neck, and the anal scute is undivided. The dorsal scales are smooth and usually are arranged in twenty-one rows at midbody. In dorsal coloration, this subspecies is gray to tan with twenty-four to fifty-four (with an average of thirty-six) black-bordered dark gray or brownish blotches (brick red in juveniles) that usually terminate laterally on the third or fourth scale row (counting from ventral scutes). One or two rows of ventrolateral, irregularly shaped, dark blotches are found on each side, alternating with the dorsal blotches. The first body blotch usually is connected to dark markings on the head creating a blotch that encircles a medial round, oblong, Y- or V-shaped light area on the back of the head or nape of the neck. The venter has rectangular black markings separated by light markings, often creating a checkerboard pattern.

**ALABAMA DISTRIBUTION** This subspecies is confined to Lookout Mountain in the northeastern corner of the state. Specimens are known from DeKalb and Jackson Counties. We consider specimens from Morgan County to be intergradient with *L. t. syspila*.

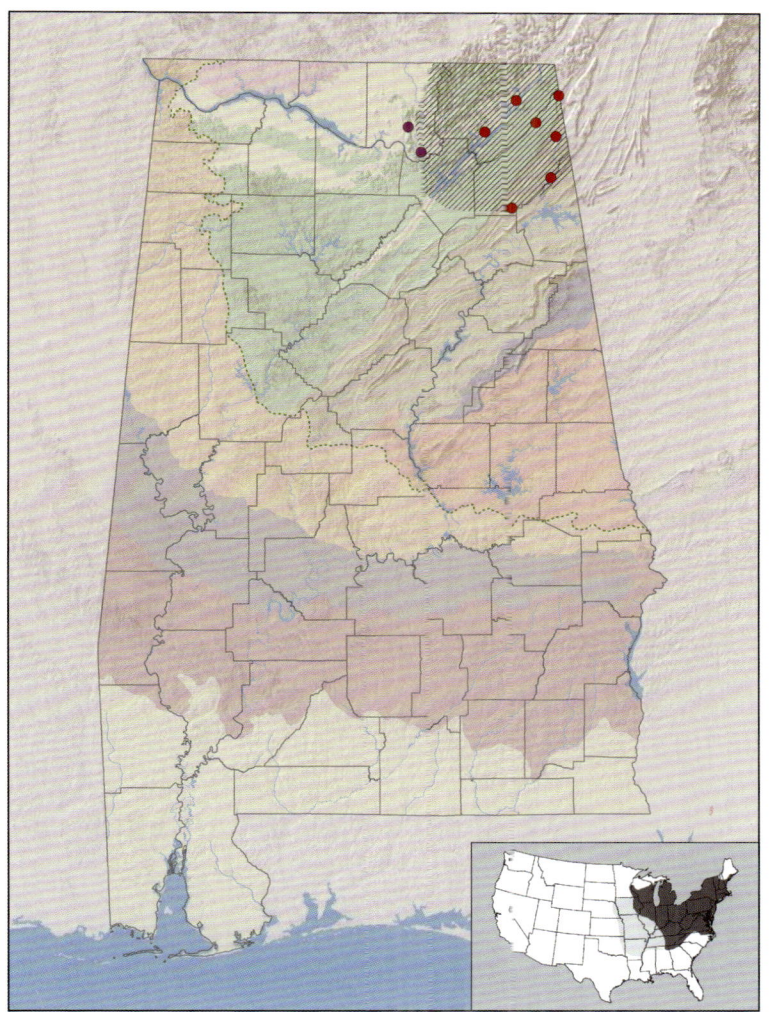

Distribution of Eastern Milksnake, *Lampropeltis triangulum triangulum*. The presumed range of the subspecies in Alabama is indicated by hatching. Solid purple dots indicate intergrades between *L. t. triangulum* and *L. t. syspila*. Inset map depicts approximate range in the United States with dark shading indicating the range of *L. t. triangulum* and light shading indicating the range of all other subspecies.

**HABITS** The Eastern Milksnake is secretive, spending much of its time under rocks, boards, logs, and in other protected places. In some parts of its range, forested areas appear to be the favored habitats, while in other parts, prairies are said to be frequently inhabited. Some seasonal variation in habitat preference is indicated, with snakes tending to frequent higher elevations during fall, winter, and spring months and moving to low, mesic areas during the summer. Specimens from Alabama are all from sandstone or limestone cliffs, where they have been found at night or at dusk in May and June. Whether these snakes hibernate as a group, as they do in northern areas, is unknown but seems likely. This subspecies breeds in early spring, and mating may take place in or near the hibernaculum as the animals emerge in spring. Females produce from two to seventeen eggs that are elliptical in shape and are deposited under rocks or rotting logs. Food is killed by constriction, and the primary dietary items are small rodents. However, birds, lizards, snakes, and eggs of these vertebrates also are consumed. The Eastern Milksnake may bite when captured but then tends to become mild mannered. The folklore that Milksnakes suck milk from cows is, of course, a fallacy.

**CONSERVATION AND MANAGEMENT** The status of this subspecies is difficult to assess because these snakes are so secretive. Eastern Milksnakes receive no special protection under state law but are tracked as being imperiled under the Alabama Natural Heritage tracking system. The subspecies is found on public lands being managed as wilderness areas (Little River Canyon; DeSoto State Park), a feature that should secure its presence in the state. Any activities, such as mining, that disturb rock outcrops are likely to disrupt hibernacula that may be used habitually by these snakes. Such sites are likely to be limiting to this and other snake species. So, care should be taken to protect these geological formations.

## *Lampropeltis calligaster* (Harlan, 1827)

**TAXONOMY** This species contains the Mole and Prairie Kingsnakes of central and eastern North America. Three subspecies are recognized, two of which are found in Alabama. Previous authors have placed this species in the genera *Ablabes, Coluber, Coronella,* and *Ophibolus.*

### KEY TO THE SUBSPECIES OF *LAMPROPELTIS CALLIGASTER* OF ALABAMA

**1a** Twenty-one or twenty-three (usually twenty-one) scale rows at mid body; fifty or fewer dark dorsal saddles; dorsum tan with reddish saddles.
   *Lampropeltis calligaster rhombomaculata*—**Mole Kingsnake** . . . .
      **page 191.**
**1b** Twenty-three or twenty-five (usually twenty-five) scale rows at mid body; fifty or more dark dorsal blotches; dorsum with faint dorsolateral stripes.
   *Lampropeltis calligaster calligaster* —**Prairie Kingsnake** . . . . **page 194.**

## Mole Kingsnake

*Lampropeltis calligaster rhombomaculata* (Holbrook, 1840)

**DESCRIPTION** Mole Kingsnakes are medium-sized serpents attaining a maximum total length of around 1,190 mm. The head is scarcely wider than the neck, and the anal scute is undivided. These snakes have smooth scales that are arranged in twenty-one or twenty-three rows at midbody. The ground color of the dorsum is light brown, yellowish brown, or pinkish brown with fifty or fewer dark brown to reddish spots extending along the back and alternating with less conspicuous dark spots along the sides. These markings fade with age, virtually disappearing in old individuals. The venter is white and is clouded or spotted with brown.

**ALABAMA DISTRIBUTION** Apparently, Mole Kingsnakes are found throughout the state, with the exception of extreme northern Alabama, where the influence of *L. c. calligaster* is found. Most Alabama specimens of *L. c. rhombomaculata* have been collected in the Fall Line Hills and Ridge and Valley regions and from the vicinity of Mobile in the Lower Coastal Plain.

Juvenile Mole Kingsnake, *Lampropeltis calligaster rhombomaculata*, Hale County, AL

Adult Mole
Kingsnake,
*Lampropeltis calli-
gaster rhombomac-
ulata*, Limestone
County, AL

**HABITS** The Mole Kingsnake is mostly fossorial, and its appearances above ground are confined largely to nighttime hours. Driving black-topped roads on warm spring and summer nights, especially during rainy weather, has proven to be the most effective collecting method. Mole Kingsnakes collected in this fashion are most often males, suggesting greater movement distances and home range areas in males than females. The subspecies may be active year round, especially in the Coastal Plain where it appears to be most common (Palmer and Braswell 1995). Little is known of reproductive habits in Alabama. Elsewhere, these snakes are known to produce clutches of three to thirteen eggs that are placed in a ground nest in late May to late June. The eggs are oblong, have leathery shells, and adhere to each other. Food consists of lizards, small rodents, and snakes, all of which are killed by constriction. Mole Kingsnakes are also known to eat bird eggs.

**CONSERVATION AND MANAGEMENT** Few data on this subspecies are available for the state of Alabama. Therefore, the size and health of populations of Mole Kingsnakes is largely unknown. Activity patterns of this subspecies suggest that males are susceptible to mortality from vehicular traffic. These snakes are most strongly associated with open, dry habitats that likely were fire maintained in the ancestral landscape. Therefore, areas with low vehicular traffic, especially at night,

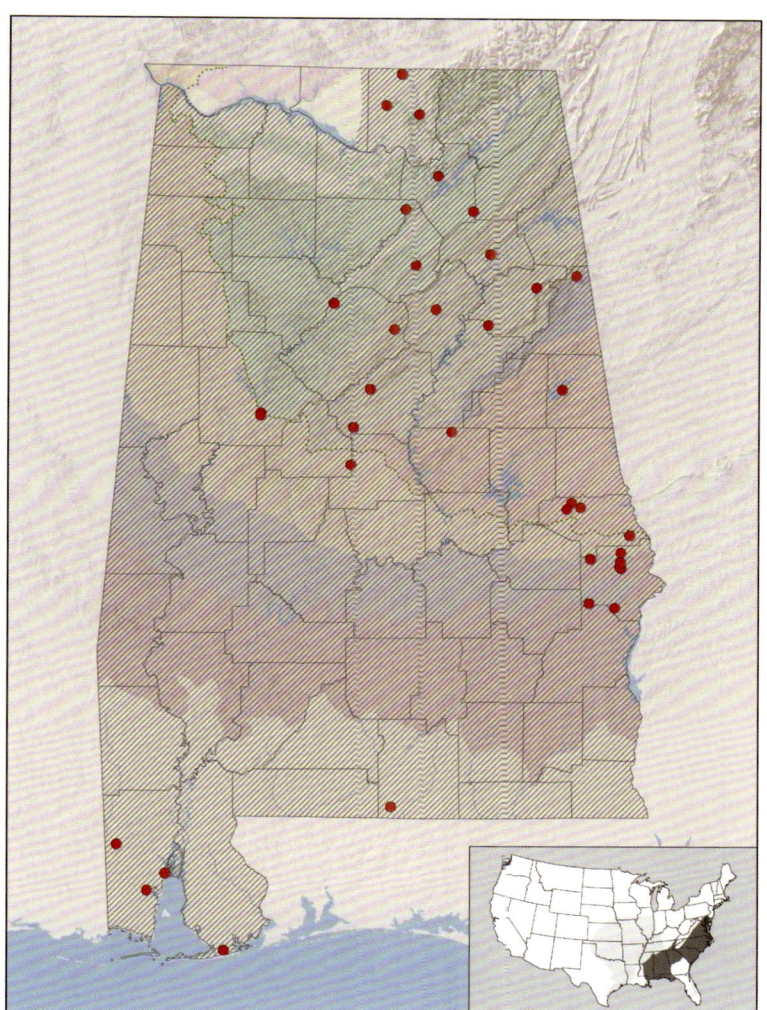

Distribution of Mole Kingsnake, *Lampropeltis calligaster rhombomaculata*. The presumed range of the subspecies in Alabama is indicated by hatching. Inset map depicts approximate range in the United States with dark shading indicating the range of *L. c. rhombomaculata* and light shading indicating the range of the other subspecies.

and managed with prescribed fire appear most likely to retain viable populations of Mole Kingsnakes. The subspecies receives no special protection under state law, but it is tracked as being vulnerable by the Alabama Natural Heritage Program.

## Prairie Kingsnake

*Lampropeltis calligaster calligaster* (Harlan, 1827)

**DESCRIPTION** Prairie Kingsnakes are medium-sized serpents attaining a maximum total length of around 1,400 mm. Their head is scarcely wider than the neck, and the anal scute is undivided. These snakes have smooth scales that are arranged in twenty-three or twenty-five rows at midbody. The ground color of the dorsum is light brown or gray with more than fifty black-edged dark brown to rusty brown spots extending down the back and alternating with less conspicuous dark spots along the sides. The markings fade with age, virtually disappearing in old individuals, which may tend to have faded light stripes. The venter is white with square brown markings.

**ALABAMA DISTRIBUTION** This subspecies is known from Madison County but likely is found in adjacent counties north of the Tennessee River.

**HABITS** This subspecies is found in grasslands, pastures, and open forests, where it may be found under surface objects or crossing roads at night. Males have home ranges that are four times larger than those of females, and the home range includes a hibernaculum, where this subspecies overwinters. Females select grasslands near roads but tend not to cross roads (Richardson et al. 2006). Males mate with females upon emergence from the hibernaculum in spring, and the female then lays six to eighteen adherent eggs in a ground nest, generally in

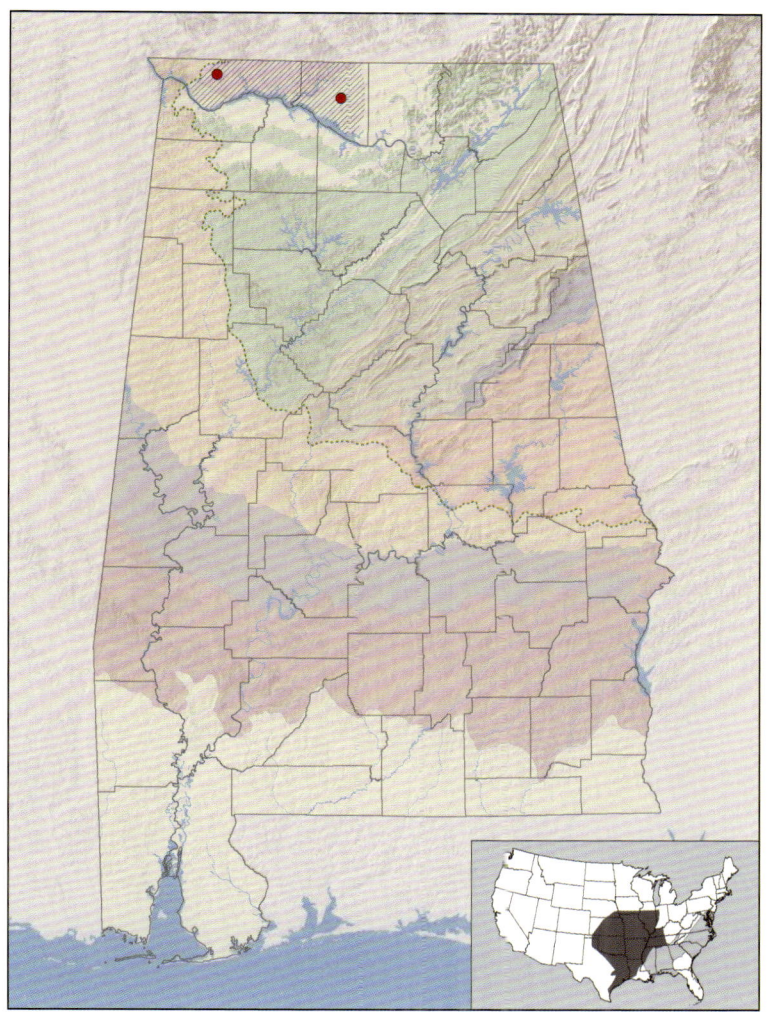

Distribution of Prairie Kingsnake, *Lampropeltis calligaster calligaster.* The presumed range of the subspecies in Alabama is indicated by hatching. Inset map depicts approximate range in the United States with dark shading indicating the range of *L. c. calligaster* and light shading indicating the range of the other subspecies.

June or July (Marion 2004). Prairie Kingsnakes are constrictors that feed on lizards, rodents, bird eggs, and other snakes.

CONSERVATION AND MANAGEMENT The Prairie Kingsnake is ranked Priority 3 (Moderate Conservation Concern) by ADCNR (Mirarchi, Bailey, Haggerty, and Best 2004), and therefore, it is unlawful to possess this subspecies of Kingsnake in the state of Alabama without a special permit. Few data are available for the state of Alabama documenting the size and health of populations of Prairie Kingsnakes. But, this subspecies is known elsewhere to decline in abundance as fire-maintained grasslands are allowed to be overtaken by hardwoods when fire is removed (Fitch 2006). Therefore, maintenance of open grasslands appears to be the best management strategy for conserving this snake subspecies. Power line rights of way may be key resources in Alabama, but if these are mowed for maintenance then mowing should be done during winter to avoid excessive mortality to Prairie Kingsnakes that might be present during the active season. These snakes seek the warmth of roads at night, where they are frequently killed by vehicular traffic. But, because Prairie Kingsnakes are found in a variety of habitats, creation of underpasses or barriers to reduce road mortality seems unlikely to succeed as a conservation strategy.

# Subfamily Dipsadinae

This subfamily is distinguished by the presence of enlarged spines along the lateral edge of the hemipenes of males and unique ornamentation on the lobes of each hemipenis (Zaher et al. 2009). However, these features are lost in the genera *Carphophis* and *Diadophis*, which have hemipenes with more simplified structural features. Several members of the subfamily have a pair of enlarged teeth on each side and at the back of the maxilla. These teeth are associated with modified salivary glands that produce venoms capable of immobilizing prey. The subfamily contains 96 genera, 5 of which occur in Alabama, and about 730 species, 7 of which occur in Alabama.

## Key to the Genera of Dipsadinae of Alabama

**1a** Prefrontal contacts orbit; tip of tail covered with a hardened spine-like scale. Go to **2**.

**1b** Prefrontal not in contact with orbit. Go to **3**.

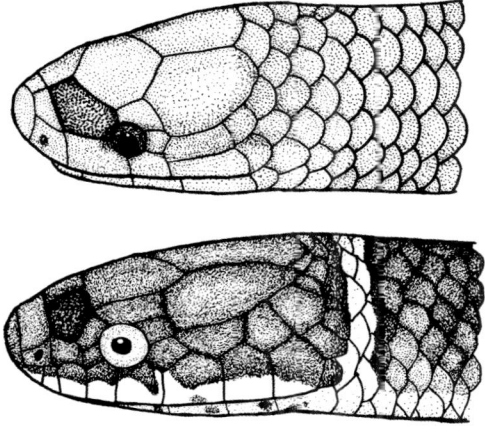

From top to bottom:

Dorsolateral view of head of a Wormsnake (*Carphophis*); based on AUM 23434

Dorsolateral view of head of a Ringnecked Snake (*Diadophis*); based on AUM 39448

**2a** Dorsum with bold markings of red and black.

> Genus *Farancia*—Mudsnakes and Rainbow Snakes . . . . page 199.

**2b** Dorsum uniform gray or brown.

> Genus *Carphophis*—North American Wormsnakes . . . . page 211.

**3a** All or most of the body scales keeled; rostral scale protruding and upturned.

    Genus *Heterodon*—**North American Hog-nosed Snakes** . . . . **page 218.**

**3b** All or most of the body scales smooth; rostral scale not protruding and upturned. Go to **4.**

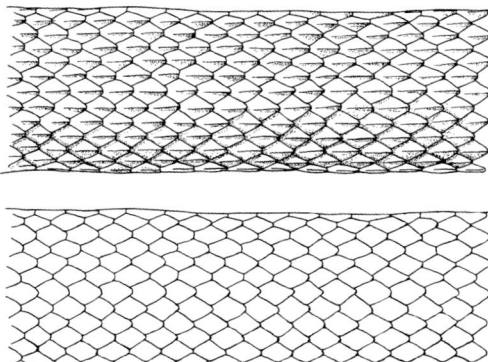

*From top to bottom:*

Dorsal view of keeled scales

Dorsal view of smooth scales

**4a** Dorsum uniform dark brown with a single light yellow (occasionally interrupted) neck ring; a dark stripe not present from tip of snout through eye to corner of mouth.

    Genus *Diadophis*—**Ring-necked Snakes** . . . . **page 226.**

**4b** Dorsum lacking light ring on neck; a dark stripe present from tip of snout through eye to corner of mouth, bordered below by light yellow upper labial scales.

    Genus *Rhadinaea*—**Littersnakes** . . . . **page 232.**

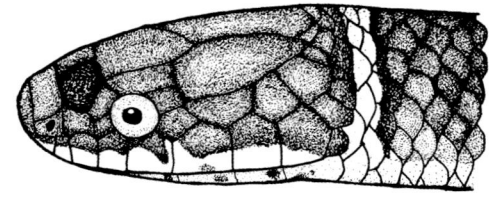

Dorsolateral view of head of a Ring-necked Snake (*Diadophis*); based on AUM 39448

# Mudsnakes and Rainbow Snakes
## Genus *Farancia* (Gray, 1842)

This genus is found only in the southeastern United States, is sister either to *Carphophis* (Pyron et al. 2011) or *Carphophis + Contia + Diadophis* (Zaher et al. 2009) and includes two species of large, aquatic or semiaquatic snakes, both of which occur in Alabama. The two species are brightly colored with red and black, are sluggish, and have thick bodies with extremely shiny, smooth scales. Because they are colored in a fashion that is vaguely reminiscent of an Eastern Coralsnake and because they have a sharp, spinelike scale at the end of the tail and, when freshly captured, will press the tip of this scale against the skin of a captor, some people believe these snakes to have a venomous bite and a stinger that is equally toxic. Although rear-fanged, neither species has been proven to have a bite that harms humans. Additionally, the tail spine is devoid of venom. Nevertheless, members of this genus are famous in folklore for the putative habit of seizing the tail in the mouth, pursuing an enemy by rolling like a hoop, and, upon overtaking the victim, administering a deadly sting. Based on this lore, stinging snake, hoop snake, thunderbolt, and lightning snake are among the names applied to them. In fact, the myth is embellished in southern lore to a full-blown tall tale in which a farmer is chased by a hoop snake that propels itself at the farmer, who ducks to avoid the snake, causing the snake to impale its stinger into a sapling tree, which then swells to giant size because of the venomous sting. The farmer then cuts down the tree, saws it into lumber, and uses the boards to build a nice addition to his house. But then, the rains come and leach the venom from the wood, causing it to shrink to its original size, leaving the farmer with only a small chicken coop attached to the house.

## Key to the Species of *Farancia* of Alabama

**1a** Dorsum and sides with longitudinal light (white, yellow, or red) stripes.

> *Farancia erytrogramma erytrogramma*—Common Rainbow Snake . . . . page 201.

**1b** Sides of body with vertical red (rarely white) bars or rings.

> *Farancia abacura* ssp. . . . . page 204.

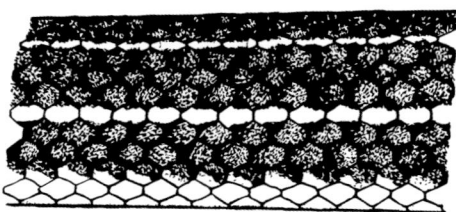

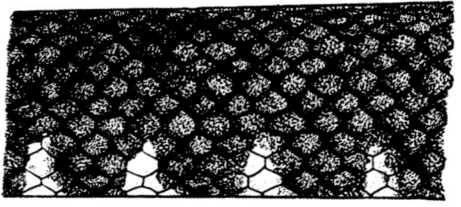

*From top to bottom:*

Lateral view of color pattern of a Rainbow Snake (*Farancia erytrogramma*)

Lateral view of color pattern of a Mudsnake (*Farancia abacura*)

## Common Rainbow Snake

*Farancia erytrogramma erytrogramma* (Palisot de Beauvois, 1801)

**DESCRIPTION** This subspecies is a large snake attaining a maximum
total length of around 1,500 mm with a stout, cylindrical body and a
short, stout tail that ends in a spine. The head and neck are approximately equal in width, and the eyes and tongue are noticeably small.
A preocular is lacking, and the anal scute usually is divided. The dorsal scales are smooth and are organized in nineteen rows at midbody
(slight keels may be present dorsally near the cloaca). The dorsum
is shiny dark blue or blue-black with three narrow, longitudinal red
stripes, one positioned mid-dorsally and the others dorsolaterally. A
yellow stripe is present on each side ventrolaterally where the dorsal
scales contact the ventral scutes. The venter is red with each ventral
scute possessing a black blotch at each side and usually with a smaller
one in the middle, these forming diffuse black ventral stripes. Subcaudals average forty-eight in males and forty in females; males have
more spots on the subcaudals than do females (Gibbons et al. 1977).

**ALABAMA DISTRIBUTION** This subspecies has a spotty distribution in
the Coastal Plain, occasionally extending slightly above the Fall Line
along major rivers.

**HABITS** The Common Rainbow Snake, one of the most beautiful of Alabama's reptiles, is also one of the least frequently encountered of our

Common Rainbow Snake, *Farancia erytrogramma erytrogramma*, Laurens County, GA

large snakes. It is found in rivers and medium- and large-sized creeks, where it is mostly active at night. Alabama specimens have been collected in piles of wet debris along the edges of streams and beneath mats of vegetation along the edges of ponds. Males reach sexual maturity in two years, and females may take three years; males reach a maximum length of about 1 m, whereas females may reach sizes of 1.6 m (Gibbons et al. 1977). Females lay eggs in nests on upland sites, and juveniles likely overwinter in the nest, entering the wetlands the following year (Gibbons et al. 1977). Common Rainbow Snakes are thought to specialize on American eels. Other fish, tadpoles, and salamanders have been reported as occasional food items. Theses snakes are extremely docile, never biting unless they mistake the handler for food.

**CONSERVATION AND MANAGEMENT** The Common Rainbow Snake is ranked Priority 1 (Highest Conservation Concern) by ADCNR (Mirarchi, Bailey, Haggerty, and Best 2004), and therefore, it is unlawful to possess this subspecies in the state of Alabama without a special permit. This subspecies requires large creeks and rivers with eels, a fish that needs migratory access to the Gulf of Mexico. Thus, dams on Alabama's major rivers undoubtedly have adversely affected Rainbow Snakes by reducing eel populations (Hughes and Nelson 2004). Rainbow Snakes are most common in the upper Mobile Bay and Tensaw River. However, they are frequently killed on major roads in

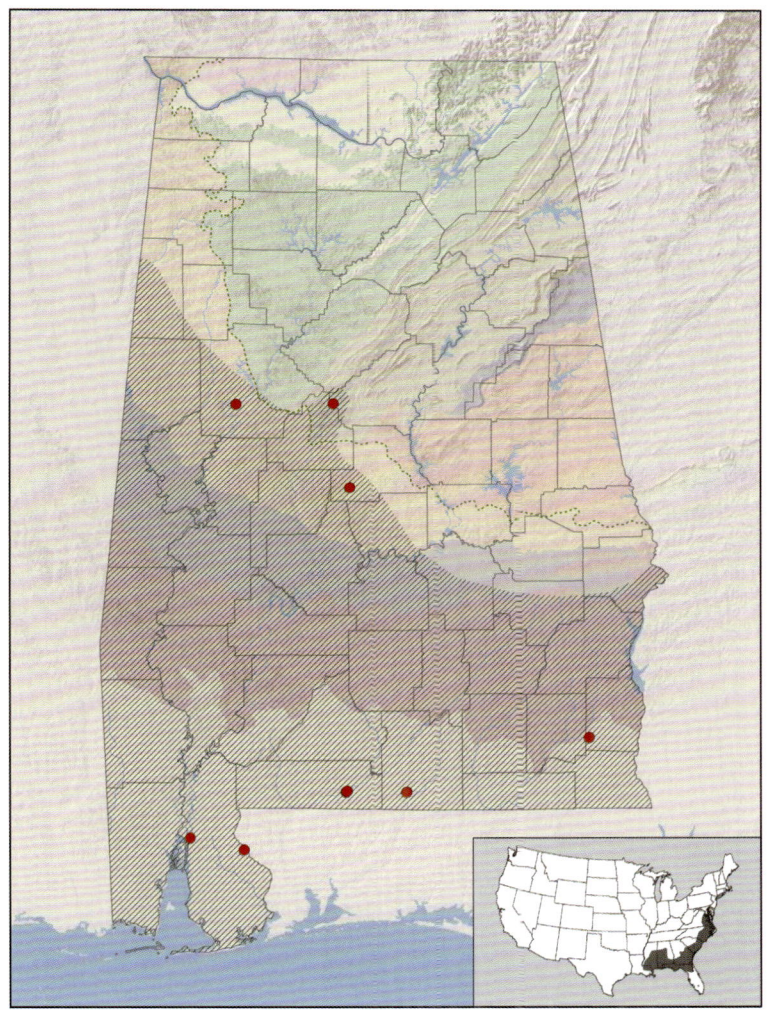

Distribution of Rainbow Snake, *Farancia erytrogramma*. The presumed range of the species in Alabama is indicated by hatching. Inset map depicts approximate range in the United States.

this region. So, carefully designed ecological passages for vertebrates are important for maintaining Rainbow Snakes in this area. Elsewhere, proper management of riparian forests along major rivers of the Coastal Plain and protection of sand bars should improve habitat for this subspecies. This subspecies is known to make long distance movements across upland areas, suggesting that primary wetland habitats must also be embedded in landscapes of well-managed uplands (Steen, Stevenson, et al. 2013).

TAXONOMY This species is sister to *F. abacura*, the only other living member of the genus. Traditionally, two subspecies of this snake are recognized. However, one, the Southern Florida Rainbow Snake, is known from a single specimen that has not been seen in repeated efforts to find additional specimens. At best, this taxon appears to be extinct. Previous authors have placed this subspecies in the genera *Abastor, Calopisma, Coluber, Helicops, Homalopsis, Hydrops,* and *Natrix*.

## *Farancia abacura* (Holbrook, 1836)

TAXONOMY This is the species to which the mudsnakes belong. They are so named because they often occur in the aquatic sites occupied by sirens and amphiumas, their primary dietary items. Two subspecies are recognized, both of which occur in Alabama. However, considerable regional variation occurs in *F. a. abacura*, including south Alabama. Two specimens from Big Creek and Cowarts Creek in Houston County, headwaters of the Chipola River, have a color pattern of thin, light rings. Specimens with similar rings toward the anterior end are found from the Perdido, Conecuh, and Yellow Rivers, suggesting a possible separate, currently unnamed lineage in this region. Previous authors have placed this species in the genera *Calopisma, Helicops,* and *Homolopsis*.

### KEY TO THE SUBSPECIES OF *FARANCIA ABACURA* OF ALABAMA

1a Upward extensions of red (rarely white) ventral color forming fifty-three or more figures with acutely angular apices.
   *Farancia abacura abacura*—**Eastern Mudsnake** . . . . **page 205.**
1b Upward extension of red ventral color forming fifty-two or fewer figures with rounded or squared apices.
   *Farancia abacura reinwardtii*—**Western Mudsnake** . . . . **page 208.**

# Eastern Mudsnake
*Farancia abacura abacura* (Holbrook, 1836)

**DESCRIPTION** This is a large snake attaining a maximum total length
of about 2,055 mm, with the tail comprising only a small portion of
that length. Its body is cylindrical as is the head, which is about the
same width as the neck. The eyes of mudsnakes are small, black, and
difficult to detect from their background, the black color of the head.
No preocular scale is present on the head, and the anal scute usually
is divided. The dorsal scales are smooth and are arranged in nineteen
rows at midbody (slight keels may be present dorsally near the cloaca),
giving the dorsum a shiny, blue-black appearance. The ventrals are
mostly red (rarely white), and this color extends upward on the sides to
form a series of bars on the body and tail that may occasionally meet
on the dorsum to form rings. The dorsal color extends onto the venter
between the red bars, creating alternating ventral regions of red and
black. In this subspecies, light body bars usually number fifty-three or
more, their apices being pointed instead of rounded (fewer than fifty
three and with flattened or rounded apices are present in *F. a. reinwardtii*). The apices of corresponding bars on the neck connect in specimens

Ringed morph of
Eastern Mudsnake,
*Farancia abacura
abacura*, Gulf
County, FL

from the Perdido, Conecuh, Yellow, and Chipola River drainages; these markings are separated by no more than three dorsal scale rows for specimens from the upper Alabama River and its tributaries.

**ALABAMA DISTRIBUTION** This subspecies is found in southern Baldwin County and Dauphin Island in Mobile County eastward across the Coastal Plain.

**HABITS** The habitats of this subspecies include swamps, floodplains, and sluggish, mud-bottomed creeks. The presence of *Amphiuma* or *Siren*, the chief components of the diet, may be required to support populations of this snake. Daylight hours are spent in the water or in burrows in the mud. On rainy nights, these snakes may be found crossing roads in swampy areas. The Mudsnake is completely harmless, refusing even to bite. Females lay eleven to fourteen oblong or elliptical eggs in a flask-shaped cavity in the ground of upland areas or in alligator nests (Hall and Meier 1993). The female snake attends the nest. During some years, eggs hatch in fall, and in other years, they overwinter, hatching the following spring. Juveniles migrate to wetlands immediately after hatching (Semlitsch et al. 1988). Food items mentioned in the literature other than sirens and amphiumas include tadpoles, frogs, small salamanders, and fish.

**CONSERVATION AND MANAGEMENT** Eastern Mudsnakes are still found extensively in Alabama, but numbers seem to have declined. This subspecies receives no special protection by state law. The keys to

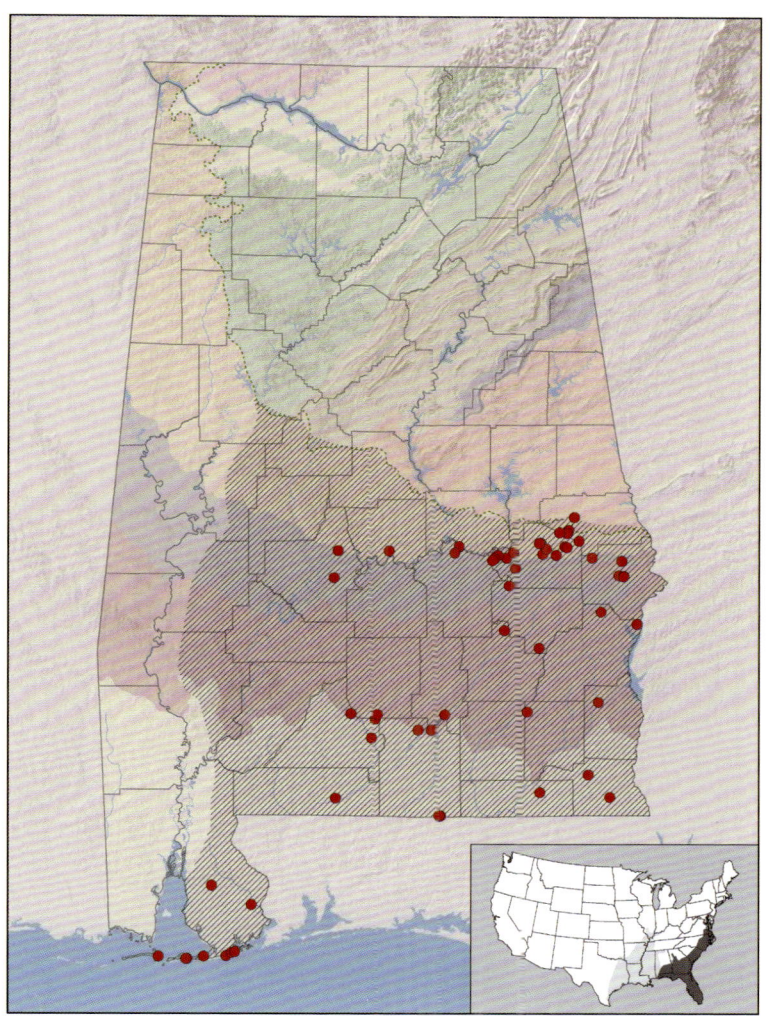

Distribution of Eastern Mudsnake, *Farancia abacura abacura*. The presumed range of the subspecies in Alabama is indicated by hatching. Inset map depicts approximate range in the United States with dark shading indicating the range of *F. a. abacura* and light shading indicating the range of the other subspecies.

maintaining this subspecies in Alabama are preservation of swampy habitats with aquatic vegetation. These features are required for nest sites and to maintain populations of food resources. This subspecies is known to make long distance movements across upland areas, suggesting that primary wetland habitats must also be embedded in landscapes of well-managed uplands (Steen, Stevenson, et al. 2013).

# Western Mudsnake

*Farancia abacura reinwardtii* (Schlegel, 1837)

**Description** This is a large snake, attaining a maximum total length of about 1,830 mm, with the tail comprising only a small portion of that length. Its body is cylindrical as is the head, which is about the same width as the neck. The eyes are small, black, and difficult to detect from the black background of the head. No preocular scale is present on the head, and the anal scute usually is divided. The dorsal scales are smooth and are arranged in nineteen rows at midbody (slight keels may be present dorsally near the cloaca), giving the dorsum a shiny, blue-black appearance. The ventrals are mostly red (rarely white), and this color extends upward on the sides to form a series of bars on the body and tail. The dorsal color extends onto the venter between the red bars, creating alternating ventral regions of red and black. In this subspecies, light body bars number fewer than fifty-three (fifty-three or more are present in *F. a. abacura*), and their apices are flattened or rounded across their tops (apices are pointed in *F. a. abacura*). In the neck region, the light bars are separated by eight or more dorsal scale rows (*F. a. abacura* has three) of dark dorsal color.

**Alabama Distribution** This subspecies is found west of the Mobile River, along the western boundary of the state. It also occurs in the Tennessee Valley. However, its presence in that region is documented by only one specimen (Madison County), and the extent of the range is not well-known.

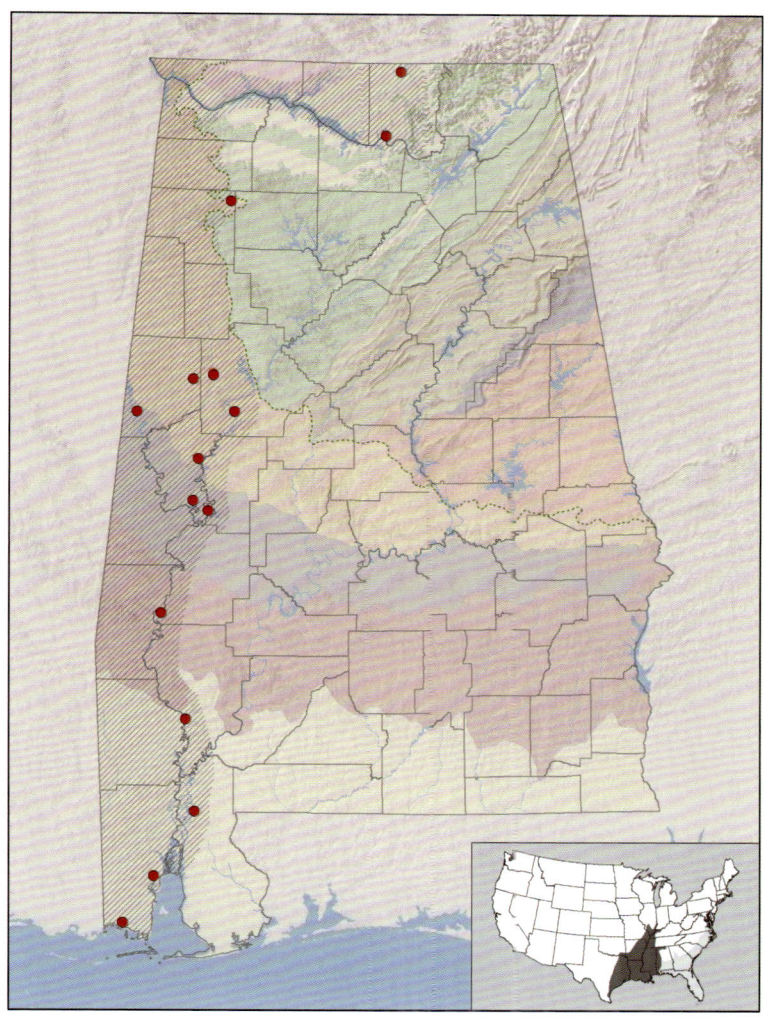

Distribution of Western Mudsnake, *Farancia abacura reinwardtii*. The presumed range of the subspecies in Alabama is indicated by hatching. Inset map depicts approximate range in the United States with dark shading indicating the range of *F. a. reinwardtii* and light shading indicating the range of the other subspecies.

**Habits** This subspecies inhabits beaver swamps, ponds, and lakes with swampy margins and abundant aquatic vegetation, floodplains, and sluggish, mud-bottomed creeks. *Amphiuma, Siren,* or possibly *Necturus,* the chief components of the diet, may be required to support these snakes. The subspecies spends most of its daylight hours in the water or in burrows in the mud. On rainy nights, it may be found crossing roads in swampy areas.

The Mudsnake is completely harmless, refusing to bite when handled. Females lay oblong or elliptical eggs in a flask-shaped cavity in the ground and remain with them, apparently, until they hatch. Clutch size ranges from eleven to fourteen eggs for the subspecies, and the eggs may overwinter, hatching the following spring.

**CONSERVATION AND MANAGEMENT** Western Mudsnakes are still found in Alabama, but the status of their populations is unknown. This subspecies receives no special protection by state law. The keys to maintaining it in Alabama are to retain nest sites and food resources. Both of these require managing for old growth trees along riparian areas and native aquatic vegetation along the margins. This subspecies is known to make long distance movements across upland areas, suggesting that primary wetland habitats must also be embedded in landscapes of well-managed uplands (Steen, Stevenson, et al. 2013).

# North American Wormsnakes
## Genus *Carphophis* (Gervais, 1843)

This genus contains the North American Wormsnakes, so named because of their superficial resemblance to earthworms and their propensity to consume them. These are small snakes found in hardwood areas with loose soil, thick leaf litter, and rocks and rotting logs as cover. These snakes are the sister taxon either to *Farancia* (Pyron et al. 2011), the Mudsnakes and Rainbow Snakes of the Southeast, or *Contia* (Zaher et al. 2009), the sharp-tailed snakes of the West. The genus is composed of two species, one of which is found in Alabama.

### *Carphophis amoenus* (Say, 1825)

TAXONOMY The members of this species are divided into two subspecies, both of which occur in Alabama. The two subspecies have relatively distinct geographic distributions in the state, but intergrades occur to the north, especially along the Tennessee River. Their distribution pattern in the state suggests that they may represent separate species. Previous authors have placed this species in the genera *Calmaria*, *Celata*, *Celuta*, and *Coluber*.

### KEY TO THE SUBSPECIES OF *CARPHOPHIS AMOENUS* OF ALABAMA

**1a** Prefrontal and internasal scales fused.
> *Carphophis amoenus helenae*—Midwestern Wormsnake . . . . **page 212.**

**1b** Prefrontals and internasals separate.
> *Carphophis amoenus amoenus*—Eastern Wormsnake . . . . **page 215.**

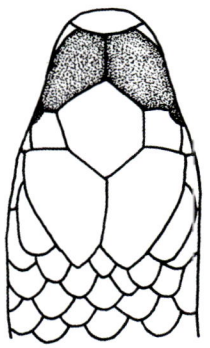

*From left to right:*

Dorsal view of head of a Midwestern Wormsnake (*Carphophis amoenus helenae*); based on AUM 23434

Dorsal view of head of an Eastern Wormsnake (*Carphophis amoenus amoenus*); based on AUM 13754

## Midwestern Wormsnake
*Carphophis amoenus helenae* (Kennicot, 1859)

**DESCRIPTION** These small snakes attain a maximum total length of about 350 mm and possess a tail that is rather short and has a spine at its tip. The head is small, flattened, and about the same width as the neck. The dorsal scales are smooth, shiny, and arranged in thirteen rows at midbody. The anal scute is divided. Dorsally, the ground color is brown, gray-brown, or pinkish brown and is otherwise unmarked. Ventrally, the belly is pink, and this color extends upward onto the first or second lateral scale rows. There is a sharp delineation between the light belly color and the dark dorsal color, and this feature is particularly evident in hatchlings. Additionally, the upper labials are bright pink ventrally, changing abruptly to brown dorsally, markings that align with the lateral light and dark markings. In this subspecies, the internasals and prefrontals are fused (approximately 95 percent of individuals; they are separate in *C. a. amoenus*). Subcaudals average thirty-five in males and twenty-seven in females, and females reach larger body sizes than do males (Russell and Hanlin 1999).

**ALABAMA DISTRIBUTION** The Midwestern Wormsnake is known to occur in the central and eastern portions of the Fall Line Hills, Black Belt, Red Hills, and Lower Coastal Plain. This subspecies is also found along the Tennessee River. Above the Fall Line, *C. a. helenae* is found

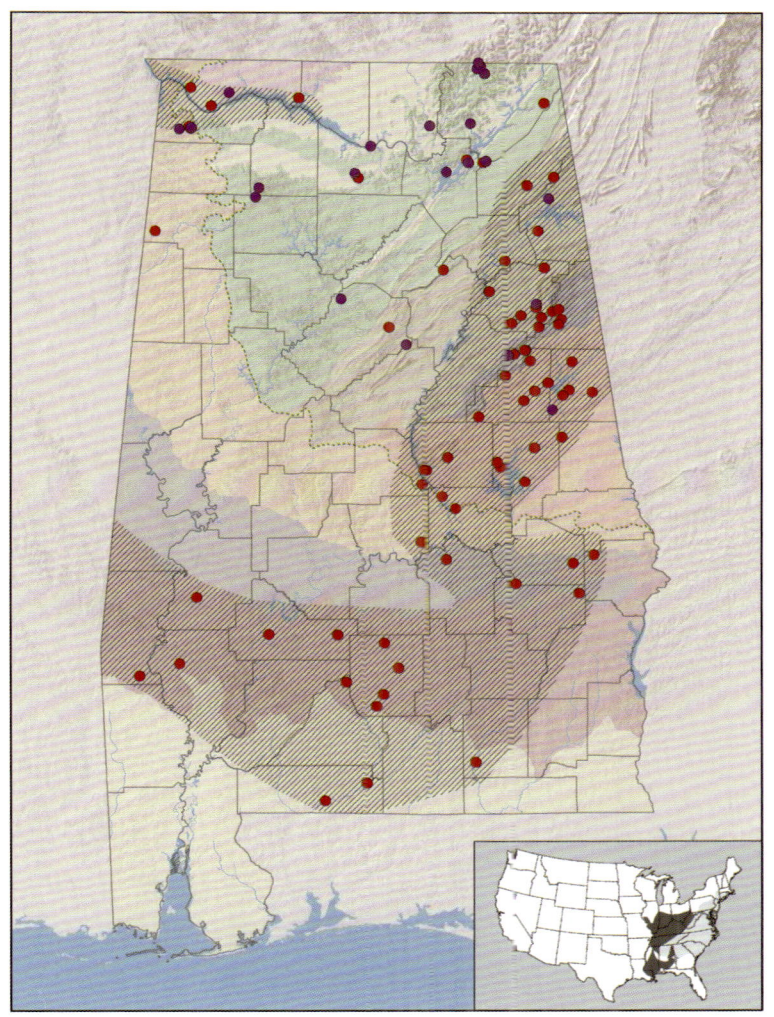

Distribution of Midwestern Wormsnake, *Carphophis amoenus helenae*. The presumed range of the subspecies in Alabama is indicated by hatching. Solid purple dots indicate intergrades between *C. a. helenae* and *C. a. amoenus*. Inset map depicts approximate range in the United States with dark shading indicating the range of *C. a. helenae* and light shading indicating the range of the other subspecies.

east of the Coosa River in the Piedmont, Talladega Upland, and portions of the Ridge and Valley. Except in a few areas of the Red Hills and Fall Line Hills, *C. a. helenae* is infrequently encountered in the Coastal Plain; above the Fall Line, this subspecies can be abundant. Intergrades between *C. a. amoenus* and *C. a. helenae* occur in the Appalachian Plateaus, especially in areas draining into the Tennessee River.

**HABITS** Optimum habitat for Midwestern Wormsnakes is mesic hardwood forests with abundant leaf litter and humus. These snakes are secretive and are usually found beneath rocks, logs, and debris. They will not bite. Wormsnakes are oviparous, laying two to eight (usually five) eggs in rotting logs, stumps, or sawdust piles. Earthworms constitute the bulk of the diet, but slugs and larval insects also are consumed.

**CONSERVATION AND MANAGEMENT** This subspecies receives no special protection from state law because it is so common. These serpents can survive in urban areas where trees are allowed to reach mature ages. Management activities that retain old-growth trees along riparian zones should enhance habitat for these snakes.

# Eastern Wormsnake

*Carphophis amoenus amoenus* (Say, 1825)

**DESCRIPTION** These small snakes attain a maximum total length of
about 350 mm and possess a tail that is rather short and has a spine
at its tip. The head is small, flattened, and about the same width as
the neck. The dorsal scales are smooth, shiny, and arranged in thirteen
rows at midbody. The anal scute is divided. Dorsally, the ground color
is brown, gray-brown, or pinkish brown and is otherwise unmarked.
Ventrally, the belly is pink, and this color extends upward onto the first
or second lateral scale rows. There is a sharp delineation between the
light belly color and the dark dorsal color, and this feature is particularly
evident in hatchlings. But, the division between light ventral and dark
dorsal coloration does not extend anteriorly to the labials. In this subspecies, the internasals and prefrontals typically are separate (approximately 95 percent of individuals; fused in *C. a. helenae*). Subcaudals
average thirty-five in males and twenty-seven in females, and females
reach larger body sizes than do males (Russell and Hanlin 1999).

**ALABAMA DISTRIBUTION** The range of this subspecies enters Alabama
from the north through the Highland Rim, broadens southward to
include most of the Appalachian Plateaus except for the northeastern
portion, extends westward into the Fall Line Hills in western Alabama,

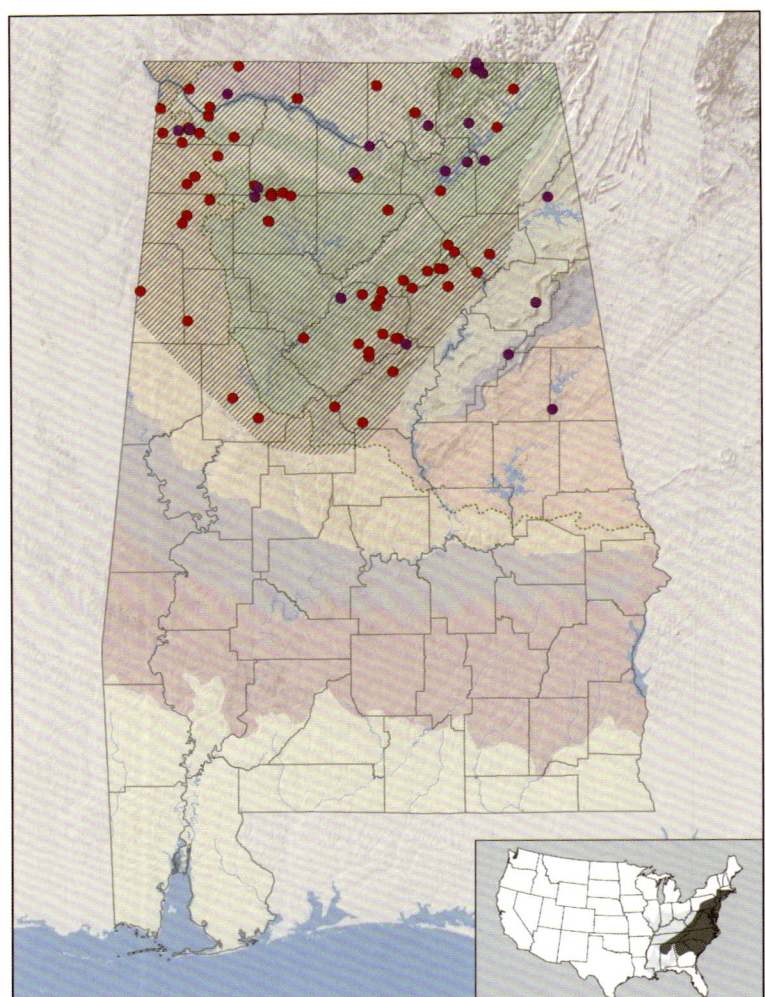

Distribution of Eastern Wormsnake, *Carphophis amoenus amoenus.* The presumed range of the subspecies in Alabama is indicated by hatching. Solid purple dots indicate intergrades between *C. a. amoenus* and *C. a. helenae.* Inset map depicts approximate range in the United States with dark shading indicating the range of *C. a. amoenus* and light shading indicating the range of the other subspecies.

and extends southward to include the lower part of the Ridge and Valley region. This subspecies gives way to *C. a. helenae* along the Tennessee River, with intergradients of *C. a. amoenus* and *C. a. helenae* occurring in the Appalachian Plateaus, especially in areas draining into that river.

**HABITS** Optimum habitats for Eastern Wormsnakes are mesic hardwood forests with abundant leaf litter and humus. These secretive snakes are usually found beneath rocks, logs, and debris. They will not bite. Activity of adults peaks in the fall, presumably because this is the mating season (Russell and Hanlin 1999). Males are captured more frequently than females, likely because males seek mates. Females lay from two to eight (usually five) eggs in rotting logs, stumps, or sawdust piles during the fall; these eggs hatch the following spring (Russell and Hanlin 1999). Earthworms constitute the bulk of the diet, but slugs and larval insects also are consumed.

**CONSERVATION AND MANAGEMENT** This subspecies receives no special protection from state law because it is so common. These serpents can survive in urban areas where trees are allowed to reach mature ages. Management activities that retain old-growth trees along riparian zones should enhance habitat for these snakes.

# North American Hog-nosed Snakes
## Genus *Heterodon* (Latreille, 1801)

This genus contains the North American Hog-nosed Snakes, a distinctive radiation of stout-bodied serpents with an enlarged and upturned rostral scale and enlarged, ungrooved teeth at the rear of the maxilla. These rear fangs are associated with modified salivary glands that produce venoms that immobilize struggling prey and cause pain, swelling, and discoloration in humans (Averill-Murray 2006). However, because the species essentially does not bite, even when handled, one need not worry about bites from these snakes. These snakes are either the sister taxon to the Sharp-tailed Snakes (*Contia*) of the western United States (Pyron et al. 2013) or are the basal member of a clade including the North American dipsadines (*Carphophis* + *Contia* + *Diadophis* + *Farancia*; Zaher et al. 2009). Three species are included within this genus. The range includes much of the United States and portions of Mexico and Canada. Two species occur in Alabama.

### Key to the Species of *Heterodon* of Alabama

**1a** Rostral scale shovel-shaped, sharply upturned; undersurface of tail usually not conspicuously lighter than belly.

   *Heterodon simus*—**Southern Hog-nosed Snake** . . . . page 219.

**1b** Rostral scale pointed, not sharply upturned; undersurface of tail usually conspicuously lighter than belly.

   *Heterodon platirhinos*—**Eastern Hog-nosed Snake** . . . . page 222.

*From left to right:*

Dorsolateral view of head of a Southern Hog-nosed Snake (*Heterodon simus*); based on AUM 34214

Dorsolateral view of head of an Eastern Hog-nosed Snake (*Heterodon platirhinos*); based on AUM 12401

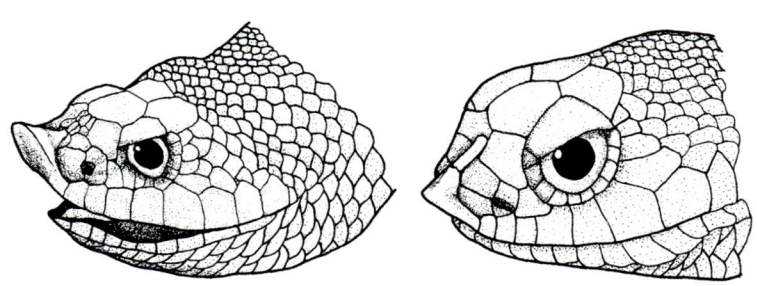

## Southern Hog-nosed Snake
*Heterodon simus* (Linnaeus, 1766)

**DESCRIPTION** Southern Hog-nosed Snakes are short, stout serpents with short tails. This species attains a maximum total length of around 610 mm. The species has a rostral scale that is greatly enlarged and sharply upturned (pointed and not sharply upturned in *H. platirhinos*). The conformation of the rostral scale of *H. simus* allows small scales to separate the prefrontals (prefrontals contact in *H. platirhinos*). Finally, the underside of the tail and belly are both light in *H. simus* (venter of tail light and rest of body dark in *H. platirhinos*). The color pattern of Southern Hog-nosed Snakes is more uniform than in *H. platirhinos*, consisting of a tan or gray ground color with a series of mid-dorsal dark brown blotches alternating with smaller dorsolateral blotches on each side. Melanistic individuals are unknown in *H. simus* but are common in *H. platirhinos*. Subcaudals average forty-two in males and thirty-three in females.

**ALABAMA DISTRIBUTION** The localities from which this species has been collected indicate a spotty distribution in the Coastal Plain and Ridge and Valley regions. Apparently, it never occurs in the Black Belt.

**HABITS** This species is semifossorial but otherwise is poorly known. Most specimens in Alabama have been collected in sandy, relatively open habitats, often crossing roads. These snakes display the same flattening of the head and neck and death-feigning behaviors described in the account for *H. platirhinos*. Natural nests are unknown, but captives

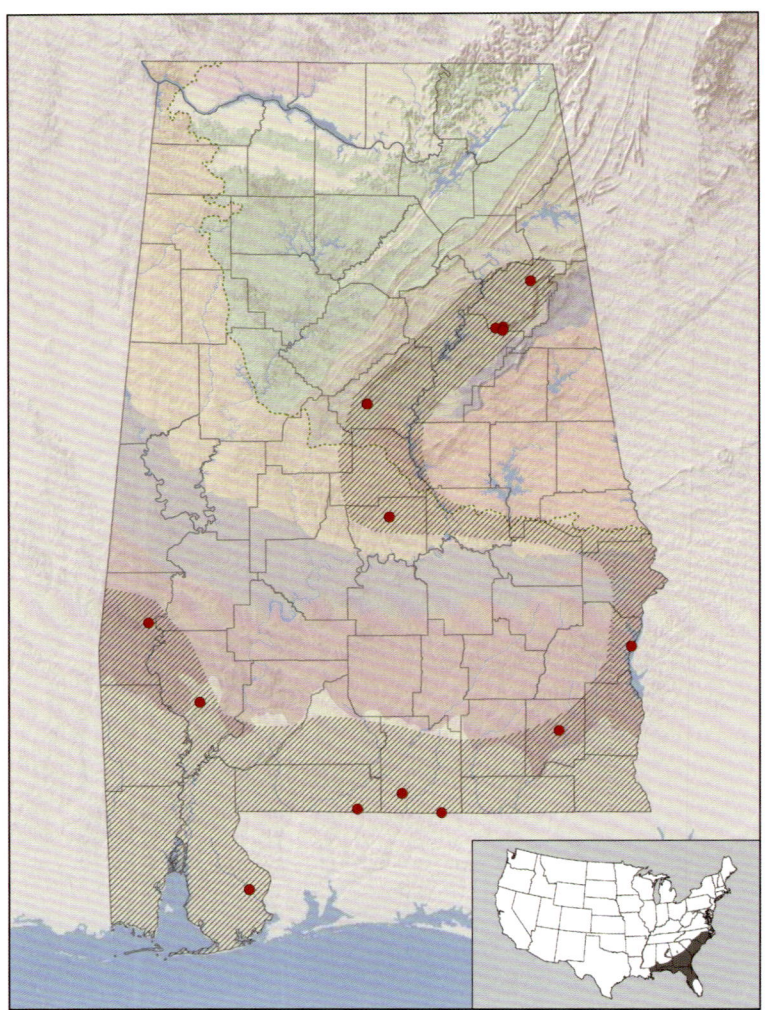

Distribution of Southern Hog-nosed Snake, *Heterodon simus*. The presumed range of the species in Alabama is indicated by hatching. Inset map depicts approximate range in the United States.

have been observed to lay six to ten eggs during mid-July to October. The Southern Hog-nosed Snake apparently feeds almost exclusively on toads (*Anaxyrus* sp.) and Spadefoot Toads (*Scaphiopus holbrooki*).

**CONSERVATION AND MANAGEMENT** The Southern Hog-nosed Snake is rated Priority I (Highest Conservation Concern), and therefore, it is unlawful to possess this species in the state of Alabama without a special permit. This species apparently has always been rare but also apparently has experienced a range-wide decline (Tuberville et al. 2000).

No known populations remain in Alabama (Jensen 2004b), and the last documented occurrence was in 1970. Therefore, a comprehensive survey for this species is needed (SWAP 2005). Its apparent disappearance from Alabama occurred at the same time that the Red Imported Fire Ant (*Solenopsis invicta*) invaded South Alabama. For this reason, increased egg mortality caused by these invasive ants is one hypothesis for the extirpation of Southern Hog-nosed Snakes from the state. Habitat loss and fragmentation undoubtedly also caused population reductions. In particular, the loss of flatwoods and xeric longleaf sandhill habitats has negatively impacted these snakes. Repatriation efforts likely will be required to return Southern Hog-nosed Snakes to the state of Alabama, but such efforts are likely to fail without simultaneous efforts to reduce fire ant densities. Sites suitable for repatriation should be thinned and burned, preferably in the growing season (Jensen 2004b). Mature pines will be required to shed needles that fuel frequent fires, the most important tool for maintaining the open habitat favored by *H. simus*. Control of invasive plants, especially cogongrass (*Imperata cylindrica*), will be vital for maintaining quality habitat for these snakes (SWAP 2005). The Conecuh National Forest, Geneva State Forest, Chattahoochee State Forest, Fort Rucker Military Installation, Perdido River Longleaf-Hills Tract, and Little River State Forest are known to have had or likely contained populations of this species and provide ample opportunities to return this species to the state's herpetofauna.

TAXONOMY The Southern Hog-nosed Snake is the sister taxon to the Western Hog-nosed Snake, *H. nasicus*, sharing with that taxon an abruptly upturned rostral scale. No subspecies are recognized within this species. Previous authors have placed this species in the genus *Coluber*.

## Eastern Hog-nosed Snake
*Heterodon platirhinos* (Latreille, 1801)

**DESCRIPTION** This medium-sized, moderately stout snake attains a maximum total length of about 1,155 mm with a short tail. The dorsal scales are keeled and arranged in twenty-three or twenty-five rows at midbody. The anal scute is divided. The rostral is keeled and pointed, projecting upward but not sharply upturned as in *H. simus*. The prefrontals are in contact with each other in Eastern Hog-nosed Snakes. The dorsal color is extremely variable. Juveniles are yellowish or tan with dark blotches. But adults may be uniform black or reddish, gray, or brown with dark dorsal blotches. The ventral surface of the tail usually is conspicuously lighter than the belly. Subcaudals average fifty in males and forty-five in females.

**ALABAMA DISTRIBUTION** This species occurs throughout the state.

**HABITS** The Eastern Hog-nosed Snake, often called spreading adder, is usually found in fields and areas of broken terrain (Plummer and Mills 2000). Hog-nosed Snakes are diurnal in activity, are relatively sluggish in their movements, and do not bite when handled. Home ranges may be as large as 50 ha in area, and movements are frequent except during the hottest months of summer (Plummer and Mills

Eastern Hog-nosed Snake, *Heterodon platirhinos*, Liberty County, FL

2000). A molested individual snake's first line of defense is a spectacular bluff. Hissing and jerking, it flattens its head and neck and inflates its lung. Sounds are produced both by inflating and deflating the lungs, with the nasal cavity serving as an echo chamber. The formidable aspect thus presented has led many to assume the Hog-nosed Snake to be dangerous. This aspect may include striking by the snake at an intruder, but this is done without opening the mouth and frequently involves apparently purposeful diversion of the strike from contacting the intruder. Should this behavior fail to intimidate the molester effectively, the snake begins to writhe as if in extreme agony. Mouth agape and tongue lolling, it rolls over, belly up. During this ordeal it may disgorge a recent meal from the stomach and defecate material from the large intestine. Then, the snake becomes still and limp, taking on every aspect of death—except one; if turned right side up, the snake immediately rolls over again.

Breeding generally occurs in April and May but may happen as early as February (Steen, Becker, Smith 2005). Between four and sixty-nine eggs are laid in sandy or gravelly soil usually at a depth of about four inches. The diet consists mostly of toads. In fact, chemical cues associated with the cloacal opening of toads appear to trigger the swallowing reflex in Eastern Hog-nosed Snakes. Additionally,

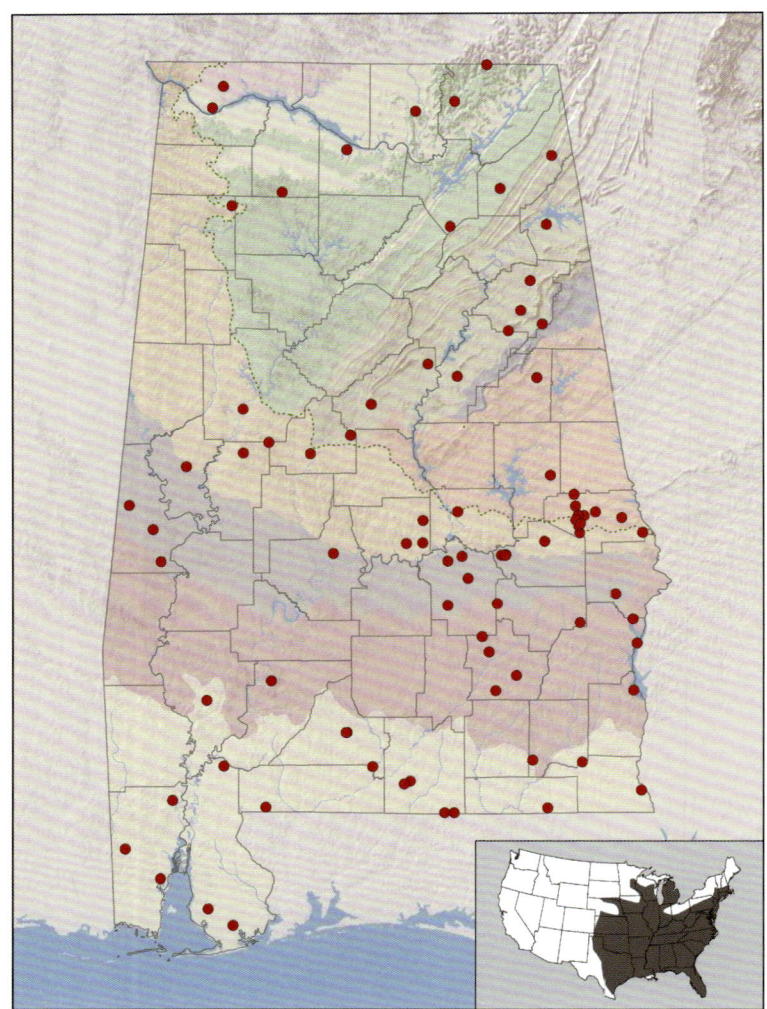

Distribution of Eastern Hog-nosed Snake, *Heterodon platirhinos*. This species is assumed to occur throughout Alabama. Inset map depicts approximate range in the United States.

these snakes are attracted more strongly to chemical stimuli from toads than to chemicals from other infrequently consumed food items (other frogs, lizards, snakes, insects; Cooper and Secor 2007).

**CONSERVATION AND MANAGEMENT** Eastern Hog-nosed Snakes are encountered infrequently, having declined in abundance since the publication of Mount (1975). Toads, their primary dietary item, are still abundant, so food appears to be plentiful. However, this species places eggs in a ground nest that may cause them to be vulnerable to predation by imported fire ants (*Solenopsis invicta*). Maintenance of pasture lands adjacent to forests, especially in areas with loose soils, is a key habitat feature for maintaining this species

**TAXONOMY** This is the basal species within the genus *Heterodon* (Pyron et al. 2013). No subspecies are recognized within this species. Previous authors have placed this species in the genus *Coluber*.

# Ring-necked Snakes
## Genus *Diadophis* (Baird and Girard, 1853)

Members of the genus *Diadophis* are referred to as Ring-necked Snakes because of the bright yellow ring just behind the head. These small snakes are found in hardwood forests where they live under logs and rocks. They possess an enlarged pair of teeth, one on each side, at the back of the maxilla. These teeth are associated with specialized salivary glands that produce venomous proteins that immobilize prey. The genus is sister to *Contia* + *Carphophis* (Zaher et al. 2009) or *Farancia* + *Carphophis* (Pyron et al. 2011). It is widely distributed in North America and consists of twelve subspecies, three of which are found in Alabama.

## Ring-necked Snake
### *Diadophis punctatus* (Linnaeus, 1766)

TAXONOMY Currently, twelve subspecies of the Ring-necked Snake are recognized based principally on color patterns. Recent examination of the mitochondrial genome recovered fourteen distinct lineages (Fontanella et al. 2008); unfortunately, these lineages do not align geographically with all the named subspecies. In their color patterns, the populations inhabiting Alabama show the influence of three subspecies (Mount 1975), and in their mitochondrial genome, three mitochondrial lineages (Cumberland Mountain, Mississippi River valley, and Piedmont) are known from the state (Fontanella et al. 2008). The degree to which these mitochondrial linages align with the color pattern of each subspecies is poorly known. However, a single Alabama specimen (LSUMZ 2187) known to be of the Cumberland Mountain lineage has the color pattern of *D. p. edwardsii*, a single Alabama specimen (FTB 888) known to be of the Mississippi River valley lineage has the color pattern of *D. p. stictogenys*, and two Alabama specimens (FTB 861 and 925) known to be of the Piedmont lineage have the color pattern of *D. p. punctatus*. Based on these limited data, we retain three subspecies within Alabama. However, the distribution of specimens referable to these three subspecies are extensive and freely overlap, suggesting either that the mitochondrial lineages within the state overlap much more than implied by Fontanella et al. (2008) or that color patterns do not align as precisely with the mitochondrial lineages as implied by our preliminary sample. For this reason, we consider all three taxa together. The presence of a fourth mitochondrial lineage (southeast Louisiana lineage) is possible in extreme southwestern Alabama. Because the type locality of *D. p. punctatus* is the Atlantic Coast of the South (Uetz 2010) and because the southeastern Louisiana, Piedmont, and Florida clades of Fontanella et al. (2008) form a monophyletic clade that includes specimens that likely include the type locality, we consider these clades to be *D. p. punctatus*. Previous authors have placed this species in the genera *Ablabes, Calamaria, Coluber, Homalosoma,* and *Natrix.*

## KEY TO THE SUBSPECIES OF *DIADOPHIS PUNCTATUS* OF ALABAMA

**1a** Venter with series of bold dark markings; seven or eight upper labials; chin with dark spots. Go to **2**.

**1b** Venter unmarked or with a single median row of small dark spots and with eight upper labials; chin usually lacking dark spots.

> *Diadophis punctatus edwardsii*—Northern **Ring-necked Snake** . . . . page 228.

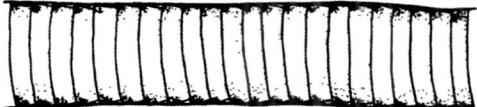

Ventral view of a Northern Ring-necked Snake (*Diadophis punctatus edwardsii*)

**2a** Venter with a single median row of conspicuous dark half moons and with eight upper labials.

> *Diadophis punctatus punctatus*—Southern **Ring-necked Snake** . . . . page 228.

**2b** Venter with paired dark spots an each scute, or irregular dark markings; seven upper labials.

> *Diadophis punctatus stictogenys*—Mississippi **Ring-necked Snake** . . . . page 228.

*From left to right:*

Ventral view of a Southern Ring-necked Snake (*Diadophis punctatus punctatus*)

Ventral view of a Mississippi Ring-necked Snake (*D. p. stictogenys*)

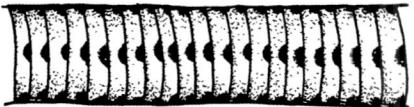 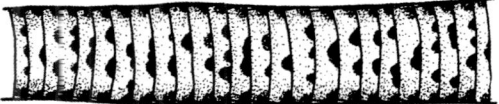

### Southern Ring-necked Snake
*Diadophis punctatus punctatus* (Linneus, 1766)
and
### Northern Ring-necked Snake
*Diadophis punctatus edwardsii* (Merrem, 1820)
and
### Mississippi Ring-necked Snake
*Diadophis punctatus stictogenys* (Cope, 1860)

DESCRIPTION These three taxa include small, slender snakes possessing a relatively long tail. In *D. p. punctatus* and *D. p. stictogenys* maximum total length is approximately 445 mm, whereas in *D. p. edwardsii* maximum total length may reach 570 mm. These snakes have small, somewhat flattened heads that are scarcely distinct from the neck. In *D. p. punctatus* and *D. p. edwardsii*, there are eight (rarely seven) upper labials, whereas in *D. p. stictogenys* there are seven (rarely eight) upper labials. In all subspecies the anal scute is divided, and the dorsal scales are smooth and are arranged in fifteen rows at midbody. The dorsal ground color is uniform slate gray to blue-black (darkest in young individuals) except for a narrow yellow or cream-colored neck ring that may be interrupted mid-dorsally, especially in *D. p. stictogenys*. The

venter is yellow or orange in all three subspecies but differs in the pattern of dark markings. In *D. p. edwardsii*, the venter has no dark marking or has a single row of very small dark spots. In *D. p. punctatus*, there is a single row of fairly large, distinct, black half-moons arranged as a stripe down the midline. Specimens from Alabama that fit this description fall into two distinct groups. In one, the midventral dark markings are very large, as are dark rectangular markings at the lateral-most portion of each ventral scute; collectively, these dark marks approach each other leaving little space for the light ground color of the venter. In the other group, the dark markings are much smaller, leaving extensive space for the light ventral ground color. Finally, the venter of *D. p. stictogenys* is characterized by irregular dark ventral spotting or a row of paired half-moon shapes, frequently partially fused, aligned down the middle of the venter. A third pattern class that we attribute to *D. p. stictogenys* involves wide dark markings that occur at the center of each ventral scute but that lack any tendency to form single or paired half-moons.

**ALABAMA DISTRIBUTION** Based on the molecular data of Fontanella et al. (2008), specimens of *D. p. punctatus* (Piedmont lineage) are known from the Chattahoochee River valley along the southeastern quarter of the state. Similarly, *D. p. edwardsii* (Cumberland lineage) is known from the extreme northeast corner of Alabama (Jackson and DeKalb Counties) and *D. p. stictogenys* (Mississippi River valley lineage) is known from Tuscaloosa County westward. Based on color pattern, specimens referable to *D. p. edwardsii* are found as far south as Elmore County (AUM 23019), and specimens showing the influence of this subspecies are found as far south as Clarke County (AUM 11134). Specimens with the color patterns of *D. p. punctatus* and *D. p. stictogenys* are throughout the state So, much additional work is needed to clarify the content and distribution of these taxa in Alabama.

**HABITS** Ring-necked Snakes occur in most terrestrial habitats but prefer those intermediate in moisture conditions. Rotting pine logs and stumps are favorable microhabitats, but this species also lives beneath rocks, in leaf litter, and in other protected places. Other small snakes frequently found in company with Ring-necked Snakes include Earthsnakes (*Virginia* spp.) and Wormsnakes (*Carphophis amoenus*).

The elongate, whitish eggs are usually laid in rotting wood within a log or stump. Clutch size is most often four or five. Food items include earthworms, insect larvae, salamanders, frogs, and small reptiles. Ring-necked Snakes almost never bite when handled, but when freshly caught, they will invariably attempt to smear the contents of the cloaca on the captor. Some individuals, when molested, will quickly curl the tail and turn it upside down, revealing the bright ventral color. This behavioral feature, shared with some other snake species, is thought to startle some potential predators and in some instances enable the snake to escape by directing the predator's attention to the tail rather than the head.

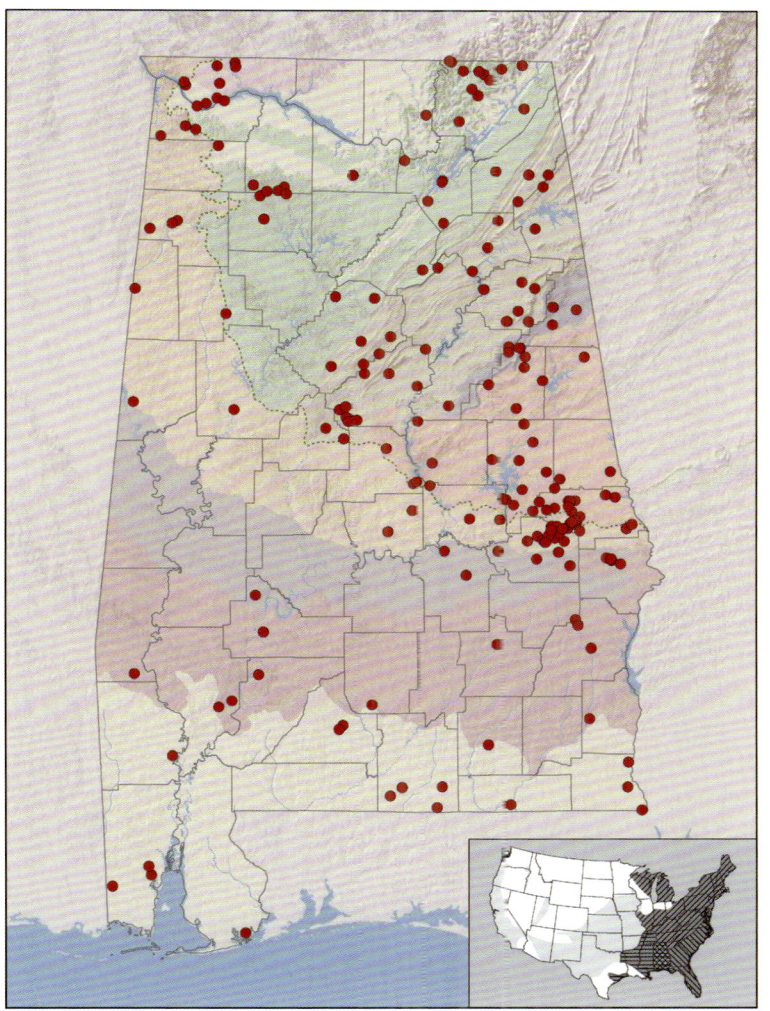

Distribution of Southern Ring-necked Snake, Northern Ring-necked Snake, and Mississippi Ring-necked Snake, *Diadophis punctatus punctatus/D. p. edwardsii/D. p. stictogenys*. This species is assumed to occur throughout Alabama. Influence of three subspecies, *D. p. edwardsii, D. p. punctatus*, and *D. p. stictogenys*, occurs within the state, but with no distinctive geographic pattern. Inset map depicts approximate range in the United States with dark shading indicating range of the three subspecies known from Alabama and light shading indicating range of all other subspecies. Right-slant hatching indicates the range of *D. p. edwardsii*; left-slant hatching indicates the range of *D. p. punctatus*; horizontal hatching indicates the range of *D. p. stictogenys*.

**CONSERVATION AND MANAGEMENT** Ring-necked Snakes are common in areas with a canopy of tree cover, thick leaf litter, and rocks or rotting logs. The species receives no special protection from state law because it is so common. These serpents can survive in urban areas where trees are allowed to reach mature ages. Management activities that retain old-growth trees along riparian zones should enhance habitat for these snakes.

## Littersnakes
### Genus *Rhadinaea* (Cope, 1863)

This is a genus of small, ground-dwelling snakes that are restricted to the New World. They possess an enlarged tooth at the posterior end of each maxilla. These teeth are associated with enlarged glands that produce venoms that immobilize prey. The genus is sister to *Coniophanes*, a Central American radiation of small snakes that live in leaf litter (Pyron et al. 2013). Most of the twenty species within *Rhadinaea* are found in tropical forests, suggesting that the single North American species is derived from a tropical origin.

Pine Woods Littersnake, *Rhadinaea flavilata*, Baldwin County, AL

## Pine Woods Littersnake (Yellow-lipped Snake)
### *Rhadinaea flavilata* (Cope, 1863)

**DESCRIPTION** This small snake attains a maximum total length of around 390 mm, a large proportion of which is the tail. The head is elongate and narrow and is slightly distinct from the neck. The dorsal scales are smooth and arranged in seventeen rows at midbody. The anal scute is divided. These snakes have a dorsum that is golden brown in ground color. A diffuse mid-dorsal stripe and a pair of lateral stripes are sometimes visible. The top of the head is darker than the body, often marked with pale vermiculations, and with a dark band that extends from the snout through the eye to the corner of the mouth. The upper labials are yellowish, some with dark spots, and are

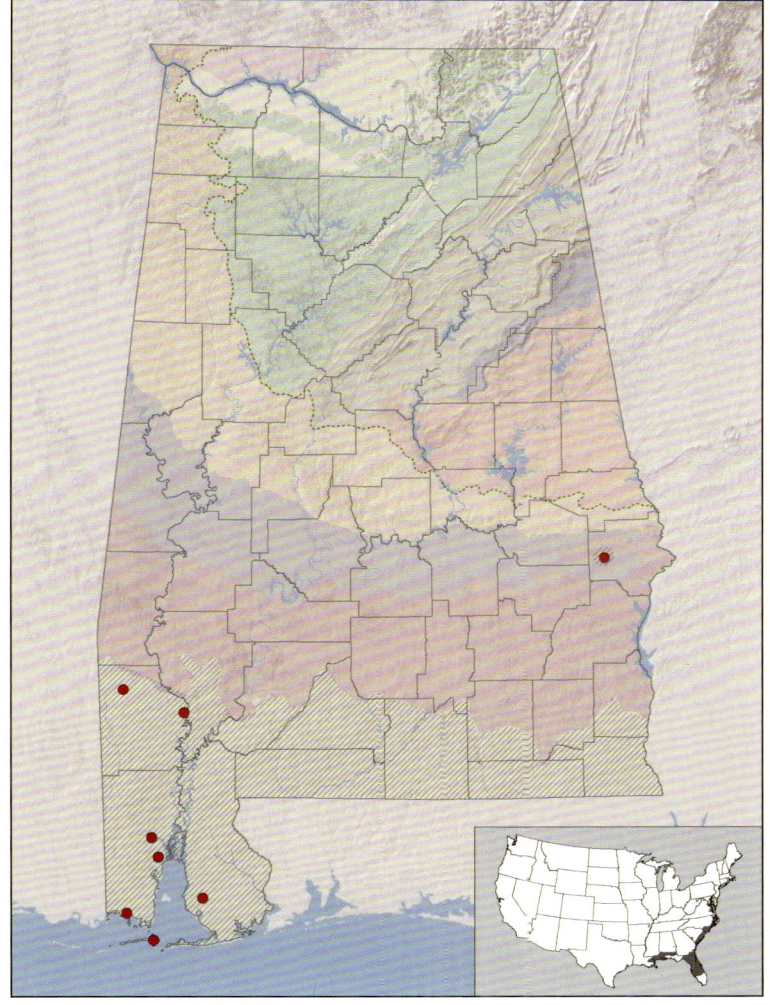

Distribution of Pine Woods Litter-snake, *Rhadinaea flavilata*. The presumed range of the species in Alabama is indicated by hatching. Inset map depicts approximate range in the United States.

strongly contrasted with the color of the rest of the head. The venter is yellow or yellowish white.

**ALABAMA DISTRIBUTION** This species has a spotty distribution in the Coastal Plain. In Alabama, it is known only from Baldwin, Mobile, Russell, and Washington Counties.

**HABITS** This poorly known, secretive species typically inhabits damp pine flatwoods or floodplains, where rotting logs and stumps are abundant. It is notably common on Dauphin Island and has on occasion been found in urban and suburban sites in the city of Mobile. It is usually encountered within or under a log, stump, or some other sheltering object during March, April, or early May, especially when the water table is high. The food of the Pine Woods Littersnake consists mostly of small amphibians and reptiles, which are swallowed alive or are killed or inactivated with mildly venomous saliva before being eaten. Enlarged maxillary teeth in the rear of the mouth assist in envenomating the prey. The snake is completely harmless to humans and makes no attempt to bite. Natural nests are unknown. Myers (1967), in his account of this species, concluded that between two and four eggs are laid and that the laying season lasts from May to August.

**CONSERVATION AND MANAGEMENT** The Pine Woods Littersnake receives no special protection under state law, but it is tracked as imperiled by the Alabama Natural Heritage Program because the species is rare in the state. It is present on Dauphin Island, where it is frequently encountered, but nowhere else in the state is this true. Gulf State Park and the Perdido River Longleaf Hills Tract, along with sites on Dauphin Island, are likely to be key public properties for conservation planning. Pine Woods Littersnakes require maintenance of pine and oak forests of the Lower Coastal Plain. Rotting logs are likely key hiding and nesting sites for this species. Activities that enhance mature pine and oak forests and have fallen logs are likely to improve habitat for Littersnakes.

**TAXONOMY** Based on the classic work of Myers (1974), we consider the Pine Woods Littersnake to be the sister taxon to *R. laureata* of Mexico, the only other member of the *flavilata* species group. No subspecies are recognized within *R. flavilata*. Previous authors have placed this species in the genera *Dromicus* and *Leimadophis*.

# Subfamily Natricinae

This subfamily consists of 32 genera within which about 220 species have been described. These snakes are widely distributed in both the Old and New Worlds, and they are the sister lineage to the South and Central American radiation Dipsadinae (Zaher et al. 2009; Pyron et al. 2013). Natricines in the New World belong to a monophyletic clade, the Thamnophini, and these likely dispersed to North America across the Bering land bridge, where they radiated to the many common species that make up the modern North American snake faunas. Within Alabama, five genera containing seventeen species are found, a rich fauna that helps make the southeastern United States a center of diversity for natricines in North America. In general, these snakes have strong ties to wet or moist places. The diagnostic characteristic of the subfamily is a unique conformation to the hemipenes (sulcus spermaticus single and highly centripetal, forming a nude region on the lobes of the hemipenes; hemipenial calyces lost; Zaher et al. 2009). However, these snakes tend to have heavily keeled scales and are diurnal in activity patterns, characters that aid in recognition of the subfamily in the field. There is also a strong tendency within the subfamily to produce live young. Most members can be found by searching along banks of wetlands or riparian zones, where these snakes frequently bask.

## KEY TO THE GENERA OF NATRICINAE OF ALABAMA

**1a** Scales smooth or weakly keeled. Go to **2**.
**1b** Scales strongly keeled. Go to **3**.

**2a** Scales noticeably shiny and with faint dark lines.
    Genus *Liodytes*—**Swampsnakes** . . . . **page 237**.
**2b** Scales dull and dorsum uniform gray or brown.
    Genus *Virginia* and *Haldea*—**North American Earthsnakes** . . . .
       **page 244**.

*From left to right:*

Dorsal view of smooth scales

Dorsal view of keeled scales

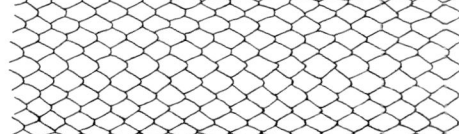

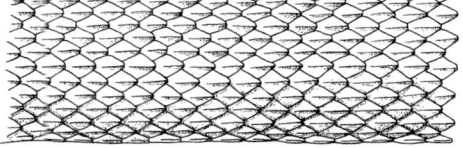

**3a** Neck with distinct light patches on sides but not joining dorsally.

    Genus *Storeria*—**North American Brownsnakes** . . . . **page 253.**

**3b** Neck without light patches. Go to **4.**

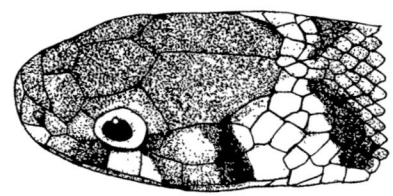

Dorsolateral view of head of a Red-bellied Snake (*Storeria occipitomaculata*)

**4a** Ventrolateral dark lines converging ventrally under neck.

    Genus *Regina*—**Queensnakes** . . . . **page 263.**

**4b** No ventrolateral dark lines. Go to **5.**

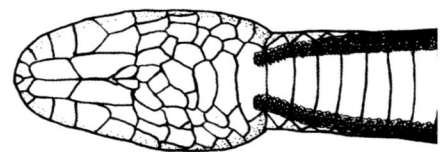

Venter of chin of a Queensnake (*Regina septem-vittata*); based on AUM 32600

**5a** Light mid-dorsal stripe but no light stripes on venter.

    Genus **Thamnophis**—**North American Gartersnakes** . . . . **page 266.**

**5b** Dorsum variously marked, but if light stripes present, then with light stripes on venter.

    Genus **Nerodia**—**North American Watersnakes** . . . . **page 272.**

# Swampsnakes
## Genus *Liodytes* (Cope 1982)

We follow McVay and Carstens (2013) in resurrecting this genus for three species that used to be placed in the genera *Regina* (*R. alleni* and *R. rigida*) and *Seminatrix* (*S. pygaea*). This action was required because, starting with Alfaro and Arnold (2001), the crayfish-eating watersnakes formerly placed in *Regina* are known to be polyphyletic. We restrict that genus to *R. grahami* and *R. septemvittata*, crayfish specialists that share a phylogenetic history with the genera *Tropidoclonion* and *Nerodia* (McVay and Carstens 2013). We use the common name Queensnake for the genus *Regina* and use the common name Swampsnake for members of the genus *Liodytes* because this common name adequately describes their habitat use and retains an association of this common name with *L. pygaea*. Members of *Liodytes* are sister to the genus *Clonophis* or to *Clonophis* + *Virginia* (Alfaro and Arnold 2001; McVay and Carstens 2013) and are confined to the southeastern United States, including the two species found in Alabama.

### KEY TO THE SPECIES OF *LIODYTES* OF ALABAMA

1a Venter cream in color with two rows of conspicuous dark spots, forming stripes, these converging anteriorly and uniting on the throat.

    *Liodytes rigida sinicola*—**Gulf Glossy Swampsnake** . . . . page 238.

1a Venter predominantly red or reddish without two rows of conspicuous ventrolateral dark spots.

    *Liodytes pygaea pygaea*—**Northern Florida Swampsnake** . . . . page 241.

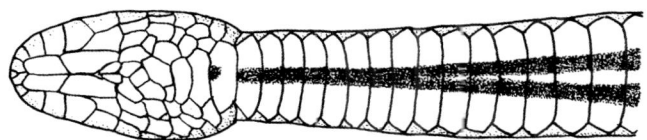

Venter of chin of a Gulf Glossy Swampsnake (*Liodytes rigida sinicola*); based on AUM 32600

## Gulf Glossy Swampsnake
*Liodytes rigida sinicola* (Huheey, 1959)

DESCRIPTION The Gulf Glossy Swampsnake is a relatively small, stout-bodied, aquatic serpent attaining a maximum total length of about 780 mm. The head is small and scarcely distinct from the neck. The eyes are large and the upper labials are swollen. The dorsal scales are shiny, weakly keeled (except for the first row on each side), and arranged in nineteen rows at midbody. The anal scute is divided. In dorsal coloration, the ground color is olive brown, becoming paler on the sides. Two faint longitudinal dark stripes extend along the dorsum, and two additional stripes occur along each side. The belly is yellowish to cream with two rows of conspicuous dark spots that usually coalesce to form stripes; these stripes converge anteriorly and unite under the chin. The underside of the tail usually has a median dark line. In this subspecies, the chin and gular region are virtually immaculate. Usually there are two preoculars on each side. Subcaudals of males average sixty-three and those of females average fifty-four. This subspecies is most similar to the North Florida Swampsnake, but that species has seventeen scale rows at midbody and lacks ventrolateral dark stripes that unite under the chin.

ALABAMA DISTRIBUTION This subspecies is found throughout the Coastal Plain, Red Hills, and Fall Line Hills, except possibly for the northwestern portion of the state.

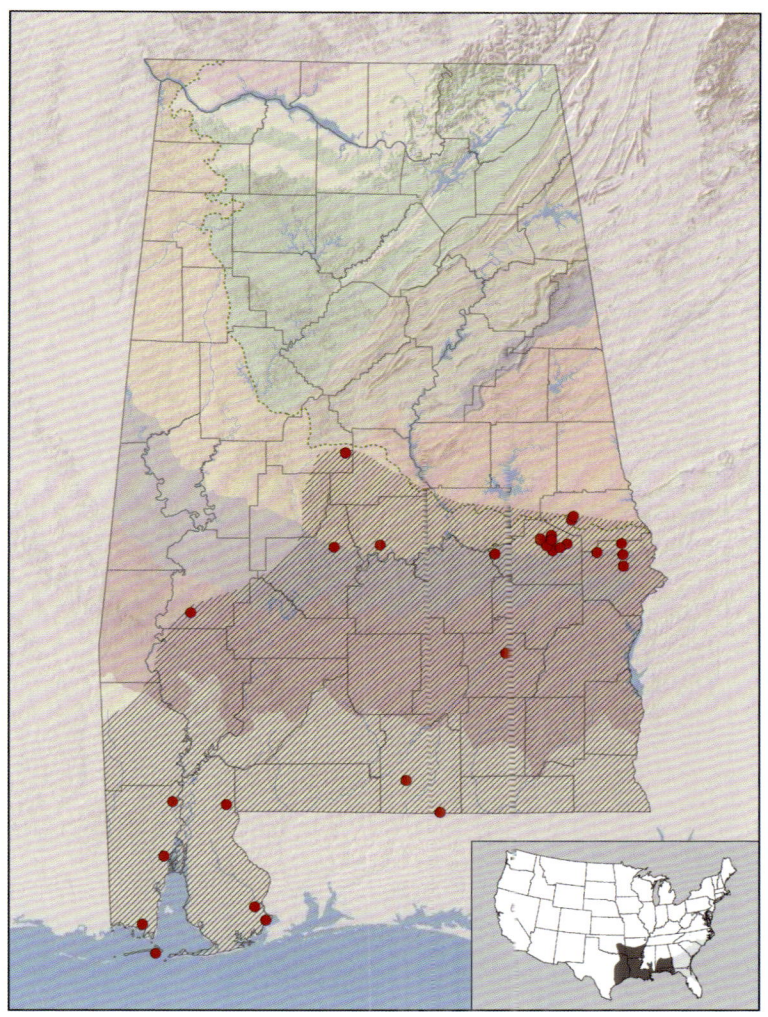

Distribution
of Gulf Glossy
Swampsnake,
*Liodytes rigida
sinicola*. The presumed range of
the subspecies in
Alabama is indicated by hatching.
Inset map depicts
approximate range
in the United States
with dark shading
indicating the range
of *L. r. sinicola*
and light shading
indicating the
range of all other
subspecies.

**HABITS** Highly secretive, the Gulf Glossy Swampsnake is seldom seen in Alabama except at night during rainy weather. Most specimens have been collected on roads traversing floodplains and other low, damp places and around the edges of ponds and swamps. The subspecies is most active in summer (Palmer and Braswell 1995), and when it is active, individuals may perch on limbs above water, frequently after heavy rains (Gibbons and Dorcas 2004). However, this subspecies does enter terrestrial traps up to about 500 m from the nearest wetland (Steen et al. 2011).

Gulf Glossy Swampsnakes feed on soft and hard crayfish, which they crush with constricting coils and consume tail first (Gibbons and Dorcas 2004). There is a strong possibility that *R. rigida* relies heavily on crayfish burrows for retreats. Additional prey items include fish, frogs, and sirens. Information on reproduction is scarce. Data from Alabama are restricted to a single female that contained eight ova. Elsewhere, this subspecies is known to give birth to six to fourteen offspring from August to September (Gibbons and Dorcas 2004).

CONSERVATION AND MANAGEMENT This subspecies is not abundant anywhere in Alabama, but it is seen persistently across its known distribution in the state. It receives no special protection by state law. Management efforts that maintain native aquatic vegetation around wetlands of the Coastal Plain and Fall Line Hills should enhance populations of this subspecies. Use of chemicals that kill crayfish would likely reduce populations of Gulf Glossy Swampsnakes. Maintenance of native upland forests for up to 500 m around each wetland may be necessary (Steen et al. 2011).

TAXONOMY This species is sister to *Liodytes pygaea* (Alfaro and Arnold 2001; McVay and Carstens 2013). Three subspecies of this species are recognized, one of which occurs in Alabama. Previous authors have placed this species in the genera *Coluber, Liodytes, Natrix, Nerodia,* and *Tropidonotus.*

Northern Florida Swampsnake, *Liodytes pygaea pygaea*, Covington County, AL

## Northern Florida Swampsnake
*Liodytes pygaea pygaea* (Cope, 1895)

**DESCRIPTION** This subspecies is a small, short-tailed snake attaining a maximum total length of about 425 mm. The head is only slightly distinct from the neck, and the anal scute usually is divided. The dorsal scales are smooth and arranged in seventeen rows at midbody, creating a shiny black dorsum. Three to four scale rows have obscure dark longitudinal lines that give the appearance that the scales are keeled. The venter is bright red to dull orange, but all ventral scales have long, narrow, curved black bars on the leading edges of the scutes. Subcaudal scutes average forty-eight in males and thirty-eight in females. This species is most similar to the Gulf Glossy Swampsnake, but that species has nineteen scale rows at midbody and dark ventrolateral stripes that unite under the chin.

**ALABAMA DISTRIBUTION** This subspecies is known from only four localities, three in Covington County and one in Escambia County, the latter being the northwestern-most extent of the subspecies' range. A shed skin that appears to be of this form was found in a weedy pond

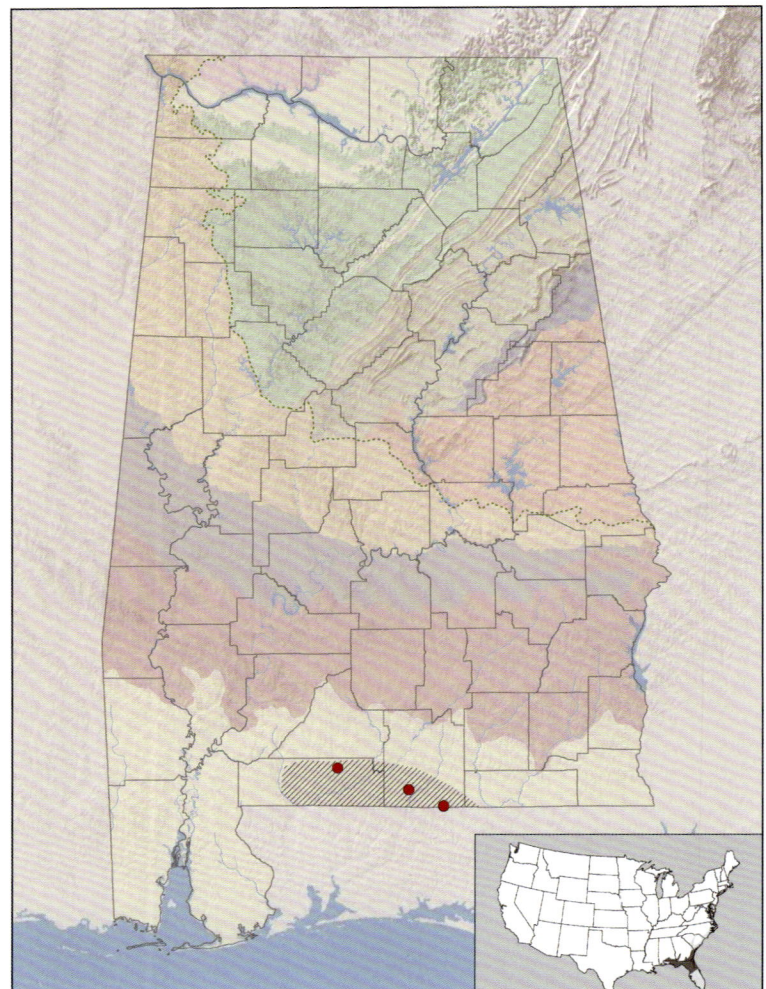

Distribution of Northern Florida Swampsnake, *Liodytes pygaea pygaea*. The presumed range of the subspecies in Alabama is indicated by hatching. Inset map depicts approximate range in the United States with dark shading indicating the range of *L. p. pygaea* and light shading indicating the range of all other subspecies.

in southern Houston County, suggesting its presence in other aquatic sites of extreme southern Alabama.

**HABITS** Preferred habitats of Northern Florida Swampsnakes are swamps, ponds, and lakes with abundant emergent vegetation. Floating mats of aquatic plants or mats of vegetation along the edge of these wetlands provide ideal habitats. In Alabama, we have collected specimens from under mats of water pennywort (*Hydrocotyle umbellata*) around the edges of ponds and lakes and in pitfall traps near an ephemeral pond. These snakes appear to be strongly tied to aquatic sites because of high evaporative water loss (Winne et al. 2001). Activity of this subspecies peaks during summer and individuals appear to be capable of migrating to sites with more water during droughts (Dodd 1993). Reproduction of this subspecies in Alabama is known from a single female from Covington County that gave birth to seven offspring (Mount 1975). Elsewhere, the subspecies produces litters ranging from four to fifteen offspring (Winne et al. 2006). Females reproduce annually during wet years (Winne et al. 2005) but may skip reproduction during drought years (Winne et al. 2006).

CONSERVATION AND MANAGEMENT The Northern Florida Swampsnake was ranked Priority 3 (Low Conservation Concern) by ADCNR in 2012. Nevertheless, it is still illegal to possess this subspecies in the state of Alabama without a scientific collecting permit. The distribution of *L. p. pygaea* barely includes Alabama, so any conservation efforts for this subspecies are likely to be concentrated elsewhere. Nevertheless, maintenance of shallow wetlands is vital for conserving this subspecies. The edges of these wetlands need to have lush herbaceous growth with thick mats of decaying vegetation covering moist soil. The Conecuh National Forest is the only public lands known to harbor this subspecies, but Gulf State Park, Perdido River Longleaf Hills Tract, Geneva State Forest, Fort Rucker Military Reserve, and Chattahoochee State Park should be examined for the presence of this subspecies and managed to retain it if it is found.

TAXONOMY This species is sister to *Liodytes rigida* (Alfaro and Arnold 2001; McVay and Carstens 2013). Three subspecies of this small, secretive snake are recognized. The range of one of these barely extends into southern Alabama. Previous authors have placed this species in the genera *Contia, Seminatrix,* and *Tropidonotus.*

# North American Earthsnakes
## Genus *Virginia* (Baird and Girard, 1853)
and
## Genus *Haldea* (Baird and Girard, 1853)

These two genera contain small, drab snakes that are found in loose soils and that are related to the genera *Clonophis*, *Liodytes*, and *Storeria* (McVay and Carstens 2013), within a larger radiation of small, semi-fossorial snakes (Alfaro and Arnold 2001). We consider them together in a single account because their similar morphology and ecology were used to retain them in a single genus *Virginia* containing two species (e.g., Wallach et al. 2014). Based on nuclear data, McVay and Carstens (2013) found the two species to be polyphyletic and returned one of them (*V. striatula*) to the genus *Haldea*, leaving a single species (*V. valeriae*) in the genus *Virginia*. We follow these designations. The two species are confined to the eastern United States and both are found in Alabama.

### Key to the Species of Earthsnakes of Alabama

1a Five upper labial scales; one postorbital scale.

      *Haldea striatula*—**Rough Earthsnake** . . . . page 245.

1a Six upper labial scales; two postorbital scales.

      *Virginia valeriae*—**Smooth Earthsnake** . . . . page 247.

*From left to right:*

Lateral view of head of a Rough Earthsnake (*Haldea striatula*)

Lateral view of head of a Smooth Earthsnake (*Virginia valeriae*); based on AUM 485

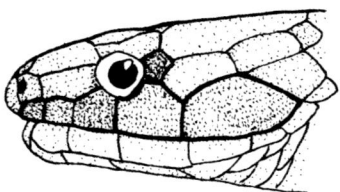 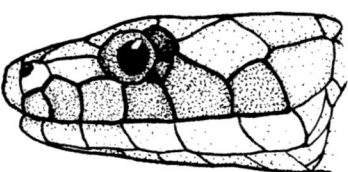

Rough Earthsnake,
*Haldea striatula*,
Escambia County, AL

## Rough Earthsnake
*Haldea striatula* (Linnaeus, 1766)

**DESCRIPTION** Rough Earthsnakes are small, attaining a maximum total length of around 320 mm, and possess short tails. The head is small, has a pointed snout, and is distinct from the neck. The anal scute is divided. There are five upper labials (six in *Virginia valeriae*) and one postorbital scale (two in *V. valeriae*). Dorsal scales are small, keeled, and arranged in eleven rows at midbody. The dorsal ground color is light brownish to gray (generally darker in *V. valeriae*). Most juveniles and some adults have a light band at the back of the head. The venter is cream to yellow. Subcaudals average forty in males and thirty-four in females.

**ALABAMA DISTRIBUTION** This species is found in most of the Coastal Plain and lower portions of the Piedmont, Talladega Upland, and Ridge and Valley regions. It possibly occurs in other upland regions, but records are lacking.

**HABITS** This small, secretive snake is most abundant in relatively dry Coastal Plain woodlands. It is usually encountered when rocks or logs are overturned or when rotting pine logs and stumps are pulled apart. The Eastern Crowned Snake, *Tantilla coronata*, is one of its frequent associates. Reported food items include a variety of small invertebrates,

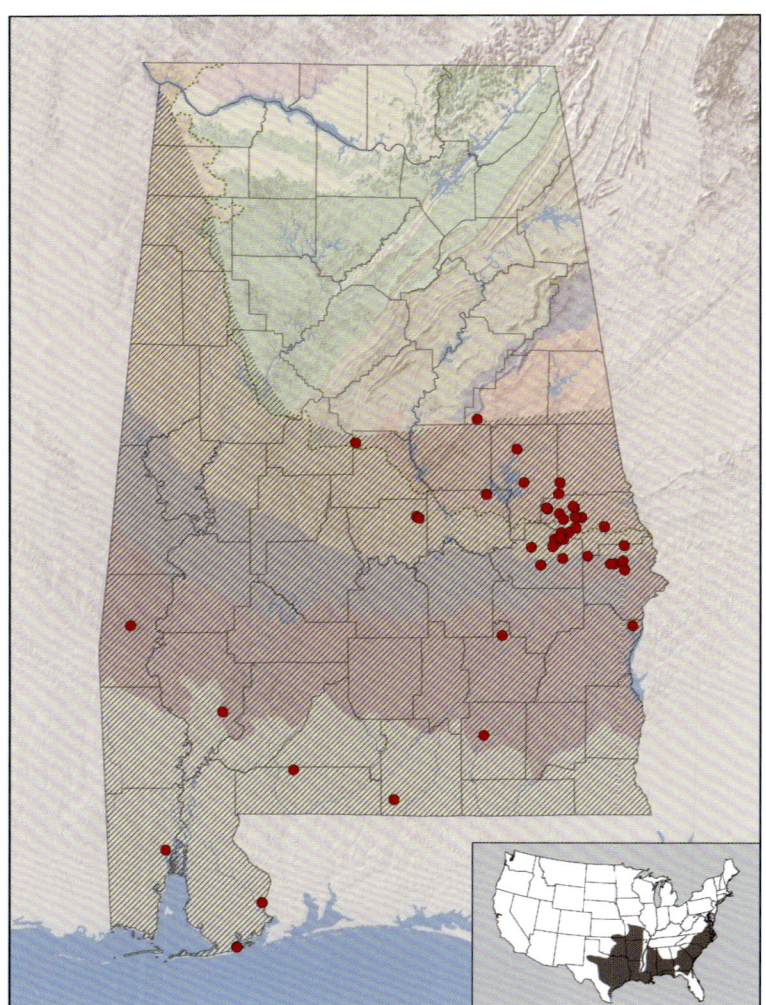

Distribution of
Rough Earthsnake,
*Haldea striatula*.
The presumed
range of the species
in Alabama is indi-
cated by hatching.
Inset map depicts
approximate range
in the United
States.

in addition to small frogs and lizards. The Rough Earthsnake is viviparous, with two to thirteen young being born in mid- or late summer.

**CONSERVATION AND MANAGEMENT** This is a common snake in urban
areas where it can be abundant under potted plants in soft garden soils.
Therefore, no special management efforts are required to maintain it
in the state of Alabama. It receives no special protection by state law.

**TAXONOMY** This species has no recognized subspecies. Previous authors have placed this species in the genera *Coluber, Conocephalus, Natrix, Falconeria, Potamophis,* and *Virginia*.

*Virginia valeriae* (Baird and Girard, 1853)

**Taxonomy** Three subspecies of this small snake are recognized, two of which are found in Alabama. Previous authors have placed this species in the genera *Carphophis* and *Haldea*.

## Key to the Subspecies of *Virginia valeriae* of Alabama

**1a** Dorsal scales in fifteen rows at midbody, lacking keels.

> *Virginia valeriae valeriae*—**Eastern Smooth Earthsnake** . . . . **page 248.**

**1b** Dorsal scales in seventeen rows at midbody, with faint keels (visible on close observation).

> *Virginia valeriae elegans*—**Western Smooth Earthsnake** . . . . **page 250.**

Eastern Smooth
Earthsnake,
*Virginia valeriae
valeriae*, Bibb
County, AL

# Eastern Smooth Earthsnake
*Virginia valeriae valeriae* (Baird and Girard, 1853)

**Description** This subspecies is a small snake, attaining a maximum total length of about 325 mm with a short tail. The head is small and pointed, and the anal scute is divided. The scales are smooth or, rarely, very slightly keeled; they are arranged in fifteen rows at midbody (compared to seventeen rows of weakly keeled scales in *V. v. elegans*). The dorsal ground color is gray to brown or yellowish brown with widely scattered small, dark flecks. The venter is light gray to white and is not as sharply delineated from the dorsal color as it is in *Carphophis amoenus*, a species with which the Eastern Smooth Earthsnake is often confused. Additionally, this subspecies has six upper labials and two postorbitals, characteristics that distinguish it from *Haldea striatula*, a second species with which Eastern Smooth Earthsnakes can be confused. Subcaudals average thirty-five in males and twenty-six in females.

**ALABAMA DISTRIBUTION** This subspecies occurs throughout the state, with the possible exception of Sumter and Mobile Counties, in which *V. v. elegans* is known to occur.

**HABITS** This small, secretive snake occurs most abundantly in mesic, open forests, where it hides beneath rocks and logs and in piles of organic debris. It is frequently encountered in wooded residential areas of cities and towns and is one of several species of small snakes mistakenly called ground rattlers. With the approach of cold weather, Eastern Smooth Earthsnakes often congregate in small numbers,

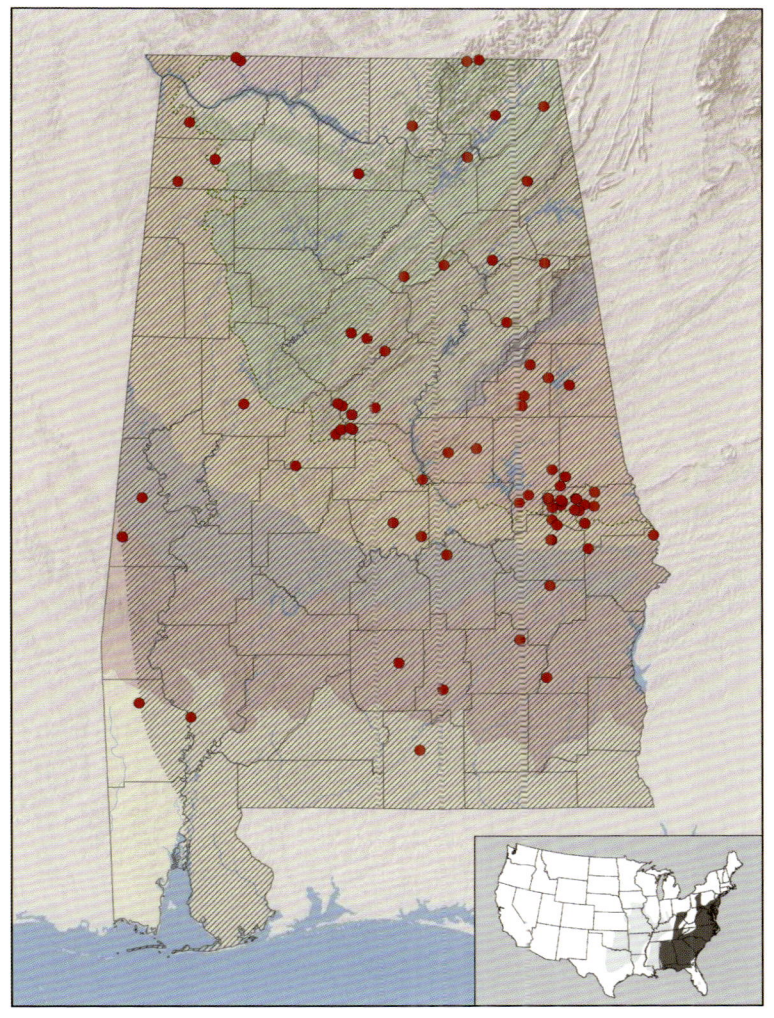

Distribution of Eastern Smooth Earthsnake, *Virginia valeriae*. The presumed range of this subspecies in Alabama is indicated by hatching. Inset map depicts approximate range in the United States with dark shading indicating the range of *V. v. valeriae* and light shading indicating the range of all other subspecies.

sometimes in company with individuals of *H. striatula*, and often overwinter under large rocks. Eastern Smooth Earthsnakes feed predominantly on earthworms. Females give live birth to four to fourteen offspring per litter in August and September.

**CONSERVATION AND MANAGEMENT** This subspecies can be abundant in urban gardens and wooded slopes. For this reason, it receives no protection under state law. It is more abundant in thinned forests than unthinned forests and less abundant in clear-cuts with litter removed than clear-cuts with litter retained (Todd and Andrews 2008).

# Western Smooth Earthsnake

*Virginia valeriae elegans* (Kennicott, 1859)

DESCRIPTION These snakes are small, attaining a maximum total length of about 325 mm, and have a short tail. The head is small and pointed, and the anal scute is divided. The scales are arranged in seventeen rows at midbody and are weakly keeled (fifteen rows of smooth scales appear in *V. v. valeriae*). The dorsal ground color is gray to brown or yellowish brown with widely scattered, small, dark flecks. The venter is light gray to white and is not as sharply delineated from the dorsal color as it is in *Carphophis amoenus*, a species with which the Western Smooth Earthsnake is often confused. Additionally, this subspecies has six upper labials and two postorbitals, characteristics that distinguish it from *Haldea striatula*, a second species with which Western Smooth Earthsnakes can be confused. Subcaudals average thirty-five in males and twenty-six in females.

ALABAMA DISTRIBUTION This subspecies occurs throughout Mississippi and is documented for Alabama from a single female specimen (AUM 21115) and her offspring from Sumter County and a second individual observed by Mount (1975) from Mobile County. Given the distribution pattern of other taxa that approach Alabama from the west, we expect *V. v. elegans* to be present in the southern half of the western tier of counties. But, specimens from Clarke and Washington Counties are referable to *V. v. valeriae*, indicating a complex boundary between the eastern and western subspecies.

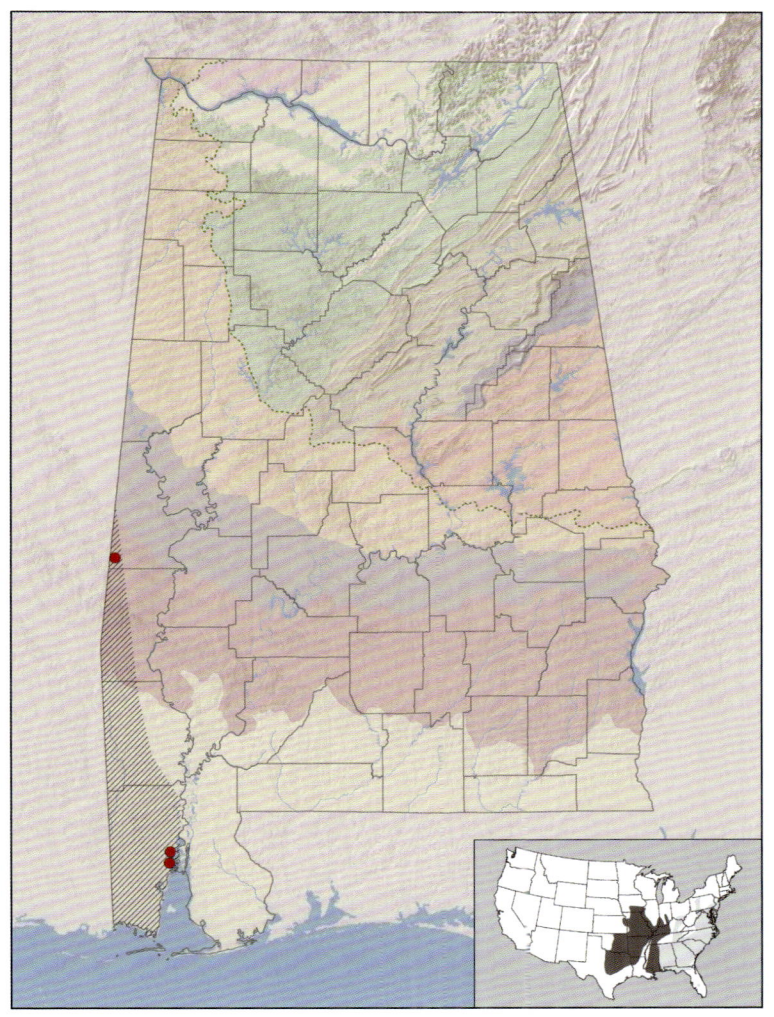

Distribution of Western Smooth Earthsnake, *Virginia valeriae elegans*. The presumed range of the subspecies in Alabama is indicated by hatching. Inset map depicts approximate range in the United States with dark shading indicating the range of *V. v. elegans* and light shading indicating the range of all other subspecies.

**HABITS** Members of this subspecies overwinter along the edges of woodlands and migrate into grasslands during the active season (Pisani 2009). Males move more frequently than females in spring, possibly for mate seeking. Once inseminated, females select warm sites during spring, allowing for rapid development of young, which are likely to be deposited in crevices in the earth or grass clumps in early August (Pisani 2009). Earthworms are the primary dietary item of *V. v. elegans*, and these prey often are eaten at night.

**CONSERVATION AND MANAGEMENT** This subspecies can be common in open pastures bordered by forests. It is likely to be common in urban areas under potted plants or in soft garden soils. No special management efforts are required to maintain it in the state of Alabama, and it receives no special protection by state law.

# North American Brownsnakes
## Genus *Storeria* (Baird and Girard 1853)

This is a genus of small, live-bearing terrestrial snakes found from the eastern United States to Guatemala and Honduras. They belong to a lineage of semifossorial snakes that includes the genera *Liodytes*, *Tropidoclonion*, and *Virginia* (Alfaro and Arnold 2001) or is the sister taxon to *Virginia* (Pyron et al. 2013). Teeth in *Storeria* are elongate and thin and associated with a well-developed Duvernoy's gland, adaptations that are thought to be associated with the primary diet of these snakes, gastropods. The genus contains four species, two of which occur in Alabama.

### KEY TO THE SPECIES OF *STORERIA* OF ALABAMA

1a Fifth upper labial scale with a light spot; labials otherwise wholly dark.
> *Storeria occipitomaculata occipitomaculata*—**Northern Red-bellied**
> **Snake** . . . . **page 254.**

1b Upper labials plain or variously marked, including possessing a black labial spot, but never with a light labial spot.
> *Storeria dekayi* spp. . . . . . **page 257.**

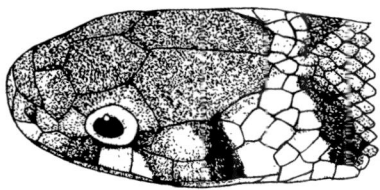

*From top to bottom:*

Head and neck of a Red-bellied Snake (*Storeria occipitomaculata*)

Head and neck of a Midland Brownsnake (*Storeria dekayi*)

Northern Red-
bellied Snake, *Store-
ria occipitomaculata
occipitomaculata*,
Bibb County, AL

## Northern Red-bellied Snake
*Storeria occipitomaculata occipitomaculata* (Storer, 1839)

**DESCRIPTION** This is a small, short-tailed snake attaining a maximum total length of about 405 mm. The head is blunt and more or less distinct from the neck, and the anal scute is divided. Keeled scales are present on the dorsum, and these are arranged in fifteen rows at midbody. The dorsum is extremely variable in ground color and pattern, with ground color ranging from gray to brown or black. There may be no other markings, or there may be a median light stripe on the dorsum. The belly is pink to orange or reddish (rarely black), usually with a longitudinal row of black spots on each side. The head is black to light brown and rarely uniform. On the nape of the neck are three pale spots, the central one being the largest. The fifth upper labial has a light spot, bordered below by black. This species is most similar to North Florida Swampsnake, but that species has seventeen scale rows at midbody, weakly keeled scales, and long, narrow, curved black bars on the leading edges of the ventral scutes.

**ALABAMA DISTRIBUTION** Apparently, the Northern Red-bellied Snake, in one form or another, occurs throughout Alabama. It is common in the northern half of the state but becomes decreasingly so southward.

Northern Red-bellied Snake, *Storeria occipito-maculata occipit-omaculata*, Jasper County, GA

**HABITS** Northern Red-bellied Snakes occur most abundantly in mesic, forested habitats where the soil is moderately heavy and the topography is hilly or mountainous. They are often found in close association with *Virginia valeriae* and, like individuals of that species, spend most of their time under logs, rocks, and other sheltering objects. *S. occipitomaculata* gives birth to young in June and July. Litter sizes of one to twenty-one have been reported and average nine young (Semlitsch and Moran 1984). The food of the Northern Red-bellied Snake consists mostly of slugs but may also include insects, snails, isopods, and other small invertebrates. The snake has a peculiar habit of curling the upper lip upward on one or both sides when freshly captured, a feature that may be an agonistic behavior or associated with eating gastropods (do Amaral 1999). The snake is, however, completely harmless and does not bite. Jordan (1970) reported death feigning in a female of this subspecies from Alabama.

**CONSERVATION AND MANAGEMENT** This snake is common in urban areas, where it can be abundant under potted plants in soft garden soils. Therefore, no special management efforts are required to maintain it in the state of Alabama. It receives no special protection by state law.

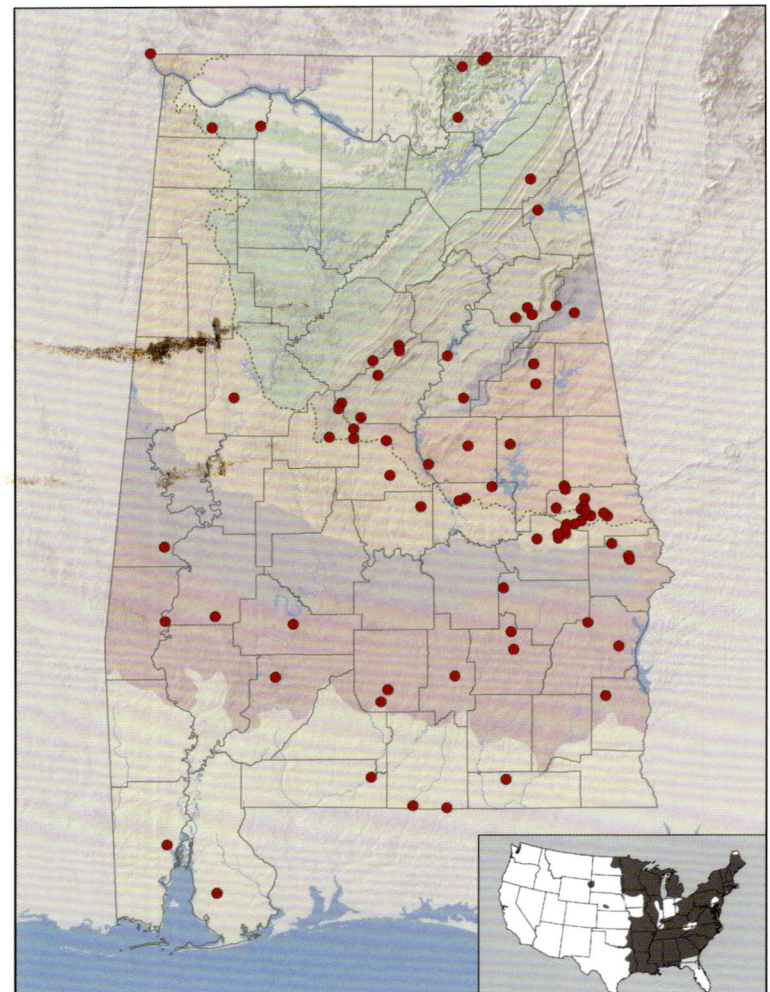

Distribution of Northern Red-bellied Snake, *Storeria occipitomaculata occipitomaculata*. This subspecies is assumed to occur throughout Alabama. Inset map depicts approximate range in the United States with dark shading indicating the range of *S. o. occipitomaculata* and light shading indicating the range of the other subspecies.

**TAXONOMY** Three subspecies of the Red-bellied Snake are recognized, one of which is present in Alabama. Influence of a second subspecies is present in some Alabama specimens (Mount 1975), but these specimens are not geographically contiguous. Therefore, we consider the state to have only the nominate subspecies. Previous authors have placed this species in the genus *Ischnognathus*.

*Storeria dekayi* (Holbrook, 1836)

**TAXONOMY** Eight subspecies of this wide-ranging species are recognized, of which two are known from Alabama. Mount (1975) considered DeKalb and Jackson Counties to have individuals showing
the influence of a third subspecies, *S. d. dekayi*. However, we follow
Christman (1982) in considering this subspecies to occur north of
Alabama. Nevertheless, Alabama specimens from north of the Tennessee River may have the light mid-dorsal stripe described in other
regions within the range of *S. d. dekayi*. South of the Tennessee River,
all specimens are referable to *S. d. wrightorum* or *S. d. limnetes*. Previous authors have placed this species in the genera *Ischnognathus* and
*Tropidonotus*.

## KEY TO THE SUBSPECIES OF *STORERIA DEKAYI* OF ALABAMA

**1a** Anterior temporal with diagonal or vertical bar; sixth and seventh upper labials with dark pigment.
> *Storeria dekayi wrightorum*—Midland Brownsnake . . . . page 258.

**1b** Anterior temporal with a horizontal bar; sixth and seventh upper labials usually unpigmented.
> *Storeria dekayi limnetes*—Marsh Brownsnake . . . page 261.

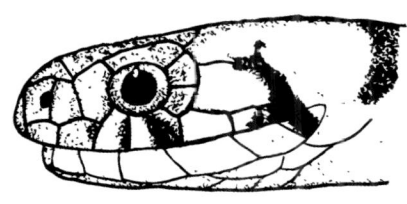

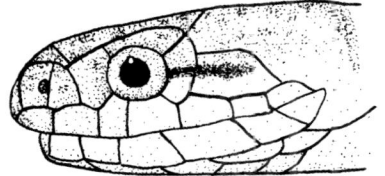

*From top to
bottom:*

Head and neck of
a Midland Brown-
snake (*Storeria
dekayi wrightorum*)

Head and neck of a
Marsh Brownsnake
(*Storeria dekayi
limnetes*)

## Midland Brownsnake
*Storeria dekayi wrightorum* (Trapido, 1944)

**DESCRIPTION** This subspecies is small, attaining a maximum total length of about 425 mm, and possesses a short tail. The head is more or less distinct from the neck, and the anal scute is divided. The dorsal scales are keeled and organized into seventeen rows at midbody. In ground color, these snakes are gray to brown with a median light stripe about four scale rows wide, outlined by a row of dark spots on each side. These dark spots possess dark projections that tend to form thin bars across the mid-dorsal light stripe. The side of the head has a vertical or diagonal dark bar involving the anterior temporal and generally the sixth and seventh upper and lower labials. This feature distinguishes *S. d. wrightorum* from *S. d. limnetes*, which has a horizontal dark bar on the side of the face. The venter of *S. d. wrightorum* is light with one or two rows of small black dots along the lateral edges. Juveniles have a yellowish collar across the neck that fades with age.

**ALABAMA DISTRIBUTION** This subspecies covers all of Alabama except for coastal regions of Mobile County. Specimens from north of the Tennessee River and the northeastern corner of the state may lack the dark crossbars expected of *D. s. wrightorum*. In parts of the Red Hills region and in the Lower Coastal Plain from Escambia County eastward, it is unaccountably rare or absent.

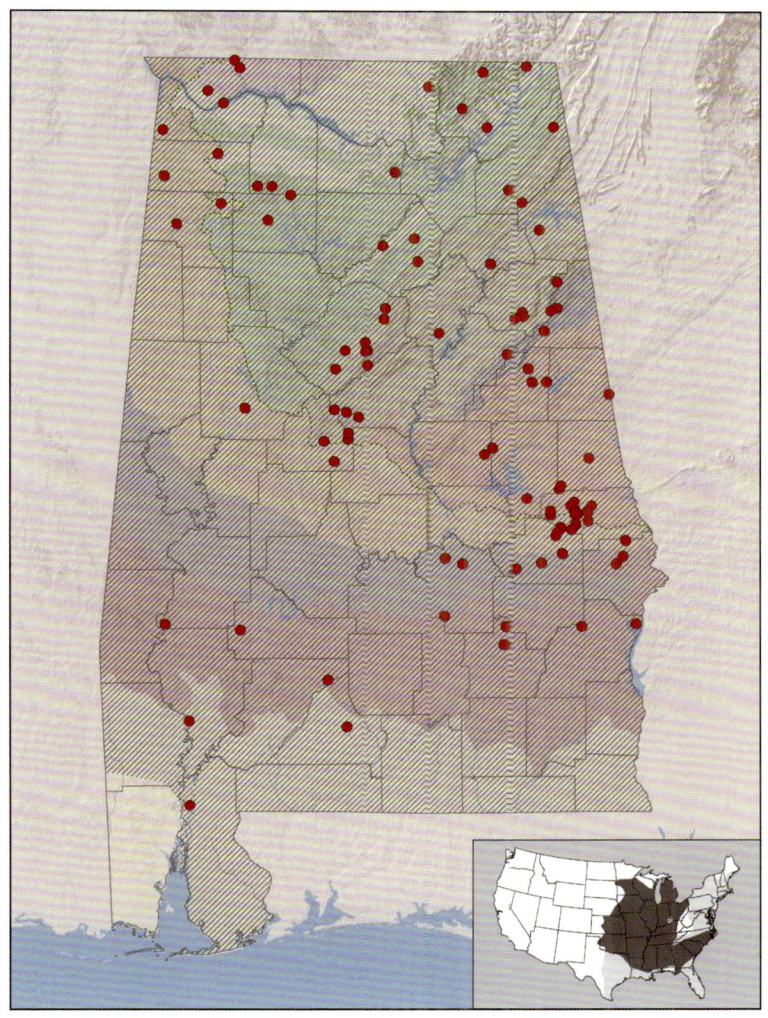

Distribution of Midland Brownsnake, *Storeria dekayi wrightorum*. The presumed range of the subspecies in Alabama is indicated by hatching. Inset map depicts approximate range in the United States with dark shading indicating the range of *S. d. wrightorum* and light shading indicating the range of all other subspecies.

**HABITS** *S. d. wrightorum* is one of the most common snakes north of the Red Hills. It thrives in urban communities as well as in rural environs. A vacant lot or golf course rough often satisfies its ecological requirements. It tends to hide under logs, pieces of tin, and other debris but is often seen abroad. The scarcity or absence of this subspecies in much of the Lower Coastal Plain is perplexing, particularly in view of the fact that it is abundant around Mobile.

*S. d. wrightorum* is docile in temperament and is not known to bite. Home ranges are small, and movement distances within the home range average 50 m (Pisani 2009). The diet includes earthworms, snails, slugs, spiders, and, on rare occasions, small amphibians. Females have one litter per year in June and July (Kofron 1979), and litter sizes vary from three to thirty-one.

**CONSERVATION AND MANAGEMENT** This is a common snake that can be abundant in garden areas. Therefore, no special management efforts are required to maintain it in the state of Alabama, and it receives no special protection by state law.

Marsh Brownsnake, *Storeria dekayi limnetes*, Brazoria County, TX

# Marsh Brownsnake

*Storeria dekayi limnetes* (Anderson, 1961)

**DESCRIPTION** This subspecies is small, attaining a maximum total length of about 440 mm, and possesses a short tail. The head is more or less distinct from the neck, and the anal scute is divided. The dorsal scales are keeled and organized into seventeen rows at midbody. These snakes are gray to brown with a median light stripe about four scale rows wide, outlined by a row of dark spots on each side. These dark spots may or may not possess dark projections that tend to form thin bars across the mid-dorsal light stripe. The side of the head has a horizontal dark bar, distinguishing it from the vertical or diagonal dark bar of *S. d. wrightorum*. The venter of *S. d. limnetes* is light with one or two rows of small black dots along the lateral edges. Juveniles have a yellowish collar across the neck that fades with age.

**ALABAMA DISTRIBUTION** Specimens referable to *S. d. limnetes* are found along the Gulf Coast as far east as Pensacola, Florida. Only a single specimen (AUM 34215), from the city of Mobile, documents the influence of this subspecies in Alabama. However, it is expected to be present in coastal regions of Baldwin and Mobile Counties.

**HABITS** This subspecies inhabits coastal marshlands. It is known from urban communities as well as in rural environs and is docile in temperament. The diet is thought to include earthworms and slugs. Females have live young that are produced in June or July.

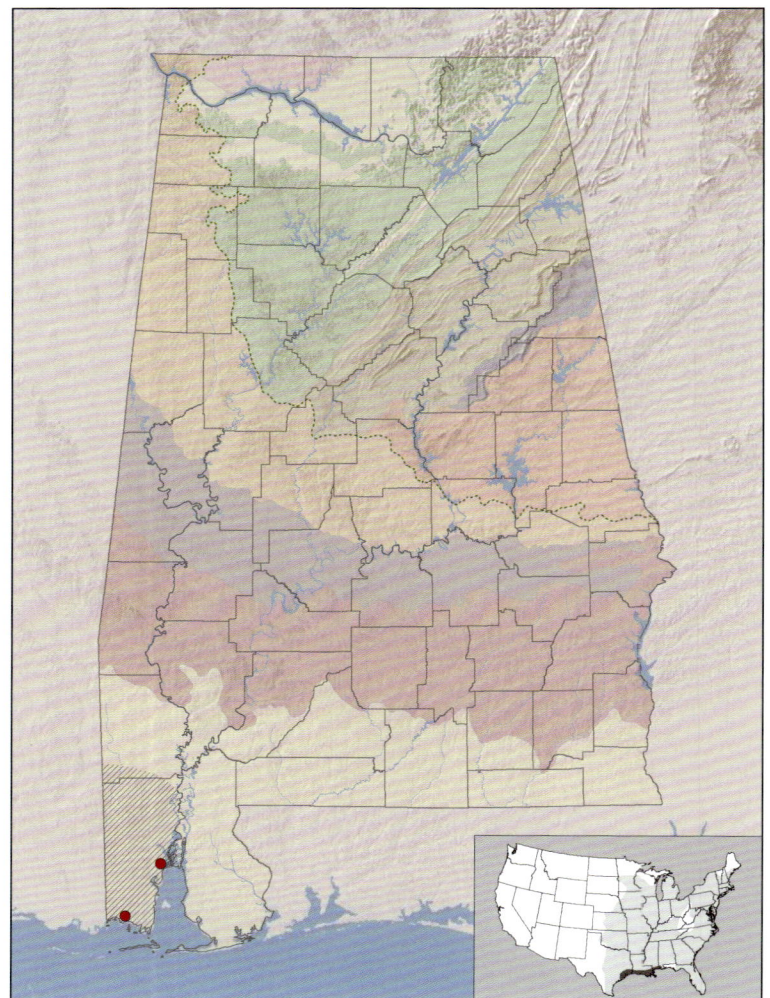

Distribution of the Marsh Brown-snake, *Storeria dekayi limnetes*. The presumed range of the subspecies in Alabama is indicated by hatching. Inset map depicts approximate range in the United States with dark shading indicating the range of *S. d. limnetes* and light shading indicating the range of all other subspecies.

**CONSERVATION AND MANAGEMENT** This is likely to be a common snake, but no series of specimens of the subspecies has been collected in Alabama to document this. Because the other subspecies is so common, Marsh Brownsnakes receive no special protection by state law. Coastal habitats protected by Bon Secour National Wildlife Refuge, Grand Bay Savannah, Gulf State Park, and Weeks Bay National Estuarine Research Reserve appear to provide sufficient public lands to maintain viable populations, but the subspecies needs to be documented from each of these sites.

# Queensnakes
## Genus *Regina* (Say 1825)

As typically conceived, this genus has long been demonstrated to be paraphyletic (Alfaro and Arnold 2001). McVay and Carstens (2013) reduced some problems of paraphyly by resurrecting the genus *Liodytes* for some former members of *Regina*. Unfortunately, the remaining two species (*R. septemvittata* and *R. grahami*) are paraphyletic relative to *Tropidoclonion* and *Nerodia* (Alfaro and Arnold 2001; Pyron et al. 2013). The simplest solution would be to place *Regina* and *Tropidoclonion* in *Nerodia*. Until that action is taken, we follow Crother et al. (2017) and retain *Regina* for *R. septemvittata* and *R. grahami* despite their paraphyly.

## Queensnake
### *Regina septemvittata* (Say, 1825)

Queensnake, *Regina septemvittata*, Liberty County, FL

**DESCRIPTION** This is a small to medium-sized aquatic snake, attaining a maximum total length of around 920 mm. The body of these snakes is slender to moderately stout with a small head that is scarcely or not at all distinct from the neck. The dorsal scales are heavily keeled and occur in nineteen rows at midbody. The anal scute is divided, and the upper labial scales are not swollen. Olive brown or black dominates the dorsal ground color, and three obscure dark stripes occasionally are visible along the dorsum. A ventrolateral light stripe begins on each side and separates the dark dorsal color from a dark ventrolateral

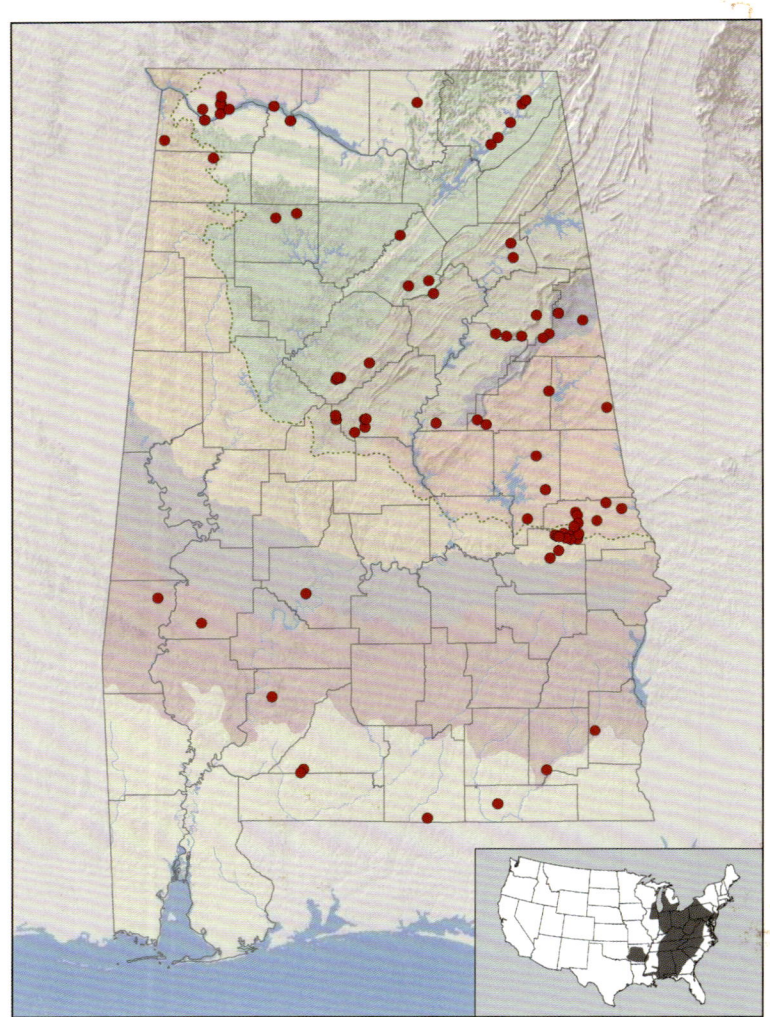

Distribution of Queensnake, *Regina septemvittata*. This species is assumed to occur throughout Alabama. Inset map depicts approximate range in the United States.

dark stripe that converges toward the midventral plane under the chin. The venter of juveniles, adult males, and young females is off-white; that of old females is predominantly dark. The ventrals show a north-to-south cline in Alabama, with populations in the northern part of the state having higher average counts than those to the south (Spangler and Mount, 1969). This species is most easily confused with the Gulf Glossy Swampsnake, but that species has shiny, weakly keeled scales, nineteen scale rows at midbody, and dark ventrolateral stripes that unite under the chin.

**ALABAMA DISTRIBUTION** This species used to be common to abundant above the Fall Line and was locally common in the Coastal Plain. However, Queensnakes have become noticeably uncommon in recent years. The limits of the range in western Alabama are not well-known.

**HABITS** Queensnakes are almost always found along streams or impoundments of streams that have crayfish. Above the Fall Line, they are common along many of the rivers as well as along smaller streams. In the Coastal Plain, they frequent small, clear creeks with sandy or rocky bottoms. Queensnakes spend much of their time basking on limbs of bushes and trees that overhang the water. Activity is largely diurnal but becomes nocturnal as well during hot summer months. Occasionally, large aggregations are found along streams during fall (Gibbons and Dorcas 2004). Queensnakes feed almost exclusively on crayfish that are small or have just finished a cycle of ecdysis and are soft shelled. Courtship behavior occurs in spring, when a male wraps his body around a female and bounces his head on the neck of the female (Ford 1982). Young are born during the summer and are deposited in litter sizes of five to twenty-three offspring (Gibbons and Dorcas 2004). Aside from the habit of discharging the cloacal contents at the time of capture, the Queensnake is inoffensive and seldom bites.

**CONSERVATION AND MANAGEMENT** This species was formerly common and currently receives no special protection by state law. However, its decline requires attention. We suspect that declines of its major prey, crayfish, may play a role. Land management that minimizes chemical pollution, retains riparian vegetation, and prevents siltation, which should enhance crayfish populations, will benefit Queensnakes.

**TAXONOMY** No subspecific variation is recognized within this species. Other authors have placed Queensnakes in the genera *Coluber, Natrix,* and *Tropidonotus.*

# North American Gartersnakes
## Genus *Thamnophis* (Fitzinger, 1843)

This genus contains the Gartersnakes and Ribbonsnakes, a radiation of aquatic serpents centered in the western United States and the sister taxon to the genus *Nerodia*, a radiation of aquatic snakes centered in the eastern United States (Alfaro and Arnold 2001). *Thamnophis* is found from Canada to Costa Rica and contains about thirty-one species, two of which occur in Alabama.

### Key to the Species of *Thamnophis* of Alabama

**1a** Dorsolateral light stripes obscure; venter with one or two rows of dark, half-moon-shaped spots along each side; tail not noticeably elongate.

> *Thamnophis sirtalis sirtalis*—**Eastern Gartersnake . . . . page 267.**

**1b** Dorsolateral light stripes bold; venter unmarked; tail noticeably elongate.

> *Thamnophis sauritus sauritus*—**Eastern Ribbonsnake . . . . page 270.**

*From top to bottom:*

Dorsal color pattern of an Eastern Gartersnake (*Thamnophis sirtalis*)

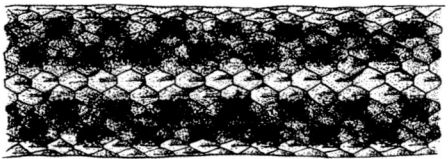

Ventral pattern of an Eastern Gartersnake (*Thamnophis sirtalis*)

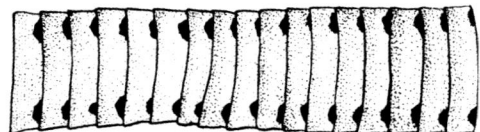

Dorsal color pattern of an Eastern Ribbonsnake (*Thamnophis sauritus*)

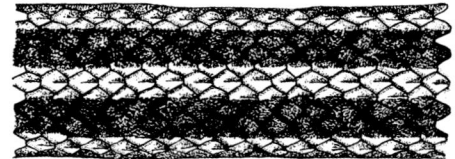

Eastern Garter-
snake, *Thamnophis
sirtalis sirtalis*, Lime-
stone County, AL

## Eastern Gartersnake
*Thamnophis sirtalis sirtalis* (Linnaeus, 1758)

**DESCRIPTION** Eastern Gartersnakes are medium-sized serpents, attaining a maximum total length of about 1,220 mm. The head is distinct
from the neck, and the anal scute is undivided. Dorsal scales are strongly
keeled and are arranged in nineteen rows at midbody. The dorsal coloration is dull greenish, brown, or bluish with a yellow, tan, or red stripe
occupying the mid-dorsal scale row and one-half of each adjacent row.
Usually, there is a light lateral stripe on each side occupying scale rows
two and three (counting up from the ventrals), but these are faded and
difficult to distinguish in most large individuals. Two rows of alternating dark spots or squarish blotches often are present between the stripes
on each side. The venter is yellowish, greenish, or bluish with one or
two rows of dark spots along the lateral-most edge of each ventral scute.
The only obvious geographic variation in this subspecies in Alabama involves the color of the mid-dorsal stripe, which is distinct in specimens
from the northern portion of the state and becomes less distinct in the
southern portion. In a majority of the live specimens from the lower
two-thirds of Baldwin and Mobile Counties, the stripe is decidedly red.
Elsewhere in the state, the stripe, when distinct, is light yellow. Subcaudal scales average seventy-four in males and sixty-nine in females.

**ALABAMA DISTRIBUTION** This subspecies is found throughout the state.

*From top to bottom:*

Eastern Garter-snake, *Thamnophis sirtalis sirtalis*, Cahaba National Wildlife Refuge, Bibb County, AL

Eastern Garter-snake, *Thamnophis sirtalis sirtalis*, Covington County, AL

**HABITS** From the standpoint of its ecological requirements, the Eastern Gartersnake is widespread and can be found in almost all terrestrial habitat types. Its food consists mostly of frogs, toads, and earthworms, but fish, salamanders, small mammals, and even snakes are eaten occasionally. These snakes give birth to young, usually seven to twenty per litter, although litters of up to eighty have been recorded. Gartersnakes tend to be somewhat ill tempered and will not hesitate to bite when captured. In their inclination to bite and smear their captor with musk, they resemble some of the species of *Nerodia*.

**CONSERVATION AND MANAGEMENT** Eastern Gartersnakes can be found in areas managed by humans, such as parks and golf courses, and are

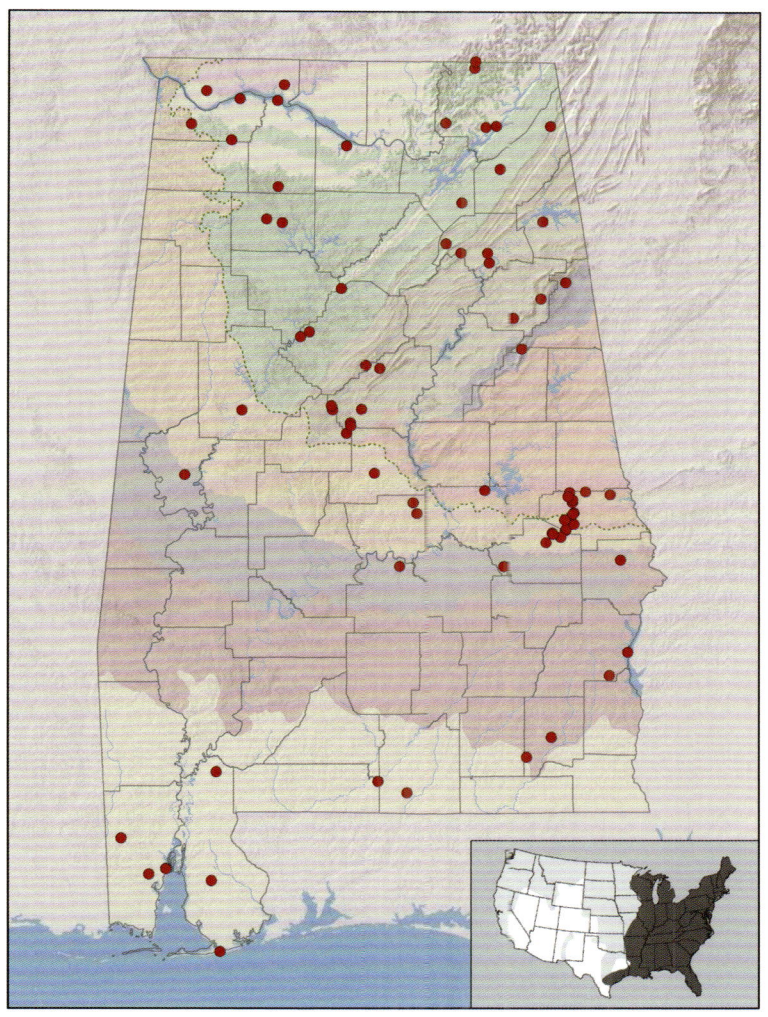

Distribution of Eastern Garter-snake, *Thamnophis sirtalis sirtalis*. This subspecies is assumed to occur throughout Alabama. Inset map depicts approximate range in the United States with dark shading indicating the range of *T. s. sirtalis* and light shading indicating the range of all other subspecies.

fairly common throughout Alabama. Therefore, the subspecies receives no special conservation protection. Management activities that create or enhance open, fish-free wetlands will increase population persistence of this species.

**TAXONOMY** This species is sister to *T. proximus* + *T. sauritus* (Pyron et al. 2011). Eleven subspecies of this snake, which ranges from the East Coast to West Coast, are recognized. Only one of these occurs in Alabama. Previous authors have placed this species in the genera *Coluber, Eutaenia, Eutainia, Eutanaeia,* and *Tropidonotus.*

## Eastern Ribbonsnake

*Thamnophis sauritus sauritus* (Linnaeus, 1766)

**DESCRIPTION** This is a medium-sized snake attaining a maximum total length of about 955 mm, with a body that is extremely slender and a tail that is exceptionally long (39 percent of total length). The anal scale is undivided. The dorsal scales are strongly keeled and are arranged in nineteen rows at midbody. The dorsum is brown to reddish brown with three conspicuous yellow stripes, a median one and a lateral one on each side. The lateral stripes involve the third and fourth scale rows on each side. Ventrally, the belly scales are yellowish and otherwise unmarked. Females attain larger sizes than males.

**ALABAMA DISTRIBUTION** This subspecies is found throughout the state but is spotty in its distribution.

**HABITS** The Eastern Ribbonsnake is semiaquatic, preferring open, damp situations. Such places include marshes, beaver swamps, weedy lake shores, stream margins, and low, wet meadows. Fish hatcheries often provide optimal habitat conditions. Fish and amphibians constitute the bulk of the diet. The Eastern Ribbonsnake is viviparous and usually gives birth in July or August, when three to twenty-six offspring are produced. This subspecies is alert, agile, and a good climber. It is nervous in temperament but does not bite.

**CONSERVATION AND MANAGEMENT** Although spotty in its distribution, the Eastern Ribbonsnake still is common in many areas of Alabama. The subspecies can be found in areas managed by humans, such as

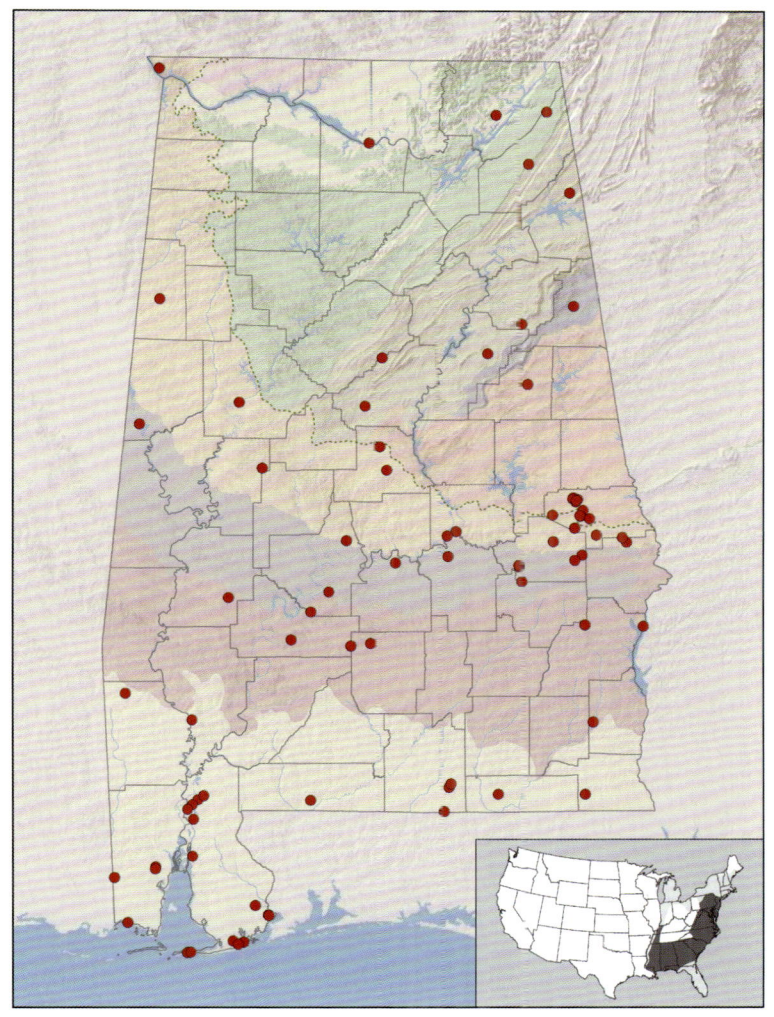

Distribution of Eastern Ribbonsnake, *Thamnophis sauritus sauritus*. This subspecies is assumed to occur throughout Alabama. Inset map depicts approximate range in the United States with dark shading indicating the range of *T s. sauritus* and light shading indicating the range of all other subspecies.

parks and golf courses. For this reason, it receives no special conservation protection by state law. But, these snakes are susceptible to road mortality in some areas (Bernardino and Dalrymple 1992).

**Taxonomy** This species is the sister taxon to *T. proximus* (Pyron et al. 2011), a wide-ranging species found in the central United States, Mexico, and Central America. Four subspecies of *T. sauritus* are recognized, one of which occurs in Alabama. Previous authors have placed this species in the genus *Coluber, Eutaenia, Leptophis,* and *Tropidonotus.*

# North American Watersnakes
## Genus *Nerodia* (Baird and Girard, 1853)

This genus, endemic to North America, contains the watersnakes, a radiation centered in the eastern United States. It is the sister taxon to the genus *Thamnophis*, a radiation of aquatic snakes centered in the western United States. Traditionally, the genus *Nerodia* has contained ten species. Nine species of *Nerodia* are found in Alabama. Many are similar in appearance to cottonmouths and frequently are killed by humans because of this similarity.

### Key to the Species of *Nerodia* of Alabama

**1a** Dorsal pattern of adults uniform or with an obscure pattern of light and dark flecks (salt and pepper appearance). Go to **2**.

**1b** Dorsal pattern of bold stripes, bands, or blotches. Go to **4**.

*From left to right:*

Lateral view of head of a Green Watersnake (*Nerodia cyclopion*); based on AUM 22105

Lateral view of head of a Brown Watersnake (*Nerodia taxispilota*); based on AUM 29277

**2a** A series of subocular scales separating upper labials from orbit. Go to **3**.

**2b** Some upper labials entering orbit.

> *Nerodia erythrogaster*—**Plain-bellied Watersnake** . . . . page 275.

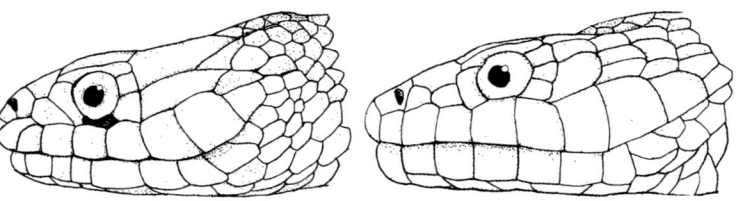

**3a** Anterior one-third of venter yellowish to white, posterior two-thirds smoky gray to brown with light half-moons.

> *Nerodia cyclopion*—**Green Watersnake** . . . . page 281.

**3b** Venter predominately light, including posterior two-thirds; may have ventral dark markings but not of half-moon shape.

> *Nerodia floridana*—**Florida Green Watersnake** . . . . page 284.

Ventral view of a Green Watersnake (*Nerodia cyclopion*); based on AUM 22105

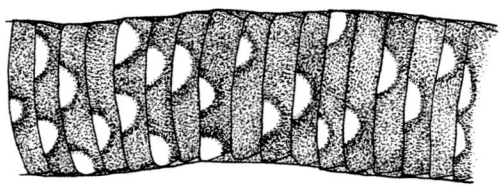

**4a** Dorsum with mid-dorsal and dorsolateral yellow stripes.

*Nerodia clarkii clarkii*—Gulf Saltmarsh Watersnake . . . . page 287.

**4b** Dorsum without longitudinal yellow stripes. Go to **5**.

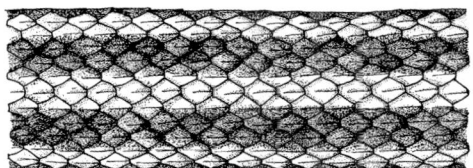

Doral color pattern of a Gulf Coast Saltmarsh Water-snake (*Nerodia clarkii*); based on AUM 5658

**5a** Parietal scutes fragmented posteriorly; body pattern of mid-dorsal dark blotches alternating with lateral ones, these not connected with chain-like dark lines.

*Nerodia taxispilota*—Brown Watersnake. . . . . page 290.

**5b** Parietal scutes not fragmented behind; body pattern with blotches, bands, or both. Go to **6**.

*From left to right:*

Head scales of a Brown Watersnake (*Nerodia taxispilota*); based on AUM 29277

Dorsal color pattern of a Brown Watersnake (*Nerodia taxispilota*)

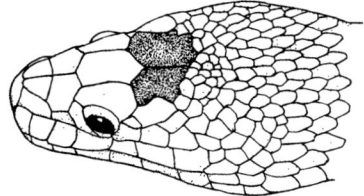
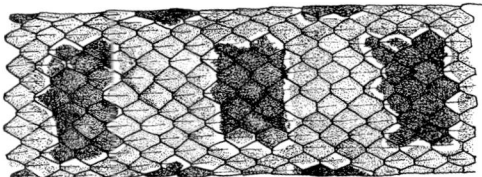

**6a** Body with mid-dorsal dark blotches connected to alternating lateral blotches by dark lines, forming a chainlike pattern.

*Nerodia rhombifer*—Diamond-backed Watersnake . . . . page 293.

**6b** Body with dark blotches or bands, but these not connecting to lateral blotches with dark lines that form a chain-like pattern. Go to **7**.

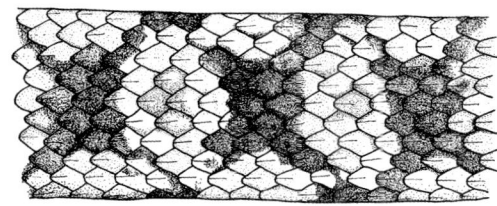

Dorsal color pattern of a Diamond-backed Watersnake (*Nerodia rhombifer*); based on AUM 12699

**7a** Body with alternating mid-dorsal and lateral dark blotches anteriorly, becoming banded posteriorly (*Nerodia sipedon pleuralis* may rarely be banded throughout its length). Go to **8.**

**7b** Body banded throughout its length.

> *Nerodia fasciata* **spp.—Banded Watersnakes. . . . . page 295.**

**8a** Venter with two rows of crescent-shaped or half-moon-shaped dark markings.

> *Nerodia sipedon pleuralis*—**Midland Watersnake . . . . page 307.**

**8b** Venter with dark transverse lines or narrow bands, one on anterior edge of each ventral scute.

> *Nerodia erythrogaster* **spp. . . . . page 275.**

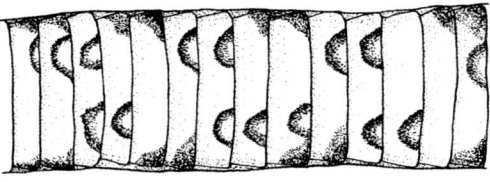

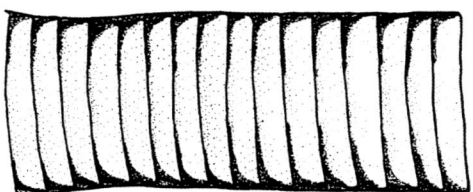

*From top to bottom:*

Ventral view of a Northern Water-snake (*Nerodia sipedon*)

Ventral view of a Plain-bellied Wa-tersnake (*Nerodia erythrogaster*)

# Plain-bellied Watersnake

*Nerodia erythrogaster* (Forster, 1771)

**TAXONOMY** This species is basal to the large-bodied watersnakes of the eastern United States (*N. harteri* + *N. sipedon* + *N. fasciata* + *N. taxispilota* + *N. rhombifer*; Pyron et al. 2011). Traditionally, four subspecies of this species have been recognized, based largely on the color of the belly. Examination of the mitochondrial genome across the range of the species identified five monophyletic lineages (Makowsky et al. 2010). A distinct divide centered on the Appalachian Mountains separates an eastern lineage from central, Louisiana, west Texas, and western lineages. The eastern lineage aligns with *N. e. erythrogaster* but includes a wider range in Alabama than implied by Mount (1975). The western lineage aligns favorably with *N. e. transversa*, a subspecies that generally has a white or pale yellow belly. But, the western lineage is distributed eastward to the Mississippi River and, along the Gulf Coast, to Mobile Bay, regions that include significant areas formerly attributed to *N. e. flavigaster*. The west Texas lineage represents substructuring of the western lineage, and the Louisiana lineage occupies geographic space formerly attributed to *N. e. flavigaster*. Finally, the central lineage includes specimens from east of the Mississippi River but west of the Appalachian Mountains, an area formerly attributed to *N. e. flavigaster*. Thus, mitochondrial data and variation in the color pattern of the venter render the traditional concept of *N. e. flavigaster* paraphyletic. To remove this problem, we retain *N. e. erythrogaster* for Red-bellied Watersnakes centered east of the Appalachian Mountains. We retain *N. e. flavigaster* for Yellow-bellied Watersnakes but expand the concept of this taxon to contain all lineages centered west of the Appalachian Mountains. Within this subspecies concept are four clades, a central lineage (representing an evolutionarily significant unit located between the Apalachicola and Mississippi drainages), a Louisiana lineage (representing an evolutionary unit located in the lower end of the Mississippi River), and the western and west Texas lineages (sister units containing specimens largely located west of the Mississippi River).

Within Alabama, yellow-bellied and red-bellied forms are present, with the red-bellied forms being more common in the southeast corner of the state, suggesting an invasion of *N. e. erythrogaster* into a state occupied largely by *N. e. flavigaster*. The degree to which these red-bellied forms align with the eastern lineage of Makowsky et al. (2010) is unknown. Similarly, the degree to which the yellow-bellied forms align with the central lineage of Makowsky et al. (2010) is unknown. What is known is that the western and central lineages are both widely distributed within Alabama, perhaps mixing across the entire state, and representatives of the western lineage have dispersed along the Gulf Coast as far east as Mobile County (Makowsky et al. 2010). Because red is not retained in many museum specimens, assignment of most specimens to the subspecies, as we view them, is not possible. Therefore, we combine both subspecies into a single account with a single map.

Other authors have placed this species in the genera *Natrix* and *Tropidonotus*.

### Key to the Subspecies of *Nerodia erythrogaster* of Alabama

**1a** Venter usually orange or reddish orange.

> *Nerodia erythrogaster erythrogaster*—**Red-bellied Watersnake** . . . .
>> **page 277.**

**1b** Venter usually cream or yellow.

> *Nerodia erythrogaster flavigaster*—**Yellow-bellied Watersnake** . . . .
>> **page 277.**

## Red-bellied Watersnake

*Nerodia erythrogaster erythrogaster* (Forster, 1771)
and

## Yellow-bellied Watersnake

*Nerodia erythrogaster flavigaster* (Conant, 1949)

**DESCRIPTION** This is a large, moderately heavy snake attaining a maximum total length of about 1,575 mm, of which the tail is a relatively short portion of total length. The head is elongate, wide, and distinct from the neck. The scales are strongly keeled and occur in twenty-three rows at midbody. In adults, the dorsum is uniform dark brown or grayish brown, becoming somewhat light on the sides. The belly color generally is red, orange, or yellow, and this color is uniform in the center of each ventral scute, becoming suffused with brownish gray along the anterior edges. Young snakes are conspicuously patterned with the anterior one-fourth to one-third of the body having distinct dark bands, these breaking up to form dark mid-dorsal saddles that alternate with vertical bars on the remainder of the body. The light interspaces separating dorsal bands and saddles are about one scale row wide, a pattern quite similar to that of *N. sipedon pleuralis*. The belly of juveniles is yellowish or cream with dark pigment on the anterior edges of the ventrals and lacks the dark half-moon-shaped markings of *N. s. pleuralis*.

*Above:* Juvenile Red-bellied Watersnake, *Nerodia erythrogaster erythrogaster*, Bullock County, AL

*Right:* Yellow-bellied Watersnake, *Nerodia erythrogaster flavigaster*, Elmore County, AL

**ALABAMA DISTRIBUTION** The distribution of the mitochondrial haplotypes constituting *N. e. erythrogaster* is poorly documented in the state. Specimens collected from near Auburn and near Tuscaloosa by Makowsky et al. (2010) are the only documented cases of this subspecies in the state. Based on the range map of Makowsky et al. (2010), we infer that the strongest influence of *N. e. erythrogaster* in Alabama occurs in the extreme southeastern corner of the state and that specimens collected elsewhere in the state might be intergrades with

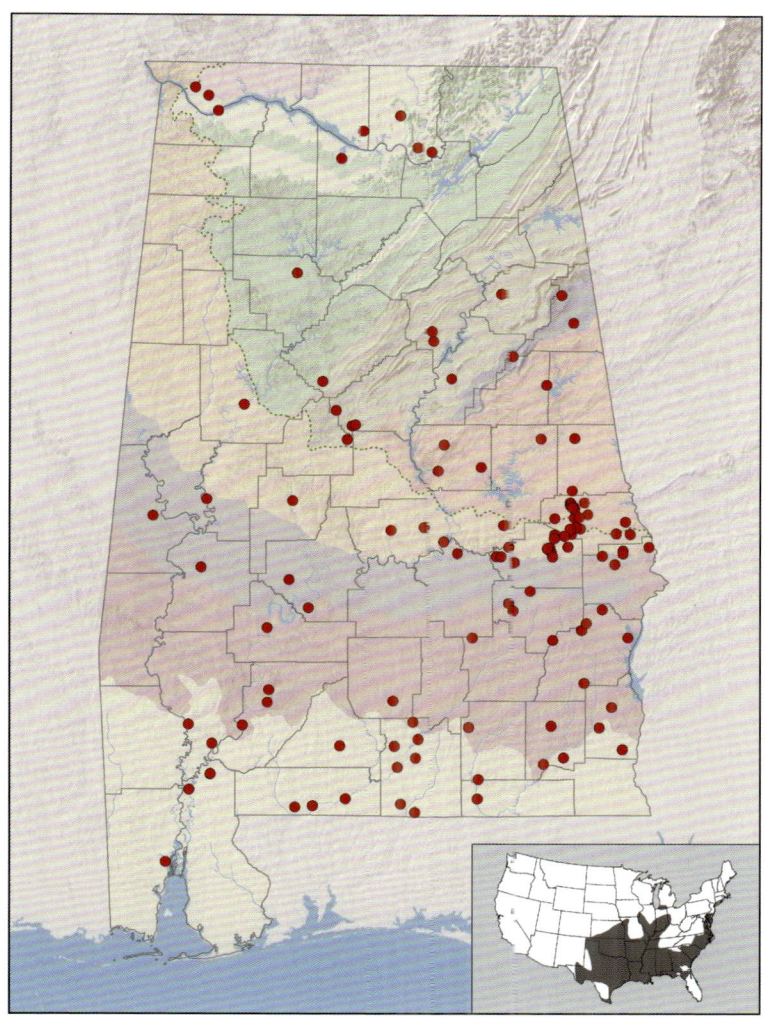

Distribution of Red-bellied Watersnake, *Nerodia erythrogaster erythrogaster,* and Yellow-bellied Watersnake, *N. e. flavigaster.* The species to which these subspecies belong is assumed to occur throughout Alabama. Inset map depicts approximate range in the United States.

*N. e. flavigaster.* However, the presence of this lineage as far north and west as Tuscaloosa, where specimens lack evidence of a red belly, indicate that our ability to distinguish two taxa in the state is limited. Influence of the central lineage of *N. e. flavigaster* is known from Jackson, Pike, and Tuscaloosa Counties and likely represents the vast majority of yellow-bellied forms in the state. Makowsky et al. (2010) document the presence of the western lineage in Alabama based on a single specimen collected in northern Mobile County. Based on distribution patterns of other taxa along the Lower Coastal Plain, we infer that this lineage is restricted to Mobile and Washington Counties and that area of Baldwin County associated with Mobile Bay.

**HABITS** Both subspecies of *N. erythrogaster* are abundant snakes in Alabama. These snakes occur in most kinds of permanently aquatic habitats, but they seem best adapted to swamps, sluggish streams, floodplain pools, and lakes and ponds with swampy margins. These subspecies are chiefly nocturnal, and on rainy nights are often found crossing roads. At such times, these snakes may move into flooded ditches and other ephemeral wetlands, where frogs congregate to breed. Driving through low country on rainy nights and investigating these places is a good way to locate and collect these subspecies, along with several other aquatic snakes. However, these snakes also wander much farther from water than do other members of the genus (Gibbons and Dorcas 2004). Unlike other members of the genus, *N. erythrogaster* remains in water rather than perching on vegetation or logs (Mushinsky et al. 1980). Frogs and fish are the main food items of these subspecies, with fish dominating the diet of juveniles and frogs dominating the diet of adults (Mushinsky et al. 1982). Mating occurs in the months of April and May. Females are viviparous, with the number of young ranging from two to fifty-five and litters being deposited in August and September (Gibbons and Dorcas 2004).

**CONSERVATION AND MANAGEMENT** These subspecies are common in wetlands and do particularly well in fish ponds, where they may reach high population densities. Because they do so well in human-altered landscapes, no specific conservation or management activities are needed. These subspecies receive no special protection by the state of Alabama.

Green Watersnake,
*Nerodia cyclopion*,
Union County,
Illinois

## Green Watersnake
*Nerodia cyclopion* (Dumeril and Bibron, 1854)

**Description** Green Watersnakes are large, heavy-bodied aquatic snakes attaining a maximum total length of about 1,270 mm, of which the tail represents only a short portion. The head is elongate and wide, followed by a distinct neck. The scales are heavily keeled, except for the first row on each side, and usually are arranged in twenty-seven rows at midbody in males and twenty-nine rows at midbody in females. The anal scute is divided and, in males, is followed by sixty-five to seventy-seven subcaudals (to distinguish this species from the seventy-three to eighty-three subcaudals of *N. floridana*); subcaudal counts of females (fifty-eight to seventy-five) do not differ from counts of female *N. floridana* (sixty-five to seventy-five). Unlike most members of the genus, Green Watersnakes have eyes that are separated from contact with the upper labials by a series of subocular scales. The ground color of the dorsum is dark olive to dark green, with a series of mid-dorsal dark blotches alternating with dark lateral bars that number from thirty-nine to forty-five (to distinguish this from the forty-six to fifty-eight bars of *N. floridana*). These markings become indistinct in old individuals. The anterior one-third of the belly is yellowish to white, and the posterior two-thirds and undersurface of the tail are smoky gray to brown with light half-moons (solid cream or yellow in *N. floridana*).

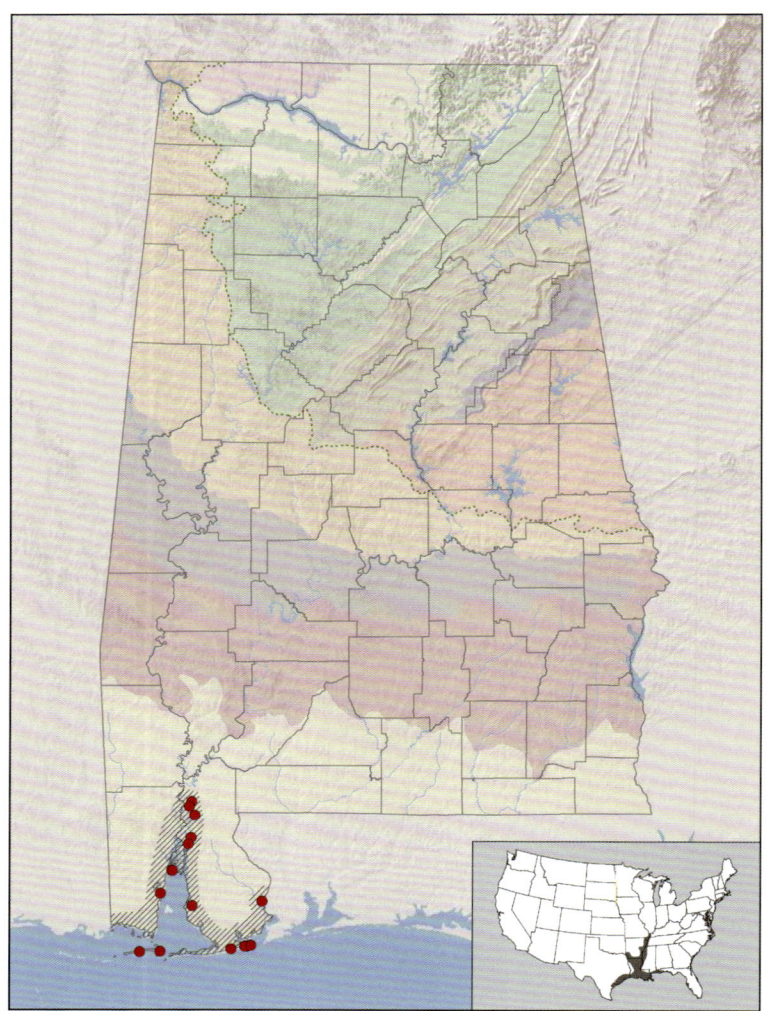

Distribution of Green Watersnake, *Nerodia cyclopion*. The presumed range of the species in Alabama is indicated by hatching. Inset map depicts approximate range in the United States.

**ALABAMA DISTRIBUTION** Specimens have been collected in swamps in lower Baldwin and Mobile Counties, in the wooded swamps and tree-lined waterways between the Mobile and Tensaw Rivers (Mount 1975), and as far east as the Perdido River (Sanderson 1993). The northern limit of the range in Alabama is undetermined but is probably the northern boundary of the Lower Coastal Plain.

**HABITS** *N. cyclopion* prefers timbered swamps, oxbows, sloughs associated with floodplains, and sluggish, tree-lined streams. In Alabama, Green Watersnakes abound in the tupelo gum-cypress swamps along the Tensaw and Mobile Rivers and their associated waterways. These snakes are fond of basking during late winter and spring, when numerous individuals may be seen, along with Cottonmouths and other aquatic snakes, in bushes and on low tree branches that overhang the water. As the weather warms in summer, these snakes spend more time in deeper water and are active mostly at night, changes that are associated with an attempt to regulate body temperature at a constant level (Mushinsky et al. 1980). The Green Watersnake is scarce or absent in much of the lower portion of the Mobile Bay delta area, where open expanses of marsh and grass flats are the prevalent habitats and *N. floridana* replaces *N. cyclopion*. Thus, habitat preferences appear to limit contact between these two species. Fish appear to be by far the most important food of Green Watersnakes, with centrachids becoming a more prevalent component of the diet as these snakes increase in size; catfish appear to be avoided (Mushinsky et al. 1982). Amphibians are eaten infrequently. Males seek mates in April with courtship apparently taking place in water. Females give birth to eleven to thirty-four young in July and August (Gibbons and Dorcas 2004).

**CONSERVATION AND MANAGEMENT** Only a very small portion of the entire range of this species occurs in Alabama and only a very small segment of Alabama has suitable habitat for this species. Fortunately, the W. L. Holland and Mobile-Tensaw Delta Wildlife Management Area and the Upper Delta Wildlife Management Areas provide exceptional habitat for this species in Alabama. As long as these areas maintain streamside management zones that retain old trees and snags, the species appears to be secure within the state. For this reason, it receives no special protection by state regulation. However, because the range of this species is so small within Alabama, it is listed as critically imperiled by the Alabama Natural Heritage system.

**TAXONOMY** Mount (1975), with reservation, considered *N. floridana* to be a subspecies within *N. cyclopion* but noted that there was no evidence of intergradation and that Mobile Bay separated the two taxa. Since that publication, the two subspecies described by Mount (1975) have been elevated to species status (Gibbons and Dorcas 2004). So, we consider *N. cyclopion* to be a species with no subspecific variation and separate it from its sister taxon, *N. floridana* (Alfaro and Arnold 2001). Other authors have placed this species in the genera *Natrix* and *Tropidonotus*.

## Florida Green Watersnake
*Nerodia floridana* (Goff, 1936)

DESCRIPTION This species is a large, heavy-bodied aquatic snake, attaining a maximum total length of about 1,270 mm, of which the tail represents only a short portion. The head is elongate, wide, and distinct from the neck. The scales are heavily keeled, except for the first row on each side, and usually are arranged in twenty-seven rows at midbody in males and twenty-nine rows at midbody in females. The anal scute is divided and, in males, is followed by seventy-three to eighty-three subcaudals (distinguishing this species from male *N. cyclopion* with sixty-five to seventy-seven subcaudals); subcaudal counts of females (sixty-five to seventy-five) do not differ from counts of female *N. cyclopion* (fifty-eight to seventy-five). Unlike most members of the genus, Florida Green Watersnakes have eyes that are separated from contact with the upper labials by a series of subocular scales. The ground color of the dorsum is dark olive to dark green, with a series of mid-dorsal dark blotches alternating with dark lateral bars that number from forty-six to fifty-eight (distinguishing this species from *N. cyclopion*, which has thirty-nine to forty-five dark bars). These markings become indistinct in old individuals. The entire belly is yellowish to white, becoming mottled with smoky gray to brown on the undersurface of the tail.

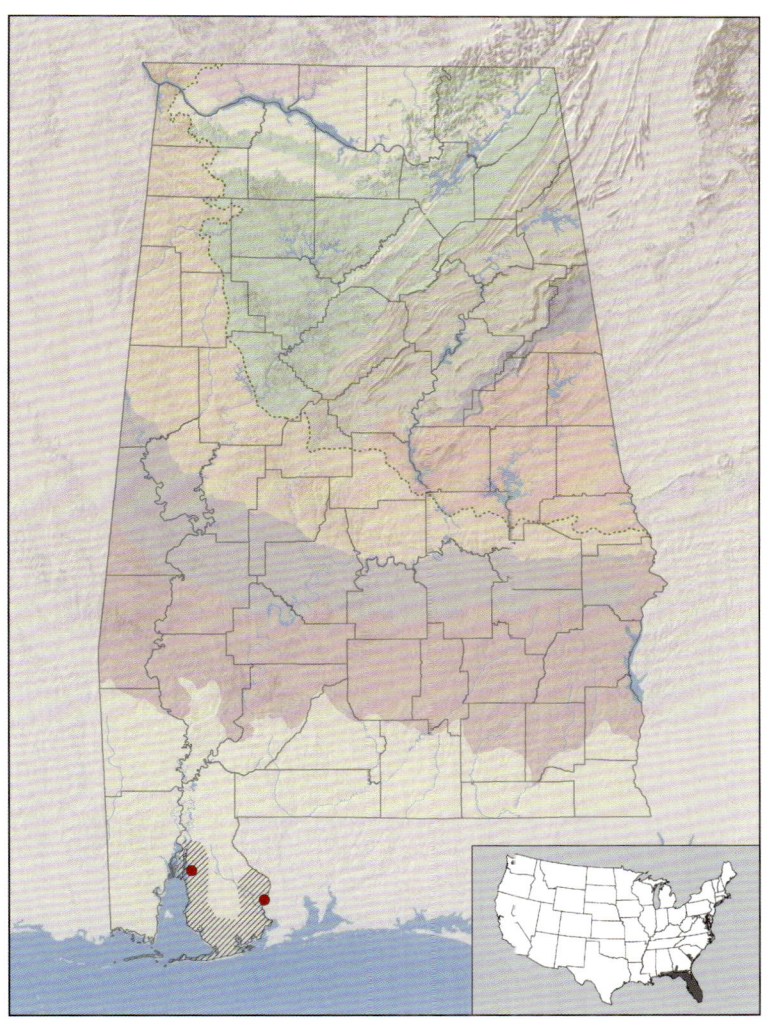

Distribution of Florida Green Watersnake, *Nerodia floridana*. The presumed range of the species in Alabama is indicated by hatching. Inset map depicts approximate range in the United States.

**ALABAMA DISTRIBUTION** This species is known only from the southern portion of Baldwin County.

**HABITS** Optimum habitat for *N. floridana* is weed-choked, freshwater marsh or wet prairie. All collecting sites in Alabama are weedy roadside ponds near the Perdido River and shallow, weedy ponds and freshwater inlets along the eastern shore of Mobile Bay. In peninsular Florida, the Florida Green Watersnake is often the most commonly encountered aquatic snake in such habitats. It is nocturnal and less inclined to bask than other members of the genus. This species is seldom found more than thirty miles from the coast, at least in the western part of its range (Alabama and the Florida Panhandle). During droughty years, populations may decline and may be slow to recover (Seigel et al. 1995). The diet of Florida Green Watersnakes is mostly frogs, perhaps explaining its largely nocturnal activity pattern. Fish also are a common component of the diet (Gibbons and Dorcas 2004). Males court females in spring (as early as February), and females produce litters of from 7 to 132 offspring during June through August (Gibbons and Dorcas 2004).

CONSERVATION AND MANAGEMENT The vast majority of the range of this species is outside of Alabama. Within the state, only a very limited area is occupied by Florida Green Watersnakes. Because of this, these snakes are considered critically imperiled by the Alabama Natural Heritage Program. Fortunately, Gulf State Park provides public protected areas occupied by this species. The Perdido River Longleaf Hills Tract may also have this species. Management practices on these sites that encourage maintenance of open marsh habitat should be sufficient to maintain populations of Florida Green Watersnakes in Alabama. The species receives no special protection by the state. It is commonly killed crossing roads, and such mortality may decrease populations; carefully constructed wildlife barriers or underpasses may be required to reduce mortality from this source in heavily populated areas of coastal Baldwin County (Smith and Dodd 2003).

TAXONOMY Mount (1975) considered *N. floridana* to be a subspecies within *N. cyclopion* but noted that there was no evidence of intergradation and that Mobile Bay separated the two taxa. Recent studies have elevated *N. floridana* to species status with its sister taxon being *N. cyclopion* (Alfaro and Arnold 2001). We consider *N. floridana* to have no identifiable subspecific variation. Other authors have placed this species in the genus *Natrix*.

Gulf Saltmarsh Watersnake, *Nerodia clarkii clarkii*, Wakulla County, FL

# Gulf Saltmarsh Watersnake
*Nerodia clarkii clarkii* (Baird and Girard, 1853)

**DESCRIPTION** This is a moderately stout aquatic snake, attaining a maximum total length of about 915 mm, with only a fairly short portion of this length being the tail. The head is elongate and wide, with swollen labials, and is distinct from the neck. The scales are heavily keeled, and the anal scute is divided. Unmistakable in appearance, the color pattern is of conspicuous mid-dorsal and dorsolateral yellowish stripes contrasting sharply against the dark gray to reddish-brown ground color. The venter is uniform black or black with tinges of yellow, having a sharply contrasting median yellow stripe that continues well onto the tail and a pair of ventrolateral yellow stripes, these involving the edges of the ventral scutes and the first row of dorsal scales. The lower labials, chin shields, and some upper labials typically are light centered.

**ALABAMA DISTRIBUTION** Coastal salt marshes of Baldwin and Mobile Counties, including Dauphin Island, are the only habitats in Alabama occupied by this subspecies.

**HABITS** The Gulf Saltmarsh Watersnake is the only snake in Alabama that habitually occupies the salt marsh habitat. It is active mostly at night, feeding on small fish and, occasionally on crabs. But individuals may be observed by day and activity may be altered by the tidal

Hybrid Gulf Saltmarsh and Banded Watersnake, *Nerodia clarkii* x *fasciata*, panhandle of FL

cycle (Gibbons and Dorcas 2004). The Mangrove Watersnake, *N. c. compressicauda*, uses lingual luring, a flickering of the tongue that attracts fish, which are then consumed (Hansknecht 2008); study of foraging strategy in *N. c. clarkii* is needed to discover whether lingual luring is part of its strategy. Data on reproduction in Alabama are lacking. Reports from other areas indicate a range in litter size of from three to forty-four offspring, with litters being deposited during July through October (Gibbons and Dorcas 2004).

**CONSERVATION AND MANAGEMENT** By statute, it is unlawful to possess a Gulf Saltmarsh Watersnake in the state of Alabama without a scientific collecting permit. Available habitat for this snake in the state has rapidly diminished as a result of dredge-and-fill operations and encroaching real estate developments. This modification and fragmentation of existing habitat has reduced to critical levels populations of all vertebrates that specialize on dune and coastal marsh habitats. The Gulf Saltmarsh Watersnake is recognized as an indicator subspecies for high-quality coastal lands remaining in the Gulf region (Kautz and Cox 2001). Populations of these snakes still thrive on Dauphin Island, Bon Secour National Wildlife Refuge, Weeks Bay National Estuarine Research Reserve, and the Fort Morgan Peninsula, suggesting that critical habitat remains sufficiently intact for this salt marsh–endemic taxon. Retention of these areas as wildlife habitat is critical to retaining Gulf Saltmarsh Watersnakes in the Alabama herpetofauna. The linear

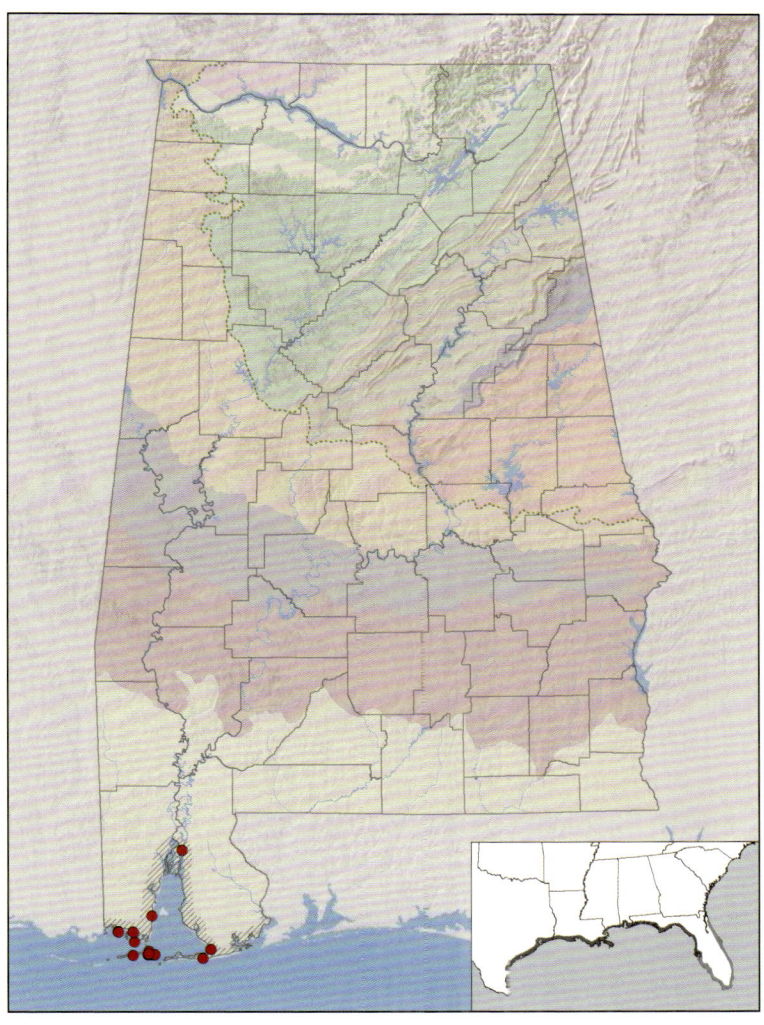

Distribution of Gulf Saltmarsh Watersnake, *Nerodia clarkii clarkii*. The presumed range of the subspecies in Alabama is indicated by hatching. Inset map depicts approximate range in the United States with dark shading indicating the range of *N. c. clarkii* and light shading indicating the range of all other subspecies.

nature of the habitat occupied by this subspecies presents management challenges for maintaining genetic diversity in the face of habitat fragmentation, as indicated by a study of the closely related Mangrove Saltmarsh Watersnake (*N. c. compressicauda*; Jansen et al. 2008).

**TAXONOMY** We follow Lawson et al. (1991) in separating this species from its sister lineage, *N. fasciata* + *N. sipedon*. Three subspecies of this species are recognized, one of which is found in Alabama. Other authors have placed this species in the genera *Natrix, Regina,* and *Tropidonotus*.

Brown Watersnake,
*Nerodia taxispilota*,
Baldwin County, AL

## Brown Watersnake
*Nerodia taxispilota* (Holbrook, 1838)

DESCRIPTION Brown Watersnakes are large, thick-bodied, short-tailed aquatic serpents, attaining a maximum total length of around 1,750 mm. The head is distinct from the neck and has unusual scutellation in that the upper labials enter the orbit and that the parietals are fragmented posteriorly. The anterior temporals are divided (80 percent of individuals). All dorsal scales are strongly keeled and occur in twenty-five to thirty-three rows, and the anal scute is divided. In dorsal coloration, these snakes are light grayish brown with a series of twenty-two to twenty-nine median, dark, square blotches alternating with a series of lateral, rectangular dark blotches. The venter is yellowish and is heavily patterned with blackish spots; the chin shields are smudged with dark pigment. Sexual size dimorphism is striking in this species; females exceed males in maximum sizes attained, and males exceed females in mean ventral and subcaudal counts.

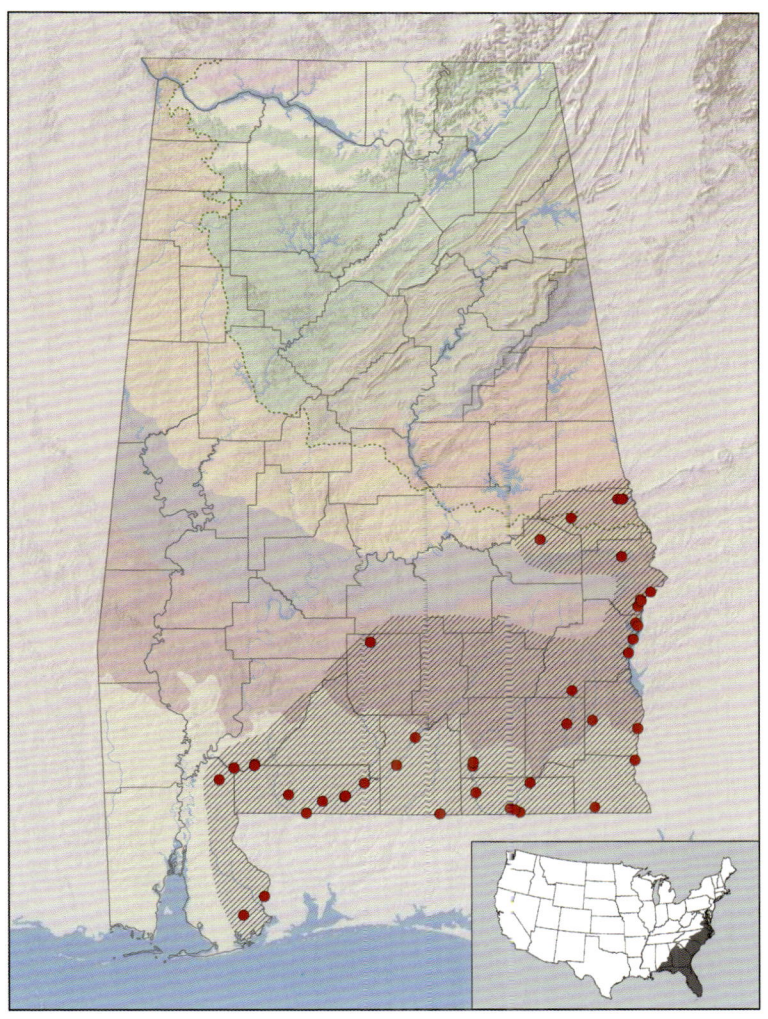

Distribution of Brown Watersnake, *Nerodia taxispilota*. The presumed range of the species in Alabama is indicated by hatching. Inset map depicts approximate range in the United States.

**ALABAMA DISTRIBUTION** This species is found in the Chattahoochee, Choctawhatchee, Conecuh, Perdido, Pea, and Yellow Rivers of the Coastal Plain and that portion of the Piedmont drained by the Chattahoochee River.

**HABITS** In Alabama, the Brown Watersnake is most frequently encountered in streams and stream impoundments. It is formidable in appearance and greatly feared by many who live within its range. The name water rattler is often applied to this snake in southern Alabama, based on the erroneous assumption that it is a Timber Rattlesnake that has adopted aquatic habits and lost its rattle. The Brown Watersnake is rather strongly diurnal and spends much of its time basking over water in trees and bushes. *N. taxispilota* puts up a vigorous defense and is capable of delivering a painful, though non-venomous, bite. Fishes (especially catfish; Tyson et al. 2008) and frogs constitute the bulk of the diet. Mating occurs from March through June, when females may be approached by multiple males (Gibbons and Dorcas 2004). Copulatory plugs, an exudate that prevents other males from inseminating a female, are known from this species, a feature also observed in mating balls of some close relatives (Gibbons and Dorcas 2004). Females give birth to four to sixty-one offspring from August through October.

CONSERVATION AND MANAGEMENT This species receives no special protection by state law but is tracked as vulnerable by the Alabama Natural Heritage Program. Brown Watersnakes can be common, being particularly abundant on Lake Eufaula, where large numbers of basking individuals can be seen each spring. However, these snakes are frequently killed by the local populace, causing it to be reduced in abundance in most areas of Alabama. The reproductive output is likely to be sufficient to overcome this source of increased mortality. Streamside management zones that retain riparian trees and snags should enhance the habitat for this species.

TAXONOMY This species is the sister taxon to *N. rhombifer* (Alfaro and Arnold 2001). No subspecific lineages have been described for this species. Other authors have placed it in the genera *Coluber, Natrix, Tropidonotus.*

# Diamond-backed Watersnake
*Nerodia rhombifer* (Hallowell, 1852)

**DESCRIPTION** Diamond-backed Watersnakes are large, thick-bodied serpents, attaining a maximum total length of around 1,600 mm and possessing a short tail. The head of this subspecies is distinct from the neck and there are conspicuous papillae under the chin. Subocular scales are absent so that the upper labials contact the orbit. The rear of each parietal is entire, a feature that distinguishes this subspecies from *N. taxispilota*, which has truncated parietals that disintegrate posteriorly into small, keeled scales. The anterior temporals are undivided, and the anal scute is divided. Dorsal scales are arranged in twenty-five to thirty-one rows and are strongly keeled. A ground color of olive to brown characterizes the dorsum, which is broken by a median series of dark blotches. The dark blotches are connected to alternating dark lateral blotches by narrow, diagonal stripes (blotches are not connected in *N. taxispilota*). The venter is yellowish with dark semilunar spots, these being more numerous posteriorly. The chin shields are immaculate, as opposed to the dark-smudged markings in *N. taxispilota*. Females exceed males in maximum size and in mean number of dorsal scale rows. Males exceed females in mean ventral count, mean subcaudal count, and mean number of dorsal body blotches (Mount 1975).

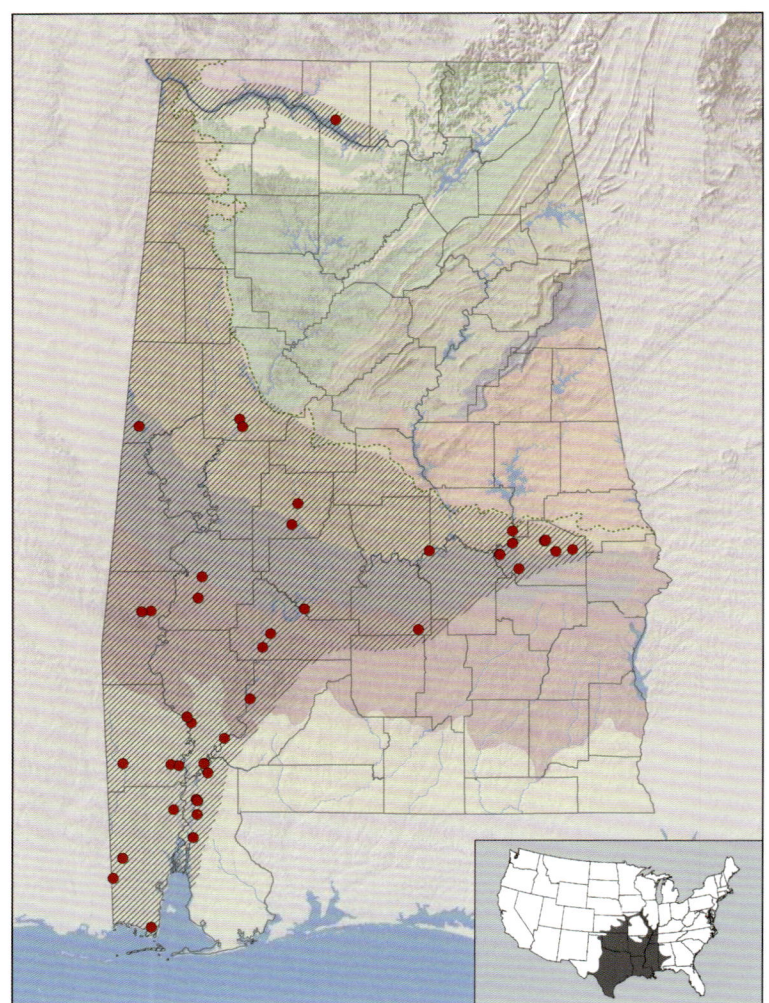

Distribution of Diamond-backed Watersnake, *Nerodia rhombifer*. The presumed range of this subspecies in Alabama is indicated by hatching. Inset map depicts approximate range in the United States.

**ALABAMA DISTRIBUTION** This subspecies is restricted almost entirely to the Coastal Plain region of the Alabama, Black Warrior, Cahaba, Coosa, Escatawpa, Tallapoosa, and Tombigbee Rivers. It is known from the Tennessee River based on a single observation from Limestone County.

**HABITS** This heavy-bodied snake is strongly aquatic and is found in greatest abundance in sloughs and lakes associated with major rivers and creeks. It is less commonly encountered in flowing water (Gibbons and Dorcas 2004). The subspecies is active day and night, differing in this respect from *N. taxispilota*, which is strongly diurnal. Also, the Diamond-backed Watersnake apparently basks less frequently than *N. taxispilota*. During the heat of summer, *N. rhombifer* becomes more nocturnal and more arboreal (Mushinsky et al. 1980). Fishes, including catfish, are dominant food items, which tend to be captured in deep water (Mushinsky et al. 1982). Frogs, salamanders, and crayfish are secondary dietary items. Females from Alabama produce litters of twenty-three to twenty-four young. Much larger broods have been reported from some other areas. Gravid females have higher body temperatures and maintain those temperatures over longer periods than do non-gravid females (Tu and Hutchinson 1994). The Diamond-backed Watersnake will bite viciously when molested.

**CONSERVATION AND MANAGEMENT** This subspecies is one of the watersnakes most often mistaken for the venomous Cottonmouth. Therefore, human persecution likely affects adult survivorship. But reproductive potential in this subspecies is large enough to overcome this source of mortality. As long as management of sloughs, ponds, and lakes along major rivers and creeks retain mature trees in the riparian zone and snags are allowed to accumulate for basking, this subspecies should maintain healthy populations within the state. Diamond-backed Watersnakes receive no special protection by state law.

**TAXONOMY** Brandley et al. (2010) examined the mitochondrial genome of this widespread taxon and determined that the Mississippi River served as a major barrier to dispersal during the Pleistocene but that gene flow has occurred since. For this reason, we follow Crother et al. (2017) in considering this to be a single species with no subspecific variation. *Nerodia rhombifer* is the sister taxon to *N. taxispilota* (Alfaro and Arnold 2001). Other authors have placed this species in the genera *Natrix* and *Tropidonotus*.

## *Nerodia fasciata* (Linnaeus, 1766)

**TAXONOMY** This species is the sister taxon to *N. sipedon* (Pyron et al. 2013), and the two taxa hybridize extensively. Currently, three subspecies are recognized with *N. fasciata*, an arrangement that has changed dramatically from the six subspecies accepted by Mount (1975). Three of those previous subspecific taxa have been allocated to *N. clarkii*

(Lawson et al. 1991). Recent range maps suggest that Alabama has either a single subspecies, *N. f. fasciata* (Dorcas and Gibbons 2004), or two subspecies, *N. f. fasciata* and *N. f. confluens* (Conant and Collins 1998). Both interpretations suggest the northern limit to the distribution of the Florida Watersnake, *N. f. pictiventris*, occurs far south of Alabama. However, we retain all three subspecies for the state's herpetofauna based on studies of geographic variation by Schwaner and Mount (1976) and Seyle (1980). This species is quite similar to *N. sipedon* and freely hybridizes with that taxon in Alabama (Schwaner and Mount 1976), Georgia (Seyle 1980), and the Carolinas (Mebert 2008). However, the pattern of introgression of genes between these two species indicates restricted gene flow across the hybrid zone, which has led to continued separation of these taxa into distinct species. The sister taxon of *N. fasciata* is *N. sipedon* + *N. harteri* (Alfaro and Arnold 2001; Pyron et al. 2011). Other authors have placed this species in the genera *Coluber, Natrix,* and *Tropidonotus.*

## KEY TO THE SUBSPECIES OF *NERODIA FASCIATA* OF ALABAMA

**1a** Venter with square dark markings along lateral edge of most ventral scutes. Go to **2.**

**1b** Venter with dark, wormlike markings along posterior edge of each ventral scute, lacking square dark markings along lateral edge of most ventral scutes.

    *Nerodia fasciata pictiventris*—Florida Watersnake . . . . page 298.

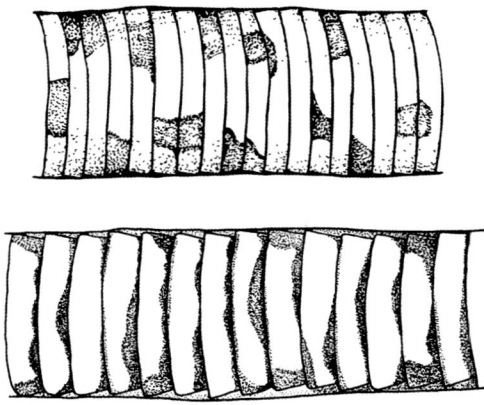

*From top to bottom:*

Ventral view of a Banded Watersnake (*Nerodia fasciata fasciata*)

Ventral view of a Florida Watersnake (*Nerodia fasciata pictiventris*)

**2a** Venter with lateral dark blotches that seldom involve more than one scute and that do not tend to merge extensively; dark dorsal bands as wide as light space between them.

*Nerodia fasciata fasciata*—**Banded Watersnake** . . . . **page 301.**

**2b** Venter with lateral dark blotches that frequently involve three or more scutes and show a tendency to merge extensively; dark dorsal bands wider than light space between them.

*Nerodia fasciata confluens*—**Broad-banded Watersnake** . . . . **page 304.**

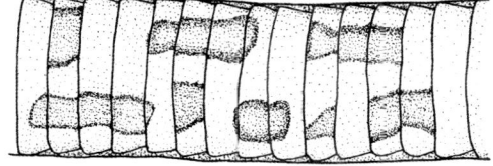

*From top to bottom:*

Ventral view of a Banded Watersnake (*Nerodia fasciata fasciata*)

Ventral view of a Broad-banded Watersnake (*Nerodia fasciata confluens*)

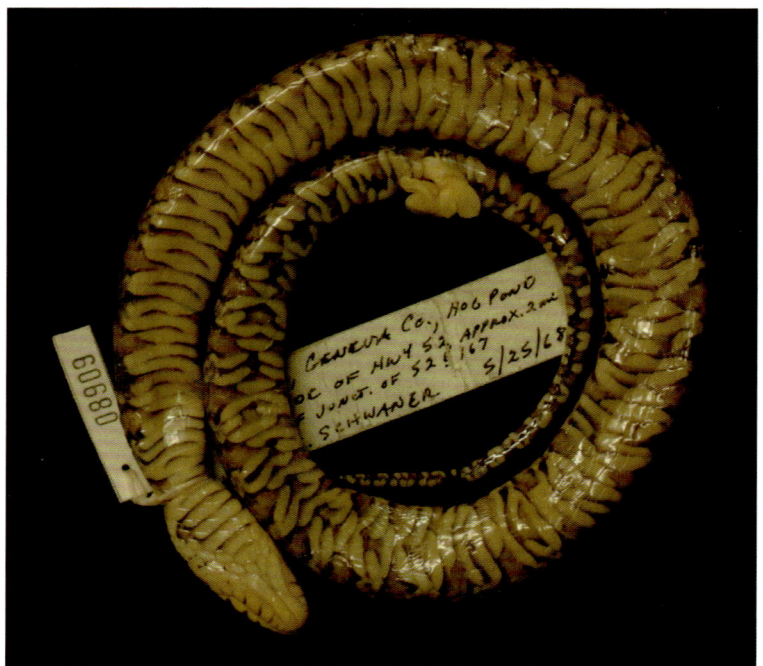

## Florida Watersnake
*Nerodia fasciata pictiventris* (Cope, 1895)

**DESCRIPTION** This subspecies is quite similar to *N. f. fasciata* but differs from that subspecies chiefly in having irregular, wormlike markings along the posterior edge of each ventral scute instead of rectangular dark blotches along the lateral margin of most ventral scutes. Additionally, this subspecies possesses secondary dark spots between the bands on the sides, a feature not present in the other two subspecies. In some adults of *N. f. pictiventris*, and less frequently in some belonging to *N. f. fasciata*, the dorsal pattern is virtually obliterated by a suffusion of melanin in the interspaces between the bands. In such individuals, indications of the bands may be indistinctly visible on the sides as bars of lighter color. This color pattern represents a curious reversal of the light-to-dark relationship that obtains in young snakes and in non-melanistic adults.

**ALABAMA DISTRIBUTION** The influence of Florida Watersnakes is found in specimens from Houston, Geneva, and Covington Counties. In the

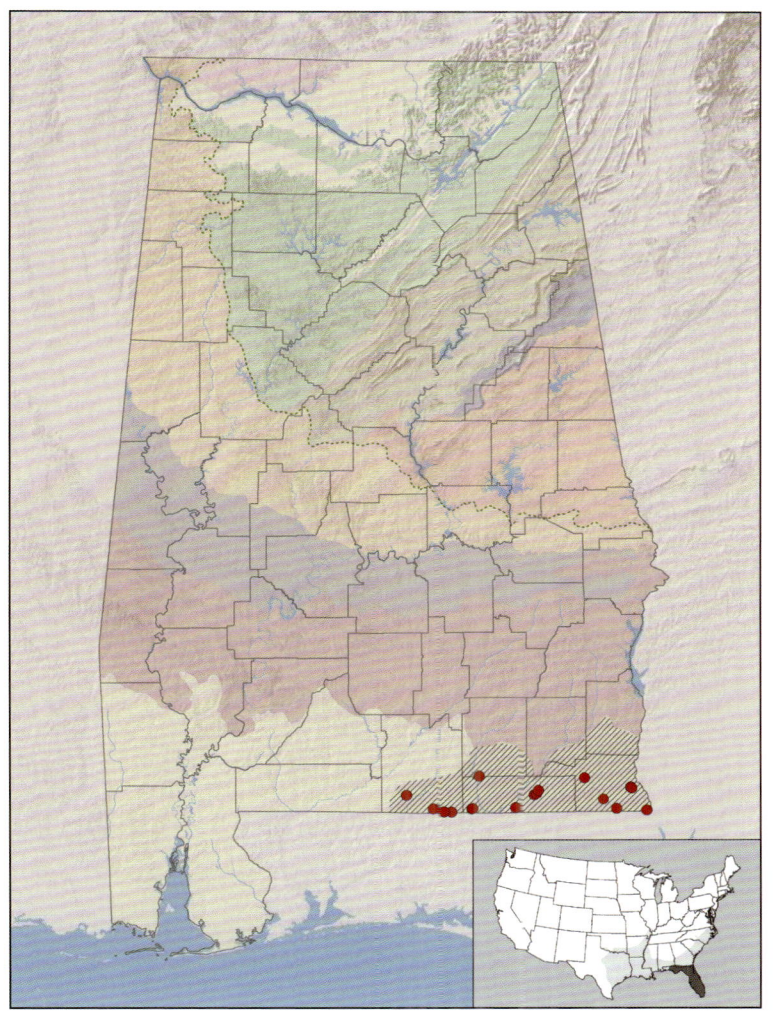

Distribution of Florida Watersnake, *Nerodia fasciata pictiventris*. The presumed range of the subspecies in Alabama is indicated by hatching. Inset map depicts approximate range in the United States with dark shading indicating the range of *N. f. pictiventris* and light shading indicating the range of all other subspecies.

Choctawhatchee and Pea Rivers, these intergradient individuals interbreed with *N. sipedon* near the northern edge of the Lower Coastal Plain.

**Habits** Florida Watersnakes occur in most kinds of permanently aquatic habitats, but shallow sinkhole lakes and ponds and swamps with abundant vegetation are especially favored. Stream habitats seldom support dense populations. These snakes are most active during the day in spring, basking along the edge of aquatic margins in morning hours and becoming more nocturnal during the summer, when these snakes may move to flooded ditches and other temporary accumulations of water where breeding frogs are abundant. Fishes and frogs are dominant prey items, supplemented by occasional salamanders and tadpoles. Mating occurs in spring when males appear to actively seek mates. *N. f. pictiventris* young are usually born in July and August, and litter size typically ranges from ten to fifty-seven.

Florida Watersnakes are commonly mistaken for the venomous cottonmouth because they share similar color patterns. Additionally, these watersnakes have the habit of flaring the quadrate and squamosal bones at the back of the skull, creating an appearance of the triangular-shaped head of a viperid snake. Florida Watersnakes are extraordinarily disagreeable and vigorously defend themselves if escape is prevented.

CONSERVATION AND MANAGEMENT Although nowhere abundant in Alabama, populations with the influence of this subspecies appear to be stable. Conservation of extensive areas of swamps and flatwoods with healthy populations of frogs is a key to retaining the influence of this subspecies in Alabama. Although persecuted by many humans, the high reproductive potential of *N. f. pictiventris* is sufficient to overcome this extra source of mortality. Because of this feature, the subspecies receives no special protection by state law.

Banded Watersnake, *Nerodia fasciata fasciata*, Covington County, AL

## Banded Watersnake
*Nerodia fasciata fasciata* (Linnaeus, 1766)

**DESCRIPTION** This is a relatively stout-bodied aquatic snake, attaining a maximum length of around 1,525 mm. The head in this subspecies is fairly short, is not distinctly widened posteriorly, possesses a wide postorbital dark stripe, has swollen labials, and is followed by a distinct neck. The dorsal scales are heavily keeled. The anal scute is divided in these snakes. In coloration, this subspecies is highly variable, with a dorsum that can be tan, brown, or reddish brown, with darker brown cross bands, which are often light edged and sometimes irregular but seldom broken. The dark dorsal cross bands usually are wider on the dorsum than on the sides. In ventral coloration, the subspecies is yellowish with reddish dark brown blotches that tend to be square or rectangular in shape and that seldom involve more than one ventral scute each.

**ALABAMA DISTRIBUTION** This subspecies is found in the Conecuh, Perdido, and Yellow Rivers of Alabama. In the Conecuh and Yellow Rivers, this subspecies interbreeds extensively with *N. sipedon*, with characteristics of the latter predominating in those populations.

**HABITS** The Banded Watersnake occurs in most kinds of permanently aquatic habitats. They are especially common in shallow sinkhole lakes and ponds and in swamps with abundant vegetation. Activity may occur year-round in these wetlands. Streams, the preferred

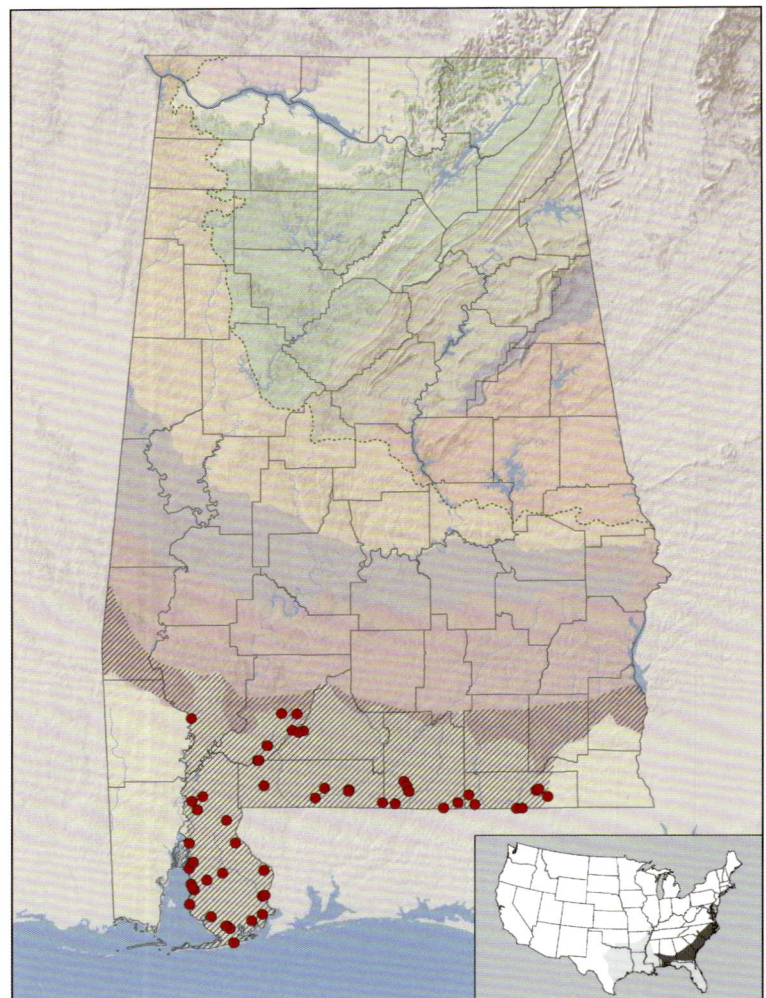

Distribution of Banded Watersnake, *Nerodia fasciata fasciata*. The presumed range of this subspecies in Alabama is indicated by hatching. Inset map depicts approximate range in the United States with dark shading indicating the range of *N. f. fasciata* and light shading indicating the range of all other subspecies.

habitat of *N. sipedon*, seldom support dense populations of *N. fasciata*. The snakes are most active during the day in spring, basking along the edge of aquatic margins in morning hours and becoming more nocturnal during the summer. During spring and summer, this subspecies tends to move out of their permanent-water habitats and into flooded ditches and other temporary accumulations of water where breeding frogs, a primary dietary item, are abundant. Fishes are a second dominant prey item, supplemented by occasional salamander and tadpole prey. Mating occurs in spring, when males appear to actively seek mates. *N. f. fasciata* young are usually born in July and August, and litter size typically ranges from ten to fifty-seven (Gibbons and Dorcas 2004). *N. f. fasciata* is commonly mistaken for the venomous Cottonmouth because of its color pattern and because of its habit of flaring the quadrate and squamosal bones at the back of the skull, creating the appearance of a triangular-shaped head. These snakes are extraordinarily disagreeable and vigorously defend themselves if escape is prevented.

**CONSERVATION AND MANAGEMENT** Although nowhere abundant in Alabama, populations of this subspecies appear to be stable, and for this reason, the subspecies receives no special protection by state law. Because frogs are a primary component of the diet, Banded Watersnakes tend to avoid sinkhole ponds that have been invaded by centrachid fishes, a frog predator that can eliminate this food resource. Although persecuted by many humans because of their similarity to Cottonmouths, the high reproductive potential of *N. f. fasciata* appears to be sufficient to overcome this source of mortality.

Broad-banded
Watersnake, *Nero-
dia fasciata con-
fluens*, Fort Bend
County, TX

## Broad-banded Watersnake
*Nerodia fasciata confluens* (Blanchard, 1923)

**DESCRIPTION** Broad-banded Watersnakes are relatively stout-bodied aquatic serpents, attaining a maximum length of around 1,525 mm. The head is fairly short, is not distinctly widened posteriorly, possesses a wide postorbital dark stripe, has swollen labials, and is followed by a distinct neck. The dorsal scales are heavily keeled, and the anal scute is divided. This subspecies has dark dorsal bands that are much wider than the interspaces between them and a venter that has extensive dark-pigmented blotches, which usually occupy three or more scutes each and often merge extensively.

**ALABAMA DISTRIBUTION** In Alabama populations, the influence of *N. f. confluens* is seen in freshwater habitats from the eastern boundary of the Mobile Bay drainage westward to the Mississippi border. Evidence of interbreeding between Broad-banded Watersnakes and Midland Watersnakes is seen in specimens collected from ponds in Monroe and Washington Counties near the northern edge of the Lower Coastal Plain (Mount 1975).

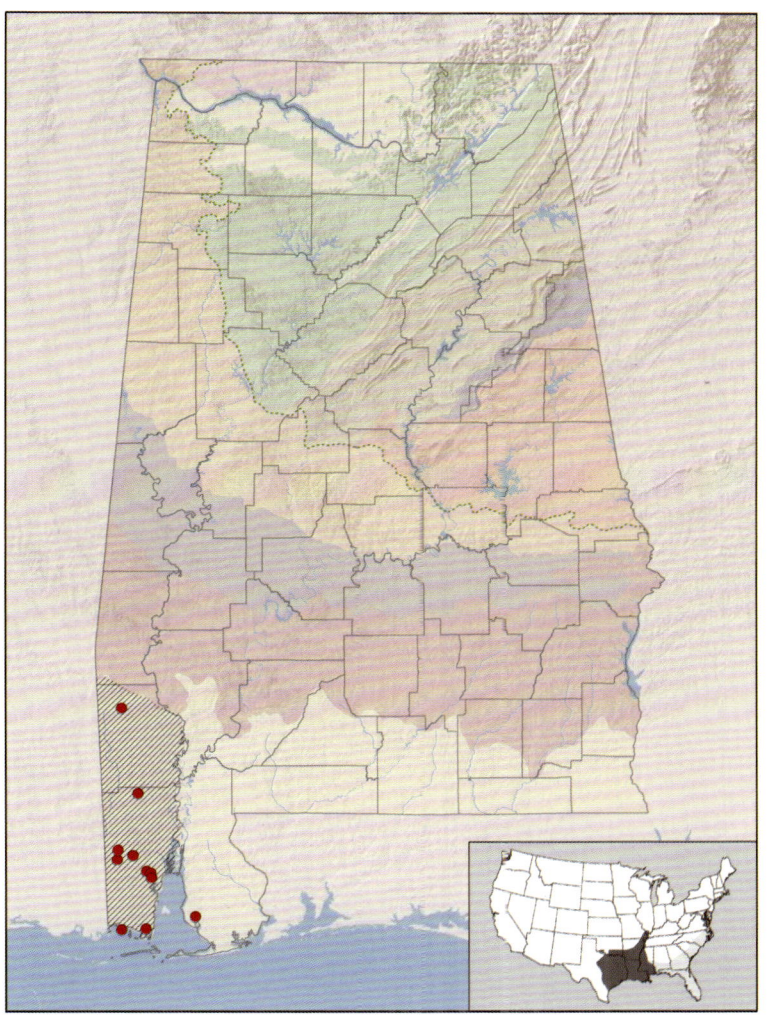

Distribution of Broad-banded Watersnake, *Nerodia fasciata confluens*. The presumed range of the subspecies in Alabama is indicated by hatching. Inset map depicts approximate range in the United States with dark shading indicating the range of *N. f. confluens* and light shading indicating the range of all other subspecies.

**HABITS** This subspecies occurs in most kinds of permanently aquatic habitats, where it is most likely to be observed in water (Mushinsky et al. 1980). During spring, Broad-banded Watersnakes are most active during the day, becoming more nocturnal during summer months (Mushinsky et al. 1980), particularly when heavy rains elicit an outburst of frog activity. Prey items are mostly fishes when the snakes are small and dominated by frogs in large snakes (Mushinsky et al. 1982). Mating occurs in spring, when males appear to actively seek mates.

Young are usually born in July and August, and litter size typically ranges from ten to fifty-seven.

As in the other two subspecies, Broad-banded Watersnakes are commonly mistaken for Cottonmouths because of similarities in color pattern and because the watersnakes can flare the quadrate and squamosal bones at the back of the skull, creating the appearance of a triangular-shaped head. These snakes are extraordinarily disagreeable and vigorously defend themselves if escape is prevented.

**CONSERVATION AND MANAGEMENT** Although nowhere abundant in Alabama, populations with the influence of this subspecies appear to be stable. Conservation of extensive areas of swamps with healthy populations of frogs is a key to retaining this subspecies in Alabama. Although persecuted by many humans, the high reproductive potential of *N. f. confluens* is sufficient to overcome this extra source of mortality. Because of this feature, the subspecies receives no special protection by state law.

Midland Watersnake, *Nerodia sipedon pleuralis*, Randolph County, AL

## Midland Watersnake
*Nerodia sipedon pleuralis* (Cope, 1892)

**DESCRIPTION** Midland Watersnakes are relatively stocky, have a fairly short tail, and attain a maximum length of about 1,310 mm. The head is distinct from the neck and has swollen labial scales. The dorsal scales are strongly keeled, and the anal scute is divided. In dorsal coloration, these snakes are gray, light brown, or reddish brown with a series of smooth-edged dark bands anteriorly and alternating dorsal and lateral blotches posteriorly. The dark markings usually are narrower than the spaces between them. The venter typically is yellow to cream colored, with a double row of crescent-shaped or half-moon-shaped red or reddish-brown markings, these extending to near the tip of the tail. Males average about seventy-five subcaudals, while females average sixty-five.

Considerable variation around this typical color pattern is observed in specimens from Alabama. A dark postocular stripe, present in *N. fasciata* but reportedly absent in *N. s. pleuralis*, is present in about 20 percent of the latter in the Alabama Coastal Plain. Between 2 and 4 percent of *N. s. pleuralis* in Alabama are banded throughout their length, as in *N. fasciata*. On large individuals, the dorsal pattern is often obscure. The ventral markings on some individuals are indistinct with indefinite boundaries. Juveniles of *N. erythrogaster* are easily mistaken for *N. s. pleuralis*.

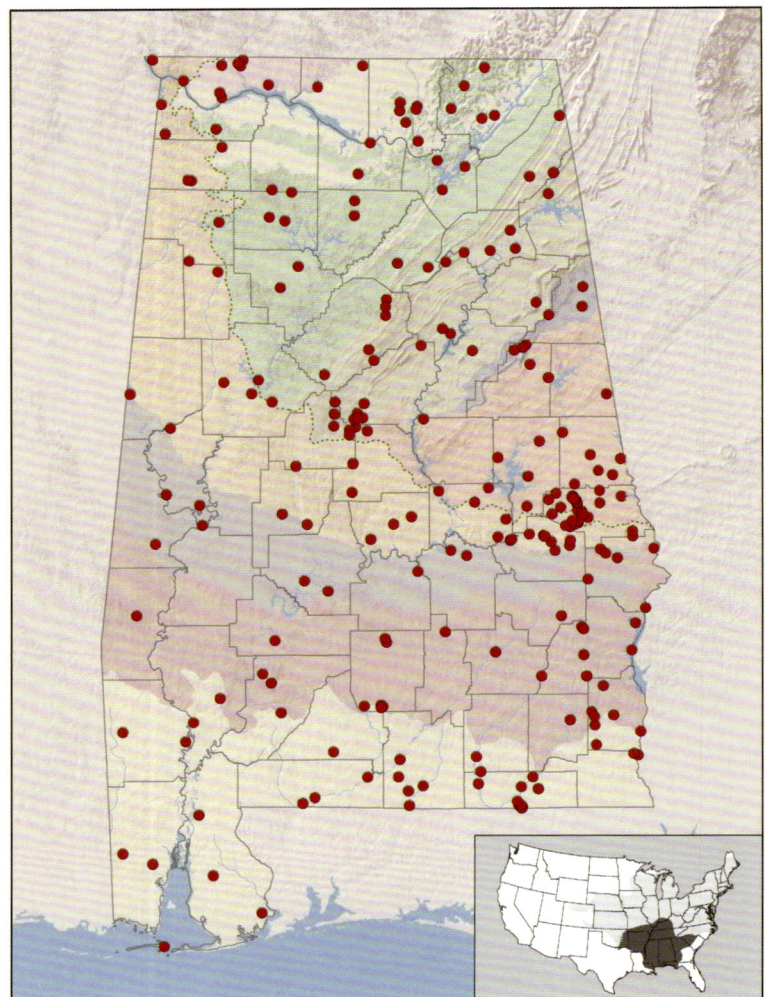

Distribution of Midland Watersnake, *Nerodia sipedon pleuralis*. This subspecies is assumed to occur throughout Alabama. Inset map depicts approximate range in the United States with dark shading indicating the range of *N. s. pleuralis* and light shading indicating the range of all other subspecies.

**ALABAMA DISTRIBUTION** This subspecies is found throughout the state, but in the Lower Coastal Plain, it is confined almost exclusively to streams that have at least some of their headwaters north of that province. Evidence of interbreeding between this form and *N. fasciata* has been found in the headwaters of the Choctawhatchee, Pea, Yellow, and Conecuh Rivers and in some ponds in Washington and Monroe Counties (Schwaner and Mount 1976).

**HABITS** Throughout much of the state, the Midland Watersnake is generally the most abundant aquatic snake. It inhabits farm ponds, lakes, streams, and most other permanently aquatic habitats. However, in small, sluggish Coastal Plain streams, beaver swamps, and other such places in that province, where the banks tend to be swampy or poorly defined, the Midland Watersnake is often greatly outnumbered by *N. erythrogaster* or one of the other members of the genus. Cottonmouths are also likely to be more common in such places. The optimal habitat for *N. s. pleuralis* seems to be a moderate-sized stream with a rock, gravel, or sand bottom and an abundance of minnows and other small fishes. When threatened with harm or capture and escape is impossible, the Midland Watersnake assumes a formidable appearance. Flattening its head and vibrating its tail, it strikes repeatedly at the offender. It is considered venomous and killed on sight by a large segment of the populace, who believe it to be a water moccasin. Fishes are a staple food of the Midland Watersnake, many of which are caught at night. Other food items include frogs, tadpoles, salamanders, and aquatic invertebrates. Males seek mates in spring. During this time, several males may attempt to mate with a single female (Mushinsky 1979). Females produce from five to fifty-nine offspring per litter, and these are born from July to early September (Gibbons and Dorcas 2004).

**CONSERVATION AND MANAGEMENT** In spite of relentless persecution, the Midland Watersnake persists in considerable abundance throughout nearly all of its range in our state. For this reason, it receives no special protection under state law. The subspecies is common in urban areas and farm ponds. So, no special management practices are necessary to maintain populations within Alabama.

**TAXONOMY** This sister species to *N. fasciata* is difficult to distinguish from that taxon and freely interbreeds with it (Mebert 2008; Seyle 1980; Schwaner and Mount 1976). Four subspecies are recognized, only one of which occurs in Alabama. Other authors have placed this species in the genera *Coluber, Natrix,* and *Tropidonotus.*

# Vipers

## Family Viperidae

This large family, cosmopolitan in distribution, contains 35 genera and about 330 species that are the sister taxa to a large radiation of advanced snakes (Pyron et al. 2011). All members of the family are dangerously venomous, producing toxic proteins in special glands derived, evolutionarily, from salivary glands. Venom is delivered by recurved, retractable, hollow fangs situated at the front of the upper jaws and acts to kill prey with bites that are administered rapidly, followed generally by an equally rapid withdrawal. Venom proteins have a variety of effects with some disrupting blood flow, others degrading tissue structure, and still others preventing transmission of nerve impulses. The overall effect of venom is to kill the prey rapidly, to cause the prey to generate a specific scent trail, and to begin the digestive process before the prey is swallowed. The snake predator then follows the chemical trail of the dying prey and consumes it. Members of the family have undivided subcaudal scutes (except for those near the tail tip), and the vast majority have elliptical pupils to the eyes. The family is divisible into 3 subfamilies, the largest of which, Crotalinae, with 21 genera and about 225 species, contains all the New World members. The crotalines, commonly called pit vipers, are also represented in Asia and are distinctive in having a heat-sensitive depression, the pit, on each side of the head between the eye and the nostril. This pit detects changes in temperature and creates an image of the heat environment in front of the head. The heat image is then interpreted in the brain by superimposing it on the image from the eyes. From this mechanism, pit vipers can detect warm objects moving against a cold background or cold objects moving against a warm background. Five of Alabama's six venomous species of snakes are pit vipers.

**1a** Tail with a rattle or button at the end. Go to **2**.

**1b** Tail without a rattle or button at the end.

> Genus *Agkistrodon*—American Moccasins . . . . page 313.

**2a** Crown of head with nine large plates; rattles very small; size of adults less than 640 mm.

> Genus *Sistrurus*—Pygmy Rattlesnake . . . . page 329.

**2b** Crown of head with many small scales, along with a few plates; rattle large; size of adults greater than 700 mm.

> Genus *Crotalus* . . . . page 338.

*From left to right:*

Dorsal view of head scales of a Pygmy Rattlesnake (*Sistrurus*); based on AUM 38771

Dorsal view of head scales of a Rattlesnake (*Crotalus*); based on AUM 39508

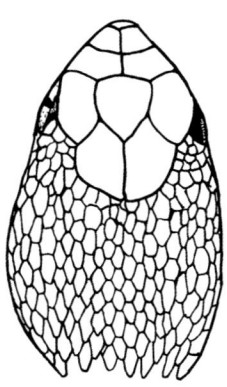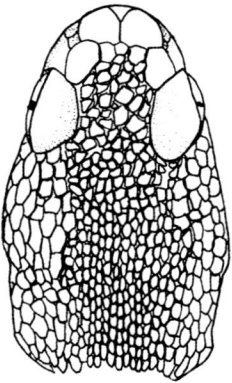

# American Moccasins
## Genus *Agkistrodon* (Palisot de Beauvois, 1799)

Six species are placed in this genus, and these snakes are found in eastern North America through Central America. *Agkistrodon* is the sister taxon to the rattlesnake genus *Crotalus* (Fyron et al. 2011), but all *Agkistrodon* lack rattles. Members of the genus are terrestrial or semi-aquatic and possess cryptic coloration that makes them difficult to detect when coiled on substrate. The genus is represented in Alabama by two species.

### Key to the Species of *Agkistrodon* of Alabama

**1a** No dark mask across side of head; dorsum typically with dumb-bell-shaped dark markings; upper labials usually separated from orbit by a row of small scales; twenty-three dorsal scale rows.

> *Agkistrodon contortrix*—**Copperhead** . . . page 314.

*From left to right:*

Color pattern of head of an Eastern Copperhead (*Agkistrodon contortrix*)

Dorsal color pattern of an Eastern Copperhead (*Agkistrodon contortrix*)

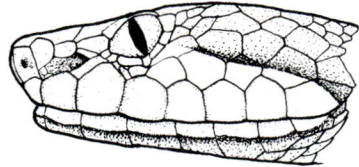 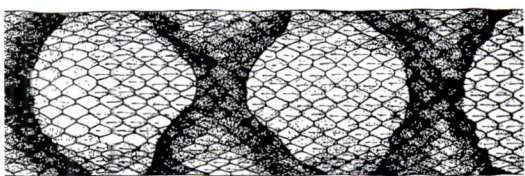

**1b** Dark mask, bordered above by light, across side of head from nostril through eye; dorsum typically with dark bands that are not dumbbell shaped; usually with at least one upper labial scale entering the orbit; twenty-five dorsal scale rows.

> *Agkistrodon piscivorus*—**Cottonmouth.** . . . . . page 318.

*From left to right:*

Color pattern of head of a Cottonmouth (*Agkistrodon piscivorus*)

Dorsal color pattern of a Cottonmouth (*Agkistrodon piscivorus*)

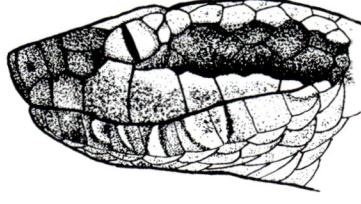 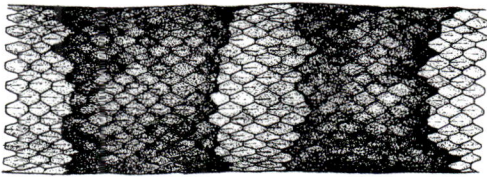

Copperhead, *Agkistrodon contortrix*, Bibb County, AL

## Copperhead
*Agkistrodon contortrix* (Linnaeus, 1766)

DESCRIPTION Copperheads are moderately large and fairly stout, attaining a maximum total length of about 1,320 mm. They have scales that are heavily keeled and occur in twenty-five rows at midbody. The anal scale is undivided, and posterior to this the subcaudals typically are undivided, except near the tip of the tail. The head is wide posteriorly and distinct from the neck, which is narrow. A facial pit or depression is present on each side of the snout between the eye and the external nares. The maxillary teeth are reduced to a pair of recurved, movable fangs, one on each side (with replacement fangs aligned behind). The pupils of the eyes are vertically elliptical, creating a cat-eyed appearance. Brown, tan, or pinkish are the basic ground colors of this species, with darker colors being associated with northern locations, becoming progressively lighter for specimens in the south. To this base color are added sixteen to twenty-one dumbbell-shaped transverse dark bands that are constricted in width at the mid-dorsum and

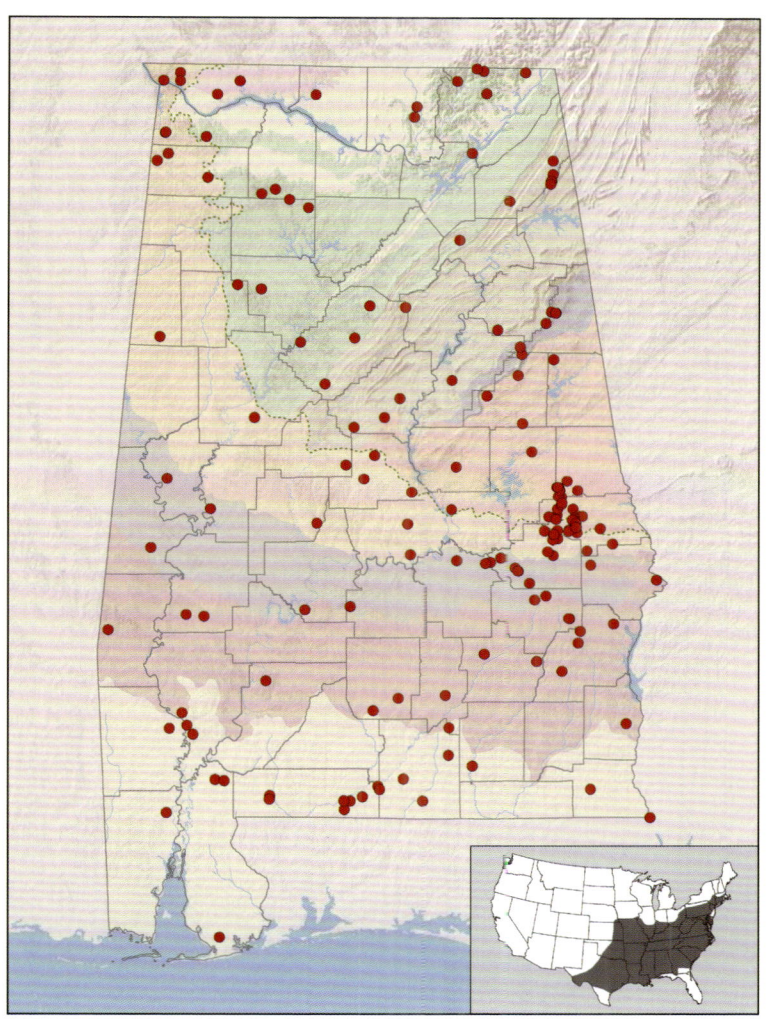

Distribution of Copperhead, *Agkistrodon contortrix*. This species is assumed to occur throughout Alabama. Inset map depicts approximate range in the United States.

separated by two to five scale rows. Generally, larger numbers of scale rows separate dark bands for snakes from the north and progressively fewer do so for snakes from the south. Occasionally, the two halves of the dark bands are offset, failing to meet dorsally. Ventrolateral dark spots are present on the sides of the body with those located opposite the body bands being pale and diffuse and those alternating with body bands being dark and well defined. The ventrolateral spots are darker and are as distinct as the dark dorsal bands for snakes in the north, becoming progressively less intense for snakes from the south.

Color pattern shows considerable individual variation in all populations, particularly in the nature of the dark cross bands. On occasional individuals, the dark pigment forms longitudinal dorsolateral stripes instead of transverse bands. Such aberrant color patterns highlight the danger of attempting to identify a snake by color pattern alone. The vertically elliptical pupil can be seen from a safe distance, and this is a reliable character for distinguishing venomous copperheads from non-venomous snakes.

**ALABAMA DISTRIBUTION** Copperheads are found throughout the state. The eastern clade (see Taxonomy section below) has the broadest distribution within the state, being found everywhere except lands west of the Tombigbee drainage, which is inhabited by members of the western clade.

**HABITS** Above the Fall Line, Copperheads occur in greatest abundance in forested areas with rocky bluffs and ravines. Preferred habitats in the Coastal Plain provinces are floodplains, edges of swamps, and hilly terrain dominated by hardwood trees. Copperheads are common in the Red Hills and Black Belt, where they are usually found along streams, hedge rows, and in places overgrown with kudzu (*Pueraria lobata*). In suitable habitat, this species is abundant but difficult to detect because it is so cryptic against a background of dead leaves. The species also is capable of relatively dramatic changes in abundance over time. In the late 1970s, Mount (1980) discovered few individuals of this species during a year-long survey of the herpetofauna of the Conecuh National Forest of south Alabama. Twenty-five years later, Guyer et al. (2007) found it to be the most abundant species in the same area. Because Mount (1980) described Eastern Kingsnakes to be abundant but declining, and Guyer et al. (2007) detected no Eastern Kingsnakes, a change in predator-prey relationships appears to have altered Copperhead abundances. This pattern is consistent with an overall negative relationship in detections of Eastern Kingsnakes and Copperheads (Steen et al. 2014).

In spring and fall, Copperheads are active during the day. During hot weather, they are mainly nocturnal, frequently absorbing heat by resting on paved roads during evening hours. The food habits of Copperheads are generalized, including small mammals, frogs, lizards, hatchling turtles, and insects, particularly emerging cicadas. The tails of newborn Copperheads are yellow and are used to lure prey by waving the tail tip in front of the juvenile's face in a way that attracts lizards and frogs. Copperheads use underground chambers to over-winter and mate shortly after emergence from brumation. The young are born during August and September and average five or six per litter. Copperheads rely on camouflage to avoid detection, and this is its first line of defense when approached by humans. However, if confrontation with a human continues, then these snakes will strike, accounting for many bites each year that require medical attention. Such bites are dangerous but infrequently cause death. However, serious problems can arise if immediate medical attention is not sought.

**CONSERVATION AND MANAGEMENT** Copperheads are abundant throughout the state and persist in high numbers in a variety of habitats. No special conservation legislation is required for them. Retention of cover objects (rock piles, boards, and roofing tin) is likely to enhance habitat for Copperheads by creating refugia for snakes as well as small rodents, a primary food item. However, in most cases, the management goal will be to rid an area of this species. We know of no product that reliably drives away these snakes. Sticky traps can be used to try to capture Copperheads in and around buildings, but this method will catch so many non-target animals that it is hard to see the value of this method in general circumstances. Maintaining excellent rodent control and a wide buffer that is cleared of debris is the best method for minimizing activity of Copperheads near human habitations.

**TAXONOMY** This is the basal species to all New World *Agkistrodon* (Pyron et al. 2011). Because the genus is thought to have invaded North America from Asia, the basal position of *A. contortrix* suggests a long association of this species with the eastern United States. Five subspecies of copperheads are traditionally recognized, based on morphology. Two of these, *A. c. mokasen* and *A. c. contortrix*, traditionally are thought to be present in Alabama (Mount 1975). Recently, Guiher and Burbrink (2008) documented three mitochondrial lineages that do not conform to the traditional subspecies designations. These authors refer to these lineages as the eastern, central, and western clades. Alabama has members of the eastern and western clades. Because most individuals from Alabama are thought to be intergradient between *A. c. mokasen* and *A. c. contortrix* (Mount 1975) and because the characteristics used to distinguish those two taxa likely represent clinal modes, we reject the use of traditional subspecies within Alabama. Instead, we use the mitochondrial clades of Guiher and Burbrink (2008) to separate the likely meaningful evolutionary units within the state. Previous authors have placed this species in the genera *Ancistrodon, Boa, Cenchris,* and *Trigonocephalus.*

## *Agkistrodon piscivorus* (Lacépède, 1789)

TAXONOMY  Three subspecies of Cottonmouths traditionally have been recognized, and all three were thought by Mount (1975) to influence specimens in Alabama. However, these subspecies are based on scale counts that are subject to such wide variation that they are unlikely to be diagnostic for the three proposed taxa. Guiher and Burbrink (2008) examined the mitochondrial genome for this species throughout its range and found evidence for only two subspecific taxa, which they referred to as a Florida clade and a continental clade. This analysis was followed with an examination of nuclear and mitochondrial genomic data analyzed in a coalescent framework that allowed gene flow between populations and that tested the hypothesis that the two mitochondrial clades represent two species (Burbrink and Guiher 2015). These authors concluded that the Florida and continental clades represent two distinct species because of extremely low levels of gene flow between the two clades. This conclusion was challenged by examination of amplified fragment length polymorphisms of the mitochondrial genome, a data set that demonstrates extensive gene flow between the two putative species (Strickland et al. 2014). Based on this evidence of recent gene flow, we retain a single species of Cottonmouths with two subspecies. Because the Florida clade conforms to *A. p. conanti,* we infer that these are synonymous and retain the common name Florida Cottonmouth for it. Although no specimen from Alabama has been demonstrated to possess the mitochondrial genome of this subspecies, we consider specimens with dark vertical rostral stripes to belong to it. We assign all other specimens from the state to the nominate subspecies, *A. p. piscivorus,* and assign the common name Northern Cottonmouth to it, as recommended by Crother et al. (2017). Previous authors have placed Cottonmouths in the genera *Acontias, Ancistrodon, Crotalus, Toxicophis,* and *Trigonocephalus.*

**1a** Rostral scute with a conspicuous dark vertical stripe on each side.

> *Agkistrodon piscivorus conanti*—Florida Cottonmouth . . . . **page 320.**

**1b** Rostral scute without a conspicuous dark vertical stripe on each side.

> *Agkistrodon piscivorus piscivorus*—Northern Cottonmouth . . . .
>
> page 323.

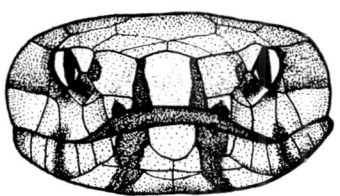

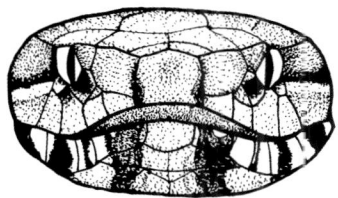

*From top to bottom:*

Anterior view of head of a Florida Cottonmouth (*Agkistrodon piscivorus conanti*)

Anterior view of head of a Northern Cottonmouth (*Agkistrodon piscivorus piscivorus*)

Florida Cotton-
mouth, *Agkistrodon
piscivorus conanti*,
Liberty County, FL

## Florida Cottonmouth
*Agkistrodon piscivorus conanti* (Gloyd, 1969)

**DESCRIPTION** This is a large, heavy-bodied aquatic snake, attaining a maximum total length of about 1,880 mm. At midbody, there are twenty-five rows of scales, and the anal scale is undivided. Posterior to the anal scale, the subcaudals are undivided except near the tip of the tail. Cottonmouths have a head that is noticeably widened posteriorly, followed by a noticeably thin neck. The head is brown, often with light patches, and has a wide black mask through the eye to the posterior edge of the jaw. This mask is bordered above by a thin light line and below by a wide white band across the upper labial scales. The tip of the snout has bold vertical dark-edged lines along the rostral scale leading to a broad dark stripe on the lower labials and chin (compared to the indistinct markings in *A. p. piscivorus*). The snout has a facial pit or depression on each side between the eye and external naris. The upper labials are white with one or two dark spots (*A. p. piscivorus* lacks spots). Teeth on the maxilla are reduced to a single, recurved, movable fang on each side (with replacement fangs aligned behind). The pupil of the eye is vertically elliptical, like a cat's. The ground color of Florida Cottonmouths is light brown to nearly black. A distinct pattern of broad

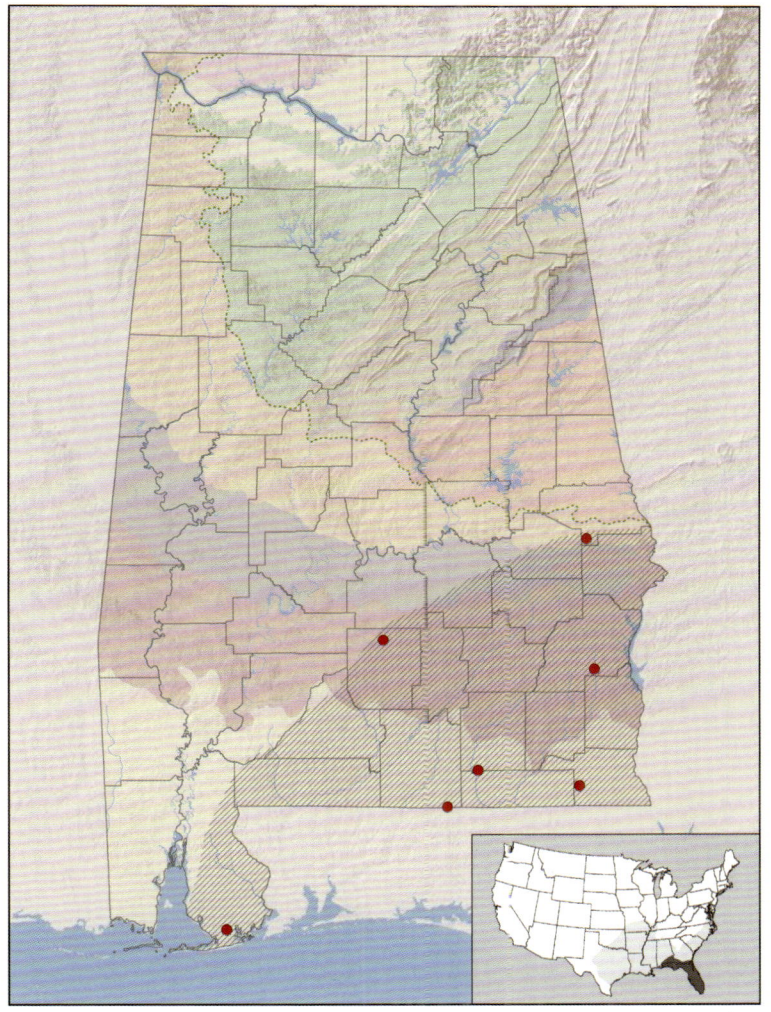

Distribution of Florida Cottonmouth, *Agkistrodon piscivorus conanti*. The presumed range of the subspecies in Alabama is indicated by hatching. Inset map depicts approximate range in the United States with dark shading indicating the range of *A. p. conanti* and light shading indicating the range of all other subspecies.

dark bands with serrate edges is present in juvenile and small adult individuals with bands being narrowest at mid-dorsum. These bands usually fade in adults, leading to a uniform dark brown or reddish color in older adults, but some large adults retain the banded pattern.

**ALABAMA DISTRIBUTION** Specimens conforming to this subspecies are found in the southeastern corner of the state, extending as far north as southern Lee County and as far west as Butler County.

**HABITS** Florida Cottonmouths occur in swamps, sloughs, bay heads, and small, slow-moving streams. They seek shelter in root mats of trees along aquatic margins during cold spells of winter but may be found on the surface during any month of the year. Food habits of the subspecies are broad, including insects, snails, fish, frogs, baby alligators, lizards, turtles, snakes, birds, bird eggs, small mammals, and carrion. Juvenile Florida Cottonmouths have a yellow tail tip that is used to lure prey to a coiled individual (Wharton 1960).

Males likely have two peaks of reproduction, one in spring and one in summer (Seigel et al. 2009). Females ovulate in spring, fertilizing eggs with sperm from a spring mating or from sperm stored from the previous summer. Females then give birth to six to seven young in August or September. Reproductive females spend the next year storing energy that allows them to begin to yolk follicles in the fall for reproduction the following year (Seigel et al. 2009). Florida Cottonmouths typically remain coiled when approached by humans and generally avoid detection when they do so. When harassed, most snakes will crawl away and hide. However, when prevented from escaping, these snakes exhibit a series of escalating aggressive behaviors. Most individuals will expel noxious-smelling chemicals from anal glands or gape to expose the light mouth linings. However, they generally will bite only when a hand approaches them (Gibbons and Dorcas 2002), contrary to the lore that these snakes will attack an unsuspecting human. Nevertheless, the bite of a Florida Cottonmouth is quite serious and requires immediate medical attention.

Several other water-inhabiting snakes in South Alabama are collectively termed water moccasins by novices and are mistaken for Florida Cottonmouths, but they seldom display the gaping habit. Another habit of cottonmouths that may aid in recognition is the tendency of these snakes to swim with the head well out of the water and the body visible above the water. This contrasts with watersnakes (*Nerodia*), which swim with the body submerged and the head barely elevated above the water.

**CONSERVATION AND MANAGEMENT** Florida Cottonmouths can be abundant and survive well in suburban as well as in rural areas. Therefore, this subspecies requires no special conservation regulation. Management activities that protect vegetation along the margins of wetlands and that limit removal of fallen logs used as basking sites are likely to be beneficial to the subspecies. We know of no product that reliably drives this subspecies from an area. Draining a wetland will rid it of these snakes, as will removal of trees and shrubs from the margins.

Adult Northern Cottonmouth, *Agkistrodon piscivorus piscivorus*, Limestone County, AL

# Northern Cottonmouth

*Agkistrodon piscivorus piscivorus* (Lacépède, 1789)

DESCRIPTION This large, heavy-bodied aquatic snake attains a maximum total length of about 1,520 mm. At midbody, there are twenty-five rows of scales, and the anal scale is undivided. Posterior to the anal scale, the subcaudals are undivided except near the tip of the tail. Northern Cottonmouths have a head that is noticeably widened posteriorly, followed by a noticeably thin neck. The head is brown, often with light patches, and has a wide black mask through the eye to the posterior edge of the jaw. This mask is bordered above by a thin, light line and below by a wide white band across the upper labial scales. The tip of the snout lacks the distinct vertical dark-edged lines along the rostral scale leading to a broad dark stripe on the lower labials and chin that are found in *A. p. conanti*. The snout has a facial pit or depression on each side between the eye and external naris. The upper labials are white and lack the dark spots of *A. p. conanti*. Teeth on the maxilla are reduced to a single, recurved, movable fang on each side (with replacement fangs aligned behind). The pupil of the eye is vertically elliptical, like a cat's. The ground color of Northern Cottonmouths is light brown to nearly black. A distinct pattern of broad dark bands with serrate edges is present in juvenile and small adult individuals with bands being narrowest at mid-dorsum. These bands usually fade in adults, leading to a uniform dark brown or reddish color in older adults, but some large adults retain the banded pattern.

Juvenile Continental Cottonmouth, *Agkistrodon piscivorus piscivorus*, unknown locality within AL

**ALABAMA DISTRIBUTION** Northern Cottonmouths are found throughout the state except for the southeastern corner.

**HABITS** Northern Cottonmouths have a wide ecological tolerance in Alabama, occurring in virtually all permanently aquatic habitats. The greatest population densities are attained in swamps, sloughs, and bay heads, but small, slow-moving streams are also preferred habitats. During winter, Northern Cottonmouths seek hibernacula that consist of holes in root mats of trees along aquatic margins. Here, they can achieve temperatures that are warmer than the temperature of water, air, and ground immediately outside a hibernaculum (Hein and Guyer 2009). Food habits of these snakes are remarkably generalized. Insects, snails, fish, frogs, baby alligators, lizards, turtles, snakes, birds, bird eggs, small mammals, and carrion are all known to be consumed. Unlike most pit vipers, Northern Cottonmouths may strike and hold on to the prey item while killing it. Juvenile Northern Cottonmouths start life with a tail tip that is sulfur yellow. When in motion, the tail resembles a worm, and this movement lures prey to a coiled individual (Wharton 1960).

Males either have one peak of courtship activity in August (Graham et al. 2008) or two peaks, one in spring and one in summer (Seigel et al. 2009). Females ovulate in spring, fertilizing eggs with sperm from a spring mating or from sperm stored from the previous summer. Females then give birth to six to seven young in August or September. Reproductive females spend the next year storing energy that allows

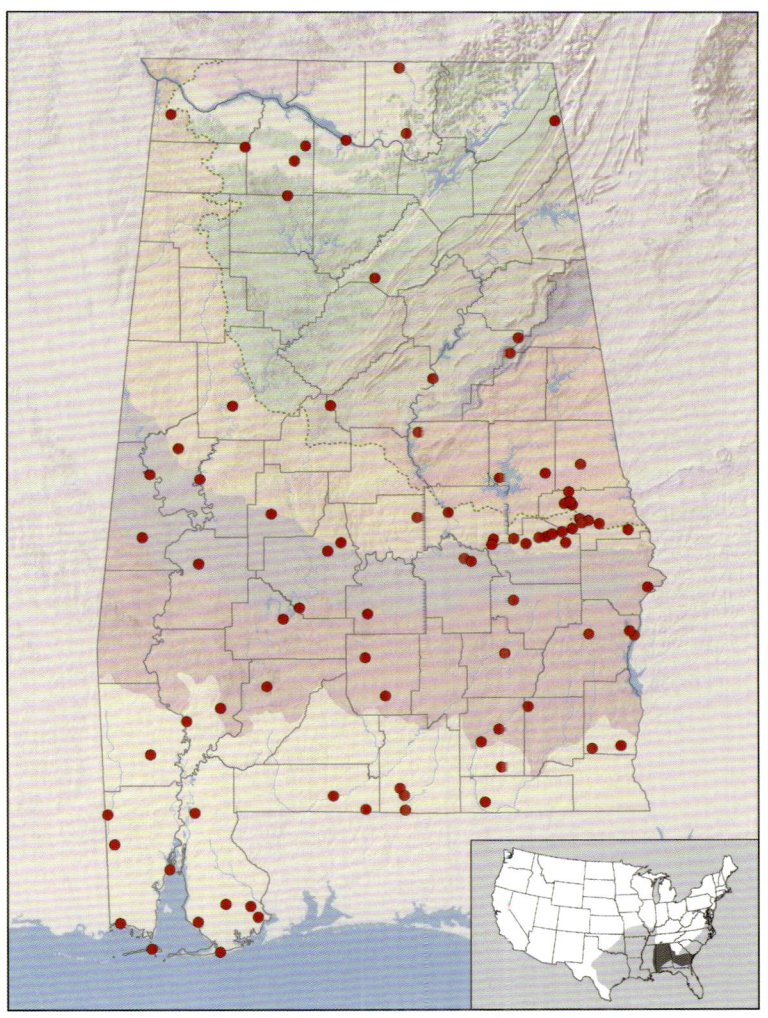

Distribution of Northern Cottomouth, *Agkistrodon piscivorus piscivorus*. This subspecies is assumed to occur throughout the state. Inset map depicts approximate range in the United States with dark shading indicating the range of *A. p. piscivorus* and light shading indicating the range of all other subspecies.

them to begin to yolk follicles in the fall for reproduction the following year (Seigel et al. 2009). Annual survival of the subspecies in Alabama is 79 percent, a remarkably high value, suggesting that this subspecies may achieve long life spans in nature (Koons et al. 2009). Northern Cottonmouths are fed upon by an assemblage of mosquitoes that specialize on amphibians and reptiles (Burkett-Cadena et al. 2008), and these snakes are capable of harboring the virus that causes Eastern Equine Encephalitis. Virus can develop titers in Northern Cottonmouths that are high enough to allow disease transmission to other organisms in wetlands occupied by these snakes. Thus, Northern

Cottonmouths may play a role in the transmission cycle of this disease by allowing the virus to survive overwinter in snakes and then infect their primary host, birds, during a subsequent season of vertebrate activity (White et al. 2011).

Northern Cottonmouths, when approached by humans, typically remain coiled and generally are undetected. If discovered and approached, most will crawl away and hide. However, when prevented from escaping, these snakes tend to be pugnacious in disposition. Often the first reaction, upon further arousal, is to expel from the anal glands a quantity of musk, which has an odor remarkably like that of a male goat. Northern Cottonmouths have the habit of gaping repeatedly at an intruder, jerking the head with each gape. This behavior reveals the light mouth linings and has given rise to the colloquial name cottonmouth. This reaction occurs frequently when humans restrain animals by lightly stepping on them. Remarkably, they generally will bite only when a hand approaches (Gibbons and Dorcas 2002). So, this subspecies does not live up to its lore of being so aggressive as to attack an unsuspecting human. However, the bite of a Northern Cottonmouth is quite serious and requires immediate medical attention. Several other water-inhabiting snakes in Alabama are collectively termed water moccasins by novices and are mistaken for Northern Cottonmouths, but they seldom display the gaping habit. Another habit of Northern Cottonmouths that may aid in recognition is the tendency of these snakes to swim with the head well out of the water and the body visible above the water line. This contrasts with watersnakes (*Nerodia*), which swim with the body submerged and the head barely elevated above the water.

**CONSERVATION AND MANAGEMENT** Northern Cottonmouths are abundant and their habitats are widespread. Additionally, they do well in parks and golf courses, where managed wetlands have adequate vegetation. Therefore, this subspecies requires no special conservation legislation. Protection of vegetation that borders wetlands and retention of fallen logs that serve as basking sites are likely to benefit the subspecies. We know of no product that reliably drives this subspecies from an area. Draining a wetland will rid it of these snakes, but obviously this will adversely affect many other species. Removing trees and shrubs from the margins of wetlands and streams generally will reduce Northern Cottonmouth populations.

# Massasauga and Pygmy Rattlesnakes
## Genus *Sistrurus* (Garman, 1883)

This genus is the second radiation of rattlesnakes, being character-ized by the presence of a tiny rattle at the tip of the tail. The rattle is so small that, from a safe distance, it is impossible to see on most specimens from Alabama. Even at close range, the rattle produces a sound that is difficult for the human ear to perceive, sounding simi-lar to an insect buzz. Nevertheless, the rattle is thought to serve as a warning signal to potential mammalian predators. Most Alabamians who know it call it a ground rattler. This leads to considerable confu-sion because many people apply that name loosely to any small snake they do not recognize. This tendency has doubtless contributed to the widespread belief that the pygmy rattler has no rattle. The bite of the pygmy rattlesnake is not likely to be serious in humans because of the small amount of venom injected, but it does require immediate medical attention and can produce several days of considerable dis-comfort. The genus is found in the United States and northern Mexico and is sister to the genus *Crotalus* (Castoe and Parkinson 2006). Three species are recognized, one of which occurs in Alabama.

## *Sistrurus miliarius* (Linnaeus, 1766)

TAXONOMY This species is sister to *S. catenatus* (Murphy et al. 2002), the Massasauga of prairie habitats from Michigan to Texas. Three sub-species of *S. miliarius* are recognized, all of which are represented in Alabama. The zones of intergradations are extensive, and their limits poorly defined. Previous authors have placed this species in the genera *Candisona*, *Crotalophorus*, and *Crotalus*.

### KEY TO THE SUBSPECIES OF *SISTRURUS MILIARIUS* IN ALABAMA

1a Dorsal ground color dark gray, heavily stippled with black; dark lateral blotches in three series; scale rows at midbody usually twenty-three; venter white with sharply contrasting dark blotches.
   *Sistrurus miliarius barbouri*—Dusky Pygmy Rattlesnake . . . . **page 329.**
1b Dorsal ground color light gray to brown, often with a pinkish cast; lateral spots in one or two series; dorsal scale rows usually twenty-one or twenty-two; venter light with dark markings but not contrasting sharply as in above. Go to **2.**

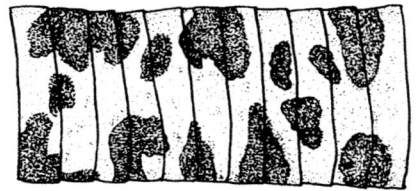

**2a** Mid-dorsal body blotches usually oval in shape, about equal longitudinally to light interspaces; mid-lateral blotches relatively round.

*Sistrurus miliarius miliarius*—Carolina Pygmy Rattlesnake . . . .
page 332.

**2b** Mid-dorsal body blotches wider than long, irregularly shaped and
narrower longitudinally than light interspaces; mid-lateral blotches
usually higher than wide.

*Sistrurus miliarius streckeri*—Western Pygmy Rattlesnake . . . .
page 335.

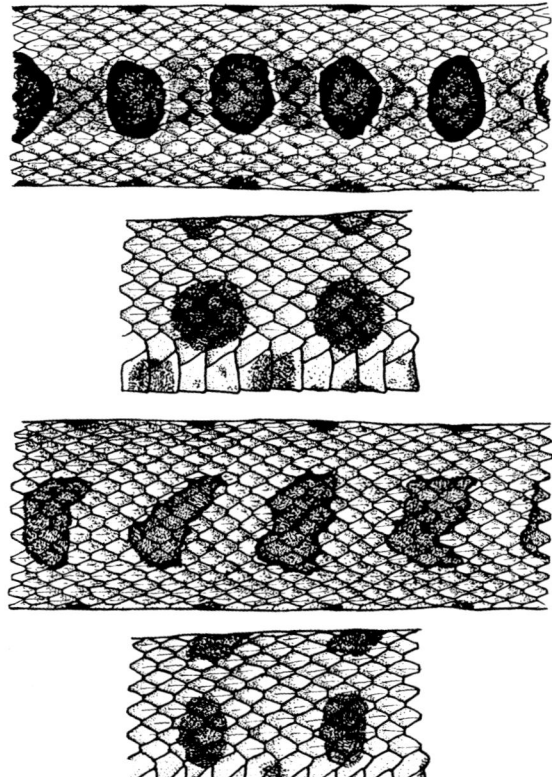

Dusky Pygmy Rattlesnake, *Sistrurus miliarius barbouri*, Levy County, FL

## Dusky Pygmy Rattlesnake
*Sistrurus miliarius barbouri* (Gloyd, 1937)

DESCRIPTION These snakes are short, stout-bodied, and attain a maximum total length of about 760 mm. The tail is attenuated at the tip and has a small rattle or button. The dorsal scales are strongly keeled and occur in twenty-three rows at midbody (twenty-one or twenty-two occur in *S. m. miliarius* and *S. m. streckeri*). The anal scute is undivided, and the head is wide posteriorly. followed by a noticeably thin neck. Like all pit vipers, there is a pit organ on each side of the head that is located between the eye and the external naris. Teeth on the maxilla are reduced to a single, movable, recurved fang on each side (followed by replacement fangs posteriorly). The top of the head is covered by nine large symmetrical plates. The dorsal ground color is dark gray that is heavily stippled with black (ground color lighter in *S. m. miliarius* and *S. m. streckeri*), frequently with a median reddish or reddish brown stripe (median stripe absent in *S. m. miliarius*). A series of twenty-five to thirty-six dark brown or black blotches are distributed along the back. These are oval in shape with regular edges and are about equal in longitudinal dimension to the light interspaces (dorsal blotches are irregular in shape and longer along mid-dorsal plane than length of light interspaces between blotches in *S. m. streckeri*). Lateral

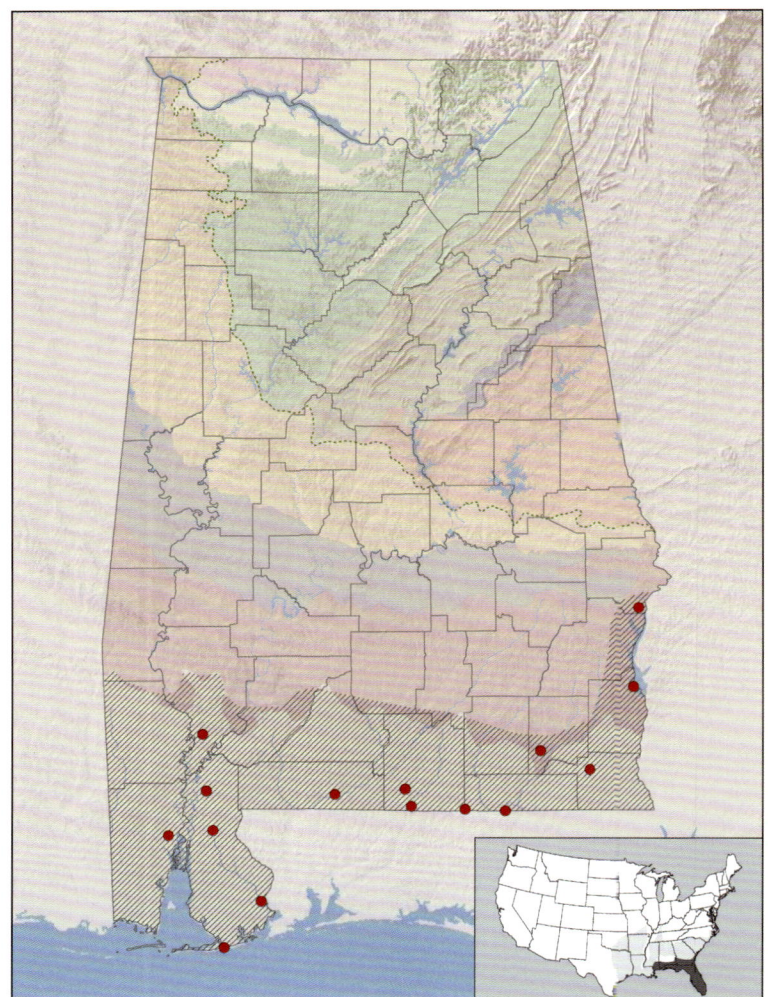

Distribution of Dusky Pygmy Rattlesnake, *Sistrurus miliarius barbouri*. The presumed range of the subspecies in Alabama is indicated by hatching. Inset map depicts approximate range in the United States with dark shading indicating the range of *S. s. barbouri* and light shading indicating the range of all other subspecies.

spots are present in three series (two are present in *S. m. miliarius* and *S. m. streckeri*). The venter is white with sharply contrasting dark brown to black blotches that tend to coalesce posteriorly (these markings are grayish brown in *S. m. miliarius*).

**ALABAMA DISTRIBUTION** This subspecies is restricted to the Lower Coastal Plain, where it is fairly common in pine flatwoods and scrubby areas. The southern tier of counties is known to be inhabited by this subspecies, and specimens that strongly conform to it are found along the eastern border of the state north to Russell County. Specimens from south central Alabama are lacking. But, based on distribution patterns of other reptiles and the Russell County specimens, we infer that this subspecies is found throughout the region east of the Alabama River.

HABITS This subspecies occurs in most terrestrial habitat types represented in the Coastal Plain, except for extensive, hardwood-dominated floodplains. These snakes are active from April through October but are seldom encountered except during late summer. Most specimens are found as they cross roads in late afternoon or at night. This small snake is pugnacious in its attitude toward a molester. Dusky Pygmy Rattlesnakes have been reported to feed on mice, lizards, frogs, insects, and spiders. The young are born between July and September and usually number four to ten individuals. Data on reproduction in Alabama are scarce but include the birth of six young on August 15 recorded by Mount (1975). This snake is reluctant to bite, with only 8 percent of provoked snakes reacting by trying to bite (Glaudas et al. 2005).

CONSERVATION AND MANAGEMENT During most times of year, this is a difficult subspecies to find. But, during late summer, it is frequently observed and seems to be relatively abundant. No conservation laws protect this subspecies and no known population declines would appear to warrant enactment of such laws. However, this is a subspecies for which population declines would be difficult to detect. Because Alabama is situated at the intersection of all three subspecies, it is important to conduct studies of gene flow among taxa within the state. Management activities that enhance pine forest habitat and wetlands associated with such habitats are likely to benefit this subspecies.

# Carolina Pygmy Rattlesnake

*Sistrurus miliarius miliarius* (Linnaeus, 1766)

DESCRIPTION These short, stout-bodied snakes attain a maximum total length of about 535 mm. The tail is attenuated at the tip and has a small rattle or button. The dorsal scales are strongly keeled and occur in twenty-one or twenty-two rows at midbody (compared to twenty-three in *S. m. barbouri*). The anal scute is undivided, and the head is wide posteriorly followed by a noticeably thin neck. Like all pit vipers, there is a pit organ on each side of the head that is located between the eye and the external naris. Teeth on the maxilla are reduced to a single, movable, recurved fang on each side (followed by replacement fangs posteriorly). The top of the head is covered by nine large symmetrical plates. The dorsum is light brown or gray, often with a pinkish cast, with a median series of twenty-five to thirty-six dark brown or black blotches (*S. m. barbouri* is dark with a reddish mid-dorsal stripe). The median blotches usually are oval in shape and have regular edges about equal in longitudinal dimension to the light interspaces (*S. m. streckeri* has irregular edges and narrower longitudinal dimension than light interspaces). Lateral spots are present in two series (three series are present in *S. m. barbouri*), the mid-lateral ones being opposite of the mid-dorsal blotches, relatively round, and with indistinct edges (these are taller than wide with distinct edges in *S. m.*

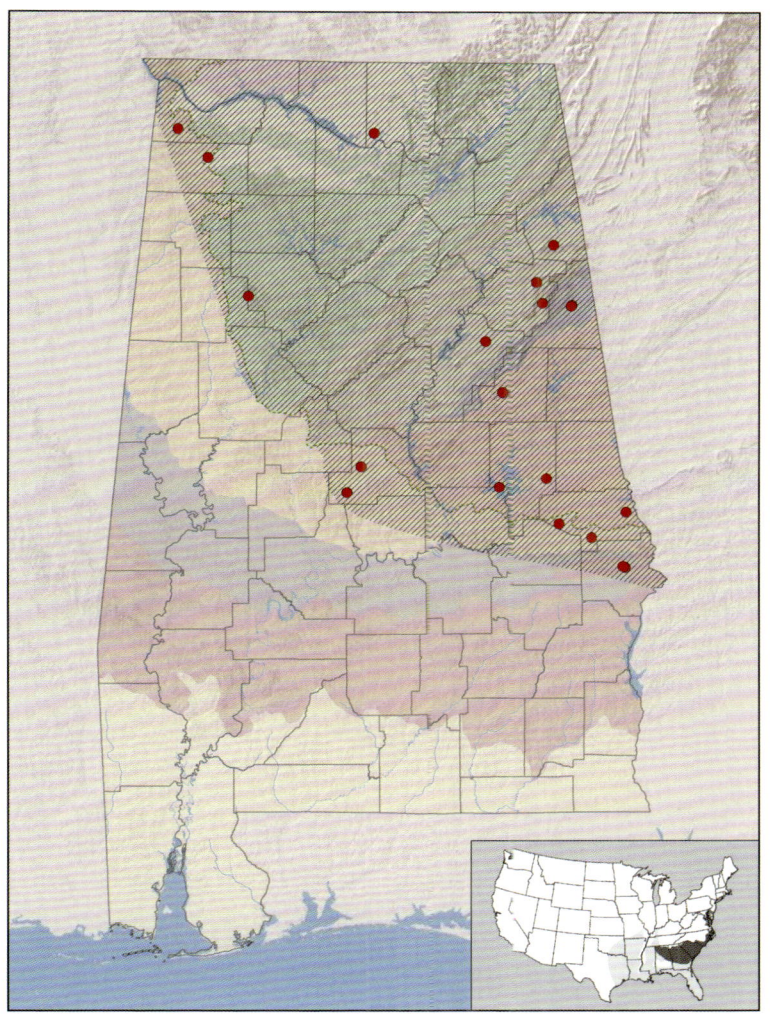

Distribution of Carolina Pygmy Rattlesnake, *Sistrurus miliarius miliarius*. The presumed range of the subspecies in Alabama is indicated by hatching. Inset map depicts approximate range in the United States with dark shading indicating the range of *S. s. miliarius* and light shading indicating the range of all other subspecies.

*streckeri*). The venter is white to cream, blotched with markings that are elliptical in shape, grayish brown in color, and equal to the width of two ventral scutes.

**ALABAMA DISTRIBUTION** This subspecies is found in the northern portion of the state above the Fall Line. Intergradient populations of *S. m. miliarius* and *S. m. streckeri* are found toward western central Alabama, and intergradient populations of *S. m. miliarius* and *S. m. barbouri* are found in the Red Hills, Black Belt, and Fall Line Hills.

**Habits** Carolina Pygmy Rattlesnakes occupy mixed pine-hardwood forests. These snakes are active from April through October but are seldom encountered except during late summer. Most specimens are found as they cross roads in late afternoon or at night. They are pugnacious in attitude toward a molester. This subspecies has been reported to feed on mice, lizards, frogs, insects, and spiders. The young are born between July and September and usually number four to ten individuals.

**Conservation and Management** No conservation regulations protect these snakes, and no known population declines would appear to demand such laws. However, this is a subspecies for which population declines would be difficult to detect. Because Alabama is situated at the intersection of all three subspecies, a study of gene flow among taxa is needed within the state. Management activities that enhance pine forest habitat and wetlands associated with such habitats are likely to benefit this subspecies.

Western Pygmy Rattlesnake, *Sistrurus miliarius streckeri*, Jasper County, MS

## Western Pygmy Rattlesnake
*Sistrurus miliarius streckeri* (Gloyd, 1935)

DESCRIPTION  These short, stout-bodied snakes attain a maximum total length of about 635 mm. The tail is attenuated at the tip and has a small rattle or button. The dorsal scales are strongly keeled and occur in twenty-one or twenty-two rows at midbody (*S. m. barbouri* has 23 rows). The anal scute is undivided, and the head is wide posteriorly followed by a noticeably thin neck. Like all pit vipers, there is a pit organ on each side of the head that is located between the eye and the external naris. Teeth on the maxilla are reduced to a single, movable, recurved fang on each side (followed by replacement fangs posteriorly). The top of the head is covered by nine large symmetrical plates. The dorsum is light brown or gray, often with a pinkish cast, with a median series of twenty-three to forty-two dark brown or black blotches. The median blotches usually are wider than long, often are irregular in shape, and typically are narrower longitudinally than the light interspaces (these are regular in shape and as wide longitudinally as the light interspaces in *S. m. miliarius*). Lateral spots are present in one or two series (three occur in *S. m. barbouri*), the mid-lateral ones being opposite of the mid-dorsal blotches and taller than wide with distinct edges (*S. m. miliarius* has round blotches with indistinct edges). The venter is light with indistinct dark blotches that are equal to the width of one ventral scute.

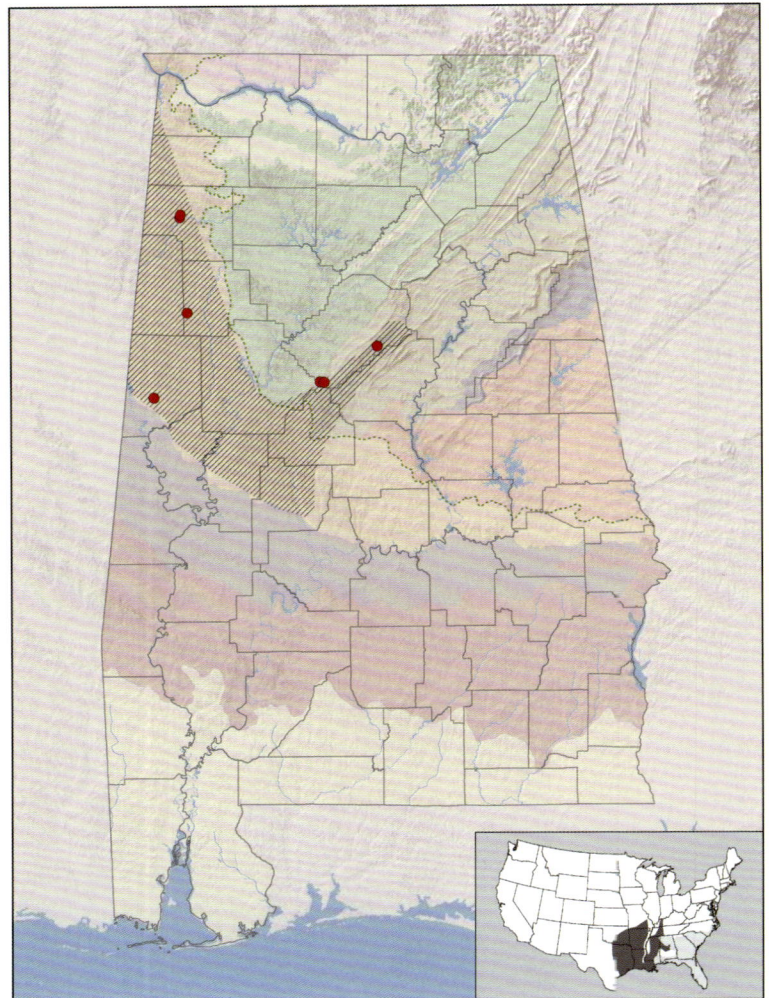

Distribution of Western Pygmy Rattlesnake, *Sistrurus miliarius streckeri*. The presumed range of the subspecies in Alabama is indicated by hatching. Inset map depicts approximate range in the United States with dark shading indicating the range of *S. s. streckeri* and light shading indicating the range of all other subspecies.

**ALABAMA DISTRIBUTION** Western Pygmy Rattlesnakes are found in the extreme western central portion of Alabama. The limits of this distribution are not precisely known, but specimens referable to this taxon are known from Pickens County and strong influence of this subspecies is seen in specimens as far east as Jefferson County.

**HABITS** This subspecies is found in forested floodplains and open, wet prairies that frequently have rocks and are near streams. These snakes are active from April through October but seldom are encountered except during late summer. They are pugnacious in their attitude toward a molester. Reportedly, they consume mice, lizards, frogs, and invertebrates. The young are born between July and September and usually number four to ten individuals.

**CONSERVATION AND MANAGEMENT** No conservation laws protect this subspecies, and no known population declines would appear to warrant enactment of such laws. However, this is a subspecies for which population declines would be difficult to detect. Because Alabama is situated at the intersection of all three subspecies, it is important for studies of gene flow among taxa to be conducted within the state. Management activities that enhance retention of forests along rivers and prairie wetlands are likely to benefit this subspecies.

# Rattlesnakes
## Genus *Crotalus* (Linnaeus, 1758)

This genus represents one of two radiations of North American snakes that have modified scales at the tip of the tail that create a rattle. In newborn snakes, the rattle consists of a single enlarged scale, the button, which is incapable of producing sound when the tail is vibrated. However, each time a rattlesnake sheds its skin, a new segment is added to the rattle; adjacent segments are joined loosely to each other creating a structure that produces a dry buzzing sound. Because rattlesnakes have extremely limited abilities to hear airborne vibrations, such as those produced by the rattle (Young and Aguiar 2002), the only known function of this structure is to warn approaching enemies. Even among approaching predators, the rattle likely only works for social mammals and birds because the bite of a rattlesnake typically is fatal, preventing that individual predator from learning to avoid the sound. However, members of a social group can learn such avoidance if one group member is observed to die from an encounter and the other observers learn to associate that death with the sound of the rattle (Mineka and Cook 1988).

Human cultures share a special fascination with rattlesnakes, creating folklore that has shaped human attitudes toward these creatures throughout the southeastern United States. In some aspects of Cherokee lore, the rattlesnake is a spirit that is prayed to because the spirit wishes people well, and in other aspects of this lore, the rattlesnake is a deceiver who tricks young boys into handling it (Langford 1987). For European settlers of Alabama and surrounding areas, the rattlesnake has been a rich source of imagery (e.g., "Rattlesnake Bit the Baby" is the name for a fiddle tune; Kuntz 2000) and of folk products (e.g., many Appalachian fiddlers place a rattlesnake rattle inside their instrument to improve its tone; Erbsen 1995). Members of the genus *Crotalus* are sister to Pygmy Rattlesnakes of the genus *Sistrurus* (Castoe and Parkinson 2006), suggesting a single origin to the rattle. The thirty-six species within this genus range from Canada to Argentina. The genus is represented in Alabama by two species.

## Key to the Species of *Crotalus* cf Alabama

**1a** Dorsum with yellow-bordered, diamond-shaped, dark markings on an olive to brown background.

> *Crotalus adamanteus*—Eastern Diamond-backed Rattlesnake . . . . page 340.

**1b** Dorsum with dark chevron-like markings on a pinkish to brown background.

> *Crotalus horridus*—Timber Rattlesnake . . . . page 345.

*From top to bottom:*

Dorsal color pattern of an Eastern Diamond-backed Rattlesnake (*Crotalus adamanteus*)

Dorsal color pattern of a Timber Rattlesnake (*Crotalus horridus*)

## Eastern Diamond-backed Rattlesnake

*Crotalus adamanteus* (Palisot de Beauvois, 1799)

DESCRIPTION This species is an extremely large, heavy-bodied snake, attaining a maximum total length of about 2,440 mm. The tail is short, stout, and contains a rattle or button at the end. It has heavily keeled scales that occur in twenty-seven to twenty-nine rows at mid-body. The anal scale is undivided. These snakes have a large head that noticeably expands posteriorly, followed by a noticeably thin neck. Like all pit vipers, there is a heat-sensitive pit on each side of the snout between the eye and the external naris. Maxillary teeth are reduced to a single large, movable, recurved fang (with replacement fangs aligned behind). The top of the head between the eyes is covered with large platelike scales, but behind the eyes are scales that are similar in size to those of the rest of the body. In ground color, the dorsum of Eastern Diamond-backed Rattlesnakes is brownish with a series of large, dark diamonds, each of which has a light center and yellow border. The head is light brown with a dark band extending obliquely through each eye to the posterior upper labial scales. These bands (one on each side of the face) are bordered by a light streak giving the snake a masked appearance. The rostral scale is outlined by yellow and vertical yellow

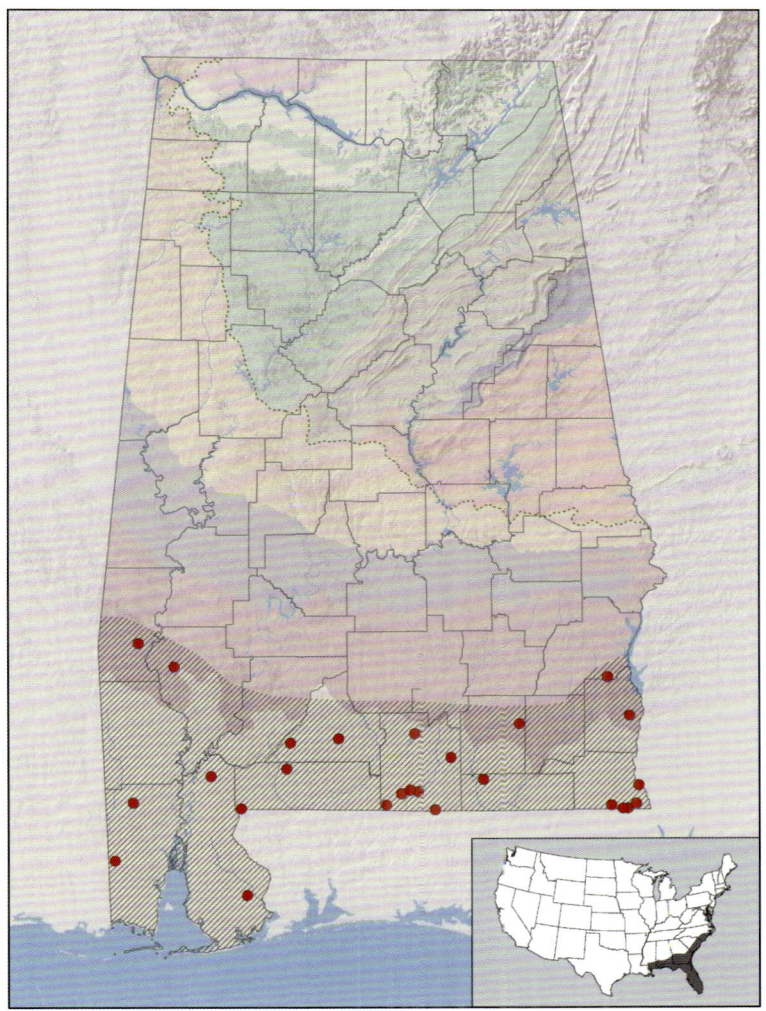

Distribution of Eastern Diamond-backed Rattlesnake, *Crotalus adamanteus*. The presumed range of the species in Alabama is indicated by hatching. Inset map depicts approximate range in the United States.

stripes emanate on each side of it in front of each external naris. The venter is yellowish to white, lightly suffused with brown laterally.

**ALABAMA DISTRIBUTION** This species occurs in the Lower Coastal Plain, where it can be relatively common. It also occurs in the Red Hills region of Alabama on ridge tops characterized by deep sandy soils and the presence of Gopher Tortoises and longleaf pines.

**HABITS** Relatively dry pine flatwoods and longleaf pine–turkey oak hills are the favored habitats during the active season of this extremely dangerous snake (Hoss et al. 2010; Waldron et al. 2006). Abandoned farmland and other such places where rabbits and small rodents abound are most likely to support sizable rattlesnake populations, but within such areas, snakes tend to place home ranges in areas lacking open fields (Hoss et al. 2011). This species has home ranges as large as 80 ha in males and 30 ha in females, but home range size is negatively associated with habitat heterogeneity (Hoss et al. 2010). Thus, movements are reduced in areas where several types of habitat interdigitate. A Gopher Tortoise burrow often serves as a refuge, particularly during cold weather. Where gopher burrows are scarce or absent, stump holes may be used. During the winter, the Diamond-backed Rattlesnake emerges on warm, sunny days but remains close to its refuge.

Mice and rats, especially Cotton Rats (*Sigmodon hispidus*), are the chief foods of small Diamond-backed Rattlesnakes, while Cotton Rats and rabbits seem to be the principal diet of large ones. Occasional food items include squirrels and ground-dwelling birds. These snakes generally move to an area with high population densities of prey, coil next to an area of high use by the prey, and then strike a prey item as it moves by. Death occurs rapidly, and the snake follows a chemical trail to find the dead food item. Besides killing the prey, the venoms start the digestive process.

Mating occurs in late summer and fall in this species (Hoss et al. 2011), when males seek receptive females with yolking follicles. Once mated, females store sperm, overwinter and then ovulate the following spring, when the eggs are fertilized with stored sperm. Pregnant females give birth to eight to fifteen young in late summer or fall. Neonates associate with the female parent until the first skin is shed; at this point, the female leaves the young (Butler et al. 1995). Each female parent then feeds through the following spring and summer in preparation for the next mating. Because of this lengthy investment in offspring, females produce litters every other year or every three years.

During encounters with humans, the Eastern Diamond-backed Rattlesnake conducts itself with poise and dignity. When not in the open, it is loath to move or rattle and will often remain perfectly still until actually touched. When touched, or when a direct confrontation is imminent, the snake assumes a defensive posture in which the body is coiled, the rattle is held erect and is buzzed, and the head is placed near the center of the coil with the neck flexed laterally. From this position, a strike of up to two-thirds of the snake's length is possible.

Once in this stance, the rattler, particularly if it is large and away from shelter, tends to hold its ground until the threat of danger has passed. Bites from this species are extremely dangerous and require immediate medical attention. Because of its large size, this species can inject a large volume of venom. Additionally, the venom components are particularly potent and include a protein that causes paralysis, generally starting with the legs. In fact, the toxicity of this venom probably is the reason that snake-handling religious groups generally have not survived in areas dominated by Eastern Diamond-backed Rattlesnakes.

**CONSERVATION AND MANAGEMENT** Although listed as Priority 2 (High Conservation Concern) by ADCNR (Mirarchi, Bailey, Haggerty, and Best 2004), Eastern Diamond-backed Rattlesnakes are not afforded the state protection enjoyed by other similarly ranked reptiles. This exemption is politically motivated, not science driven. This species has been harvested in Alabama for more than three decades in association with an annual festival operated by the Opp Jaycees. Snakes are brought in and displayed for the public. The snake hunters traditionally found rattlesnakes in burrows of Gopher Tortoises during winter months and induced them to come out of the burrow by inserting a garden hose to the bottom of the burrow, introducing a tablespoon of gasoline to the end of the hose at the mouth of the burrow, and blowing the fumes to the bottom of the burrow. These fumes irritate the snake and often cause it to crawl to the surface. Unfortunately, other inhabitants of the burrow, such as Gopher Tortoises and Florida Pinesnakes, remain in the burrow and likely are killed by the fumes (Speake and Mount 1973). The practice of gassing burrows was outlawed by the state of Alabama in 2009, making it possible for snake hunters possessing snakes that smell of gasoline to be prosecuted. Snake hunters today claim to use hooks on hoses to snag snakes and drag them from the burrow. The decades of harvest of rattlesnakes in south Alabama has been detrimental to rattlesnake populations, as evidenced by a consistent reduction in the number of snakes and the size of the largest snake processed at the Opp rattlesnake roundup (Means 2009).

There is strong evidence, especially in Alabama, that additional conservation efforts are needed to maintain Eastern Diamond-backed Rattlesnakes. This conservation effort has the obvious constraint that the species can deliver a lethal bite to humans and, therefore, is marked for eradication in many communities within its geographic range. In large rural areas, stable populations of this species can be achieved by proper maintenance of the longleaf pine community. This maintenance includes thinning of timber stands to low basal area, an action that encourages warm-season grasses. Prescribed fire at a rapid return interval (one to three years) reduces encroachment of woody shrubs and hardwoods and enhances grass coverage, which increases populations of prey consumed by rattlesnakes (Means 2004d). To maintain viable populations of Eastern Diamond-backed Rattlesnakes, tracts of 1,000 acres (400 ha) or more will be necessary.

**TAXONOMY** Eastern Diamond-backed Rattlesnakes are members of the *atrox* group of rattlesnakes, a clade that is of western desert origin (Murphy et al. 2002). Therefore, the species is likely to have invaded the southeastern Coastal Plain in relatively recent geological times (Guyer and Bailey 1993). No subspecies have been described, and no genomic lineages have been demonstrated.

Timber Rattlesnake,
*Crotalus horridus*,
Jackson County, AL

## Timber Rattlesnake (Canebrake Rattlesnake)
*Crotalus horridus* (Linnaeus, 1758)

**DESCRIPTION** Timber Rattlesnakes are large, heavy-bodied serpents, attaining a maximum total length of about 1,880 mm. They have a short, stout tail, with a rattle or button at the end. Their scales are heavily keeled and occur in twenty-three or twenty-five rows at mid-body. The anal scale is undivided in this species. Like all pit vipers, the head is noticeably wide posteriorly and is followed by a decidedly thin neck. There is a pair of pit organs, one on each side, between the eye and the external naris. Teeth on the maxilla are reduced to a single, movable, recurved fang (with replacement fangs growing behind the functional fang). The top of the head between the eyes is covered by large platelike scales, but behind the eyes, these scales are approximately the same size as the rest of the body. The ground color is yellowish, brownish, pink, gray, or occasionally nearly black. To this ground color is added a series of dark, chevron-shaped, transverse bands, some of which may be broken mid-dorsally. Along the midline of the dorsum is a tan or reddish-brown stripe, but this becomes less pronounced in specimens from the northeast corner of the state. The tail usually is black, and the belly is white to yellowish with patches of

Timber Rattlesnake,
*Crotalus horridus*,
Barbour County, AL

dark stippling. There is only a hint of a dark mask from the eye to the posterior upper labials.

**ALABAMA DISTRIBUTION** This species is found extensively throughout the state but becomes unusual along the Florida border and is absent from southern Baldwin and Mobile Counties.

**HABITS** Timber Rattlesnakes are most abundant in sparsely settled, forested areas. In the Lower Coastal Plain, where the species is sympatric with the Eastern Diamond-backed Rattlesnake, the Timber Rattlesnake is seldom found in the same habitats with its larger relative. In that region, the Eastern Diamond-backed Rattlesnake is usually found in relatively dry, sandy situations, whereas the Timber Rattlesnake is restricted mostly to swampy areas, floodplains, and upland areas where fire has been suppressed for long periods of time and hardwood encroachment is extensive (Steen, Smith, et al. 2005). Northward, the Timber Rattlesnake is much less specialized in its habitat selection, occurring in both uplands and lowlands.

Timber Rattlesnakes feed on a variety of small mammals and, infrequently, ground-dwelling birds. In its foraging mode, these snakes are classic sit-and-wait predators, coiling next to fallen logs that are

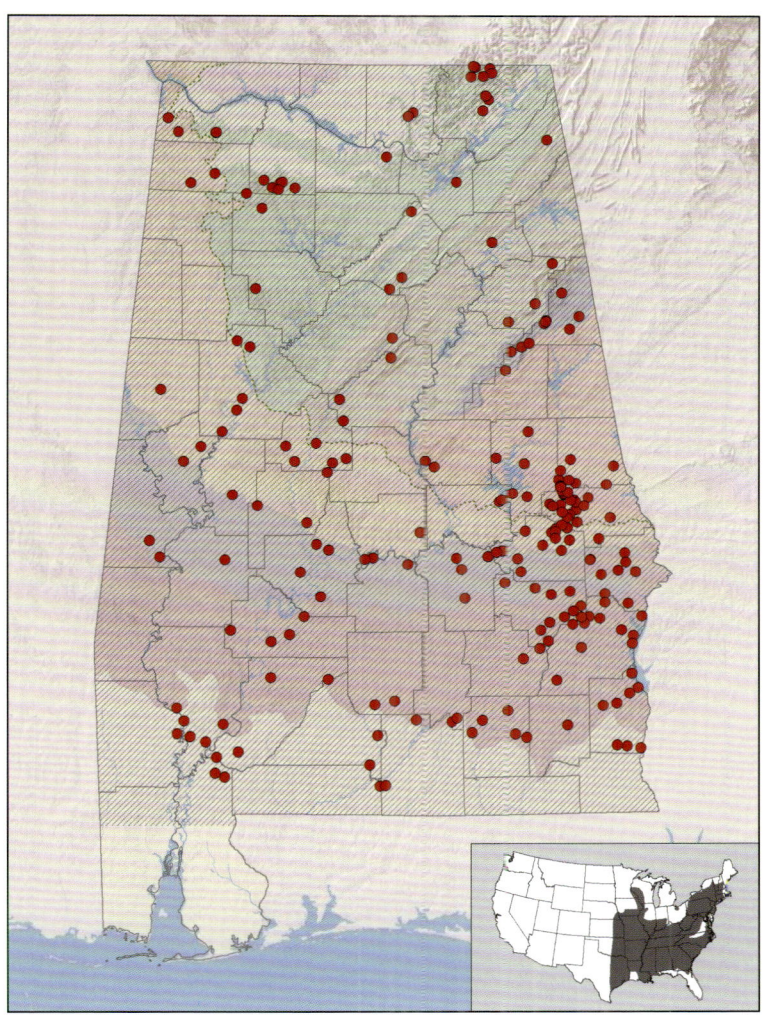

Distribution of Timber Rattlesnake, *Crotalus horridus*. The presumed range of the species in Alabama is indicated by hatching. Inset map depicts approximate range in the United States.

used as runways by rodents. A snake finds these foraging sites by following scent trails left by the rodents, with the snakes traveling to areas of highest chemical concentration along these runways. Then, the eyes and heat-sensitive pits are used to create an image of prey moving against the background. Predatory strikes are rapid, followed by an equally rapid withdrawal of the snake to its coil. An envenomated prey dies in a matter of minutes and is located by the snake from a chemical trail left by the prey.

Mating occurs in late summer and fall in Timber Rattlesnakes, when males seek receptive females with yolking follicles. Once mated, females store sperm, overwinter, and then ovulate the following spring, when the eggs are fertilized with stored sperm. Pregnant females give birth to an average of twelve young in August or September. Postpartum females feed the following year, overwinter, and begin development of eggs during the subsequent season of activity. Thus, females produce litters every three years and may have only four or five opportunities to reproduce in their lifetime (Gibbons 1972).

Timber Rattlesnakes usually remain perfectly still when approached by a human but will strike when provoked and are capable of injecting a large volume of venom. Any such bite is a serious medical emergency. Specimens from Alabama lack the neurotoxic canebrake toxin of populations from east and west of the state (Glenn et al. 1994). Therefore, the components of the venom of this species are relatively mild, and deaths from it are rare, typically being restricted to those who fail to seek medical attention until it is too late. Some Pentecostal sects display their faith in God by handling Timber Rattlesnakes during religious ceremonies, taking advantage of the general reluctance of venomous snakes to bite humans.

CONSERVATION AND MANAGEMENT Timber Rattlesnakes are declining rapidly in the northern part of the geographic range of the species (Clark et al. 2002). Habitat fragmentation and disruption of denning sites are the primary causes of these population declines. In Alabama, the species is still frequently encountered by humans. North Alabama populations use communal dens in limestone or sandstone areas, and disruption of these sites likely will cause rapid decline of the species. In the southern part of its range in Alabama, snakes overwinter individually in stump holes or under fallen debris. These populations may be more resistant to urbanization because they are less tied to specific denning sites. Because it is still fairly common, no special legislation protects this species in the state of Alabama. Management efforts that enhance hardwood forests and protect rock outcrops are likely to enhance the habitat for Timber Rattlesnakes. Because males move widely in late summer and fall in search of mates, they are frequently killed as they cross roads (Aldridge and Brown 1995). Thus, populations of this species are likely to remain viable only in areas with reduced traffic.

**Taxonomy** Traditionally, two subspecies have been recognized, a northern form, *C. h. horridus* for the Timber Rattlesnake, and a southern form, *C. h. atricaudatus*, for the Canebrake Rattlesnake. However, none of the characteristics used to separate these are diagnostic. Recent examination of the mitochondrial genome failed to document a north-south separation of intraspecific clades. Instead, an east-west split was shown, with the Appalachian and Alleghany Mountains separating the two (Clark et al. 2002). Based on this evidence, we consider this to be a single species with no subspecific variation.

# Family Elapidae

This large family of 55 genera and around 360 species contains the cobras, kraits, coralsnakes, and their allies. Elapids are the sister group to the family Lamprophiidae (Pyron et al. 2011), a group of African, Middle Eastern, European, and Southeast Asian snakes, some of which are venomous. The diagnostic feature of elapids is a pair of short, immobile fangs, one on each maxilla. These fangs are attached by a duct to a modified salivary gland that produces venoms. These venoms consist of a wide variety of low molecular weight proteins that largely affect the nervous system. Their action immobilizes prey, typically other snakes, which are then swallowed. Elapids frequently grab a prey item and chew up and down its length to distribute the venom widely. Most members of this family are tropical or subtropical in distribution, with a particularly species-rich radiation in Australia. In the New World, all species are referred to as coralsnakes because of their bright, aposematic coloration. Various combinations of red, yellow, and black rings encircle the body of most species creating a pattern that contrasts strikingly with most background colors and patterns. Two genera of elapids are found in North America, one of which is found in Alabama.

# American Coralsnakes
## Genus *Micrurus* (Wagler, 1824)

This genus contains about eighty species of brightly colored snakes distributed from the southern United States to Peru and Argentina. Referred to as coralsnakes, these serpents tend to be secretive, spending much of their time under leaf litter or other surface debris, while also searching actively for other snakes, their primary dietary items. This genus (including *Leptomicrurus*; Slowinski 1995) and its sister genus, *Micruroides*, are the only genera of the family that are found in the New World. The ancestor of these genera likely originated in the Old World and migrated from Asia to North America over the Bering land bridge. Only one species of *Micrurus* is found in the United States, and it occurs in Alabama.

Eastern Coralsnake,
*Micrurus fulvius*,
Coosa County, AL

## Eastern Coralsnake
*Micrurus fulvius* (Linnaeus, 1766)

**DESCRIPTION**  This is a medium-sized, fairly slender snake, attaining a maximum total length of about 1,090 mm. The head is short, blunt, and only slightly distinct from the neck. Each maxilla has a single,

immovable, erect fang near the front of the mouth. The scales are smooth and organized in fifteen rows at midbody, and the anal scute is divided into right and left halves. Bright yellow, jet black, and faded red are the primary colors of the body. The fading of the red rings is accomplished by black pigment found at the center of each red scale with the black spots tending to coalesce dorsally into a pair of black spots and ventrally into a single black spot. Beginning at the anterior end, the top of the head and nose are black followed by a broad yellow ring across the back of the head. Then, there is a repeated pattern of alternating red and black rings that are separated by narrow yellow rings, except for the tail, which has black and yellow rings only. The relative width of the red rings varies among individuals. Aberrant color patterns occur regularly.

**ALABAMA DISTRIBUTION** Eastern Coralsnakes generally are distributed widely across the Lower Coastal Plain, where the species is most common; the species is found infrequently in the Red Hills (in deep sands of ridge tops) and rarely elsewhere within the state. The northern limit of the range in Alabama is undetermined. Specimens have been collected above the Fall Line as far north as the vicinity of Blockton in Bibb County, in central Alabama, and Shocco Springs, Talladega County, in eastern Alabama.

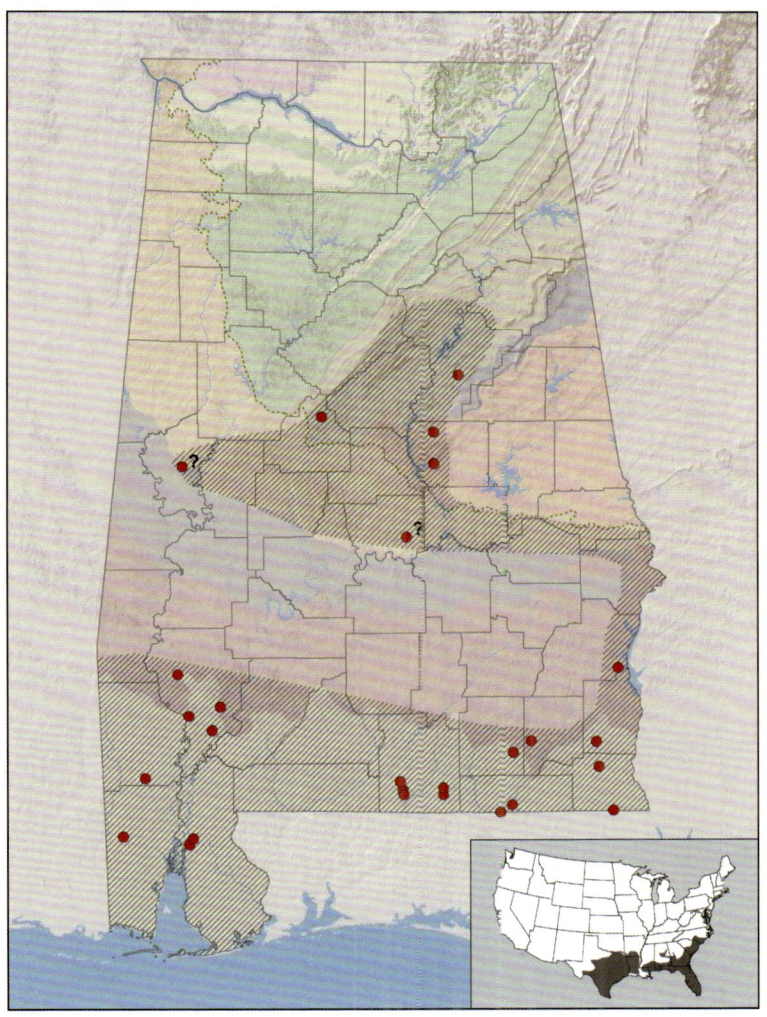

Distribution of Eastern Coralsnake, *Micrurus fulvius*. The presumed range of the species in Alabama is indicated by hatching. Inset map depicts approximate range in the United States.

**HABITS** The Eastern Coralsnake is an active diurnal species that is adept at escaping detection. Most Alabama specimens have been collected on the surface, early in the morning or late in afternoon hours. Exploited habitat types range from hardwood forests to pine flatwoods. Friable soil seems to be a requisite. The reproductive habits are poorly known. But, female snakes in north-central Florida have enlarged ovarian follicles in the spring and lay a single clutch of eggs in June or July (Jackson and Franz 1981). The eggs apparently are laid in loose soil or in rotting organic matter. Clutch size reportedly varies from three to twelve. Small snakes and lizards constitute the Eastern Coralsnake's diet.

The Eastern Coralsnake is dangerous, and its venom is fully capable of killing a human. There are many reports, most of them old, to the effect that coralsnakes are docile and inoffensive and that they may be handled with impunity. Many people who have been bitten were probably of this school of thought. In our experience, most Eastern Coralsnakes readily attempt to bite when captured or restrained. Their motions when constrained are quite rapid, with the body frequently twitching from one side to the other. If manipulated further, individual snakes often coil their tail and wave it about while hiding the head beneath coils of the body, a behavior referred to as automimicry because the tail looks like the head. This behavior appears to direct attacks to the coiled tail and protects the real head. Unlike viperids, when a Coralsnake bites a human, rarely does the snake fail to inject venom (known as a dry bite). Symptoms of envenomation are droopy eyelids, followed by respiratory failure. Symptoms frequently are not expressed until twelve hours after the bite, at which time symptoms typically emerge rapidly. Therefore, any bite from an Eastern Coralsnake requires immediate medical attention.

**CONSERVATION AND MANAGEMENT** The Eastern Coralsnake was ranked Priority 1 (Highest Conservation Concern) by ADCNR in 2012. Therefore, it is illegal to possess the species in the state of Alabama without a scientific collecting permit. It is difficult to get an understanding of population sizes of Eastern Coralsnakes in Alabama because so few of them are seen. We have only four records of the species from the state since 1987, and these have been widely separated, with two records from Covington County and one each from Barbour and Coosa Counties. Maintenance of open pine stands, managed with frequent fire, seems to be important for the species because of its relatively close association with habitats dominated by longleaf pine (Nelson 2004).

**TAXONOMY** This species is sister to *M. coralinus*, a species of South American Coralsnake, and this pair is basal to the rest of the species in the genus (Pyron et al. 2011). No subspecies of *M. fulvius* are recognized. Previous authors have placed this species in the genera *Coluber* and *Elaps*.

# Appendixes

## Appendix 1.

*Scientific and common names of non-indigenous reptiles introduced to, but not established in, Alabama.*

| Scientific Name | Common Name | Place |
|---|---|---|
| *Boa constrictor* | Red-tailed Boa | Auburn; Birmingham |
| *Eublepharis macularius* | Leopard Gecko | Auburn |
| *Gekko gecko* | Tokay Gecko | Auburn |
| *Iguana iguana* | Green Iguana | Auburn; Birmingham |
| *Lampropeltis triangulum hondurensis* | Honduran Milksnake | Loachapoka |
| *Naja naja* | Spectacled cobra | Moody |
| *Python regius* | Ball Python | Auburn |
| *Python reticulatus* | Reticulated Python | Birmingham |
| *Sphaerodactylus lineolatus* | Panama Dwarf Gecko | Mobile |

## Appendix 2.

*List of subspecies from Mount (1975) that are (A) no longer recognized as distinctive or (B) now recognized as full species.*

**A**

| Scientific Name | Source |
|---|---|
| *Agkistrodon contortrix mokasen* | Guiher and Burbrink 2008 |
| *Agkistrodon piscivorus leucostoma* | Guiher and Burbrink 2008 |
| *Sceloporus undulatus hyacinthus* | Leaché and Reeder 2002 |

**B**

| Scientific Name | Source |
|---|---|
| *Lampropeltis triangulum elapsoides* | Pyron and Burbrink 2009b |
| *Nerodia cyclopion floridana* | Lawson 1987 |
| *Nerodia fasciata clarkii* | Lawson et al. 1991 |

# Appendix 3.

*List of taxa whose eastern or western boundary abuts the Mobile-Tombigbee Rivers.*

|  | **Eastern Boundary** | **Western Boundary** |
|---|---|---|
| **Sister Taxa** | *Lampropeltis getula holbrooki* | *Lampropeltis getula getula* |
|  | *Nerodia cyclopion* | *Nerodia floridana* |
|  | *Nerodia fasciata confluens* | *Nerodia fasciata fasciata* |
|  | *Pituophis melanoleucus lodingi* | *Pituophis melanoleucus mugitus* |
|  | *Scincella lateralis* Tombigbee clade | *Scincella lateralis* Florida Panhandle clade |
| **Single Taxa** |  | *Liodytes pygaea* |
|  |  | *Ophisaurus mimicus* |
|  |  | *Plestiodon egregius* |
|  |  | *Regina septemvittata* |

# Appendix 4.

*Pairs of sister taxa whose eastern and western boundaries are associated with the Chattahoochee-Choctawhatchee-Pea Rivers.*

| **Eastern Boundary** | **Western Boundary** |
|---|---|
| *Agkistrodon piscivorus piscivorus* | *Agkistrodon piscivorus conanti* |
| *Nerodia erythrogaster flavigaster* | *Nerodia erythrogaster erythrogaster* |
| *Pantherophis obsoleta spiloides* | *Pantherophis obsoleta alleganiensis* |
| *Scincella lateralis* Florida Panhandle clade | *Scincella lateralis* East Coast clade |

# Appendix 5.

*List of species with Alabama distributions restricted within (or nearly to) the Coastal Plain.*

| Family | Species |
|--------|---------|
| **Colubridae** | *Coluber constrictor helvigularis* |
| | *Farancia abacura abacura* |
| | *Farancia abacura reinwardtii* |
| | *Lampropeltis elapsoides* |
| | *Lampropeltis getula getula* |
| | *Liodytes rigida* |
| | *Nerodia erythrogaster erythrogaster* |
| | *Nerodia rhombifer* |
| | *Nerodia taxispilota* |
| | *Pituophis melanoleucus lodingi* |
| | *Virginia valeriae elegans* |
| **Elapidae** | *Micrurus fulvius* |
| **Scincidae** | *Plestiodon anthracinus pluvialis* |
| | *Plestiodon egregius* |
| | *Scincella lateralis* East Coast clade |
| **Viperidae** | *Sistrurus miliarius barbouri* |

# Appendix 6.

*List of reptile species restricted to the Lower Coastal Plain of Alabama.*

|  | Family | Species |
|---|---|---|
| **Anguidae** | *Ophisaurus mimicus* | |
| **Colubridae** | *Drymarchon couperi* | |
| | *Farancia erytrogramma* | |
| | *Heterodon simus* | |
| | *Liodytes pygaea* | |
| | *Nerodia clarkii* | |
| | *Nerodia cyclopion* | |
| | *Nerodia fasciata confluens* | |
| | *Nerodia fasciata fasciata* | |
| | *Nerodia fasciata pictiventris* | |
| | *Nerodia floridana* | |
| | *Pituophis melanoleucus mugitus* | |
| | | *Storeria dekayi limnetes* |
| **Scincidae** | *Scincella lateralis* Mississippi East clade | |
| | *Scincella lateralis* Florida Panhandle clade | |
| **Viperidae** | *Agkistrodon piscivorus conanti* | |
| | *Crotalus adamanteus* | |
| | *Sistrurus miliarius barbouri* | |

# Appendix 7.

*Synoptic key to the subfamilies of colubrids of Alabama.*

The following sets of features characterize members of the three subfamilies of colubrids found in Alabama. Characters (columns) are scale shape (K = keeled; WK = weakly keeled; S = smooth), scale rows (count of number of scale rows at midbody), oculars (counts of numbers of preocular and postocular scales; e.g., 2/2 = 2 preoculars and 2 postoculars), prefrontal (contact = contacts orbit, no = excluded from contact with orbit by preocular), loreal (present or absent). Each row represents unique sets of character states present in some members of each subfamily.

| | Scale Shape 1 | Scale Rows 2 | Oculars | Prefrontal | Loreal |
|---|---|---|---|---|---|
| **Colubrinae (page 117)** | WK | >= 25 | 1/(2–3) | No | Present |
| | K | 17 | 1/2 | No | Present |
| | S | 19–21 | 1/2 | No | Present |
| | S | 17 | 2/2 | No | Present |
| | S | 15 | 1/2 | No | Absent |
| **Dipsadinae (page 197)** | K | 23–25 | 3/3 | No | Present |
| | S | 13 or 19 | 0/(1–2) | Contact | Present |
| | S | 17 | 1/2 | No | Present |
| | S | 15 | 2/2 | No | Present |
| **Natricinae (page 235)** | K/WK | 19–23 | 1/(2–3) | No | Present |
| | K | 15–17 | (1–2)/2 | No | Absent |
| | K | 15–17 | 0/(1–2) | Contact | Present |

# Glossary

**AMNIOTE:** A land vertebrate; any member of the group Amniota.

**AMPHISBAENIA:** Burrowing snake-like squamates frequently referred to as worm lizards because of a superficial resemblance to earthworms.

**AMPHIUMA:** Any member of the salamander family Amphiumidae; a large, entirely aquatic salamander of swampy habitat in the southeastern United States.

**ANAL SCUTE:** The plate-like scale that covers the anterior border of the cloacal opening.

**ANGUIMORPHAN:** Any member of the group Anguimorpha, a radiation of squamates with elongate bodies, forked tongues, and external ear openings.

**ANGUID:** Any member of the family Anguidae, a lineage of squamates containing the alligator lizards and legless lizards of North America.

**ANOLE:** Any member of the squamate family Dactyloidae.

**ANTIPREDATORY MECHANISM:** Behavioral or anatomical features used by squamates to avoid predators.

**APOSEMATIC:** Having bright colors, usually involving combinations of black, dark blue, red, orange, yellow, or white, on any organisms that indicate to potential predators that the organism is toxic.

**ARBOREAL:** Tree-dwelling.

**AUTOTOMIC SEPTA:** A sheet of cartilaginous tissue dividing caudal vertebrae of squamates into anterior and posterior halves that fracture when a predator attacks the tail.

**BOID:** Any member of the family Boidae, a lineage of heavy-bodied snakes that includes boas and pythons.

**BRUMATION:** A reduction of physiological activities exhibited by squamates during cold winter months, typically associated with inactivity in a burrow or hidden refuge.

**CAUDAL VERTEBRAE:** The bones of the tail of squamates.

**CENTRACHID:** Any member of the fish family Centrarchidae.

**CHEMOSENSORY:** Serving the function of detecting chemicals in the environment.

**CHERT:** A type of sedimentary rock with silica forming microcrystals, cryptocrystals, or microfibers.

**CHERT BELT:** A contiguous geological region of the southern Appalachian Mountains dominated by chert deposits.

CIRCUMORBITALS: A ring of scales that encircle the eye of squamates.

CLEAR-CUT: A method of forest management that involves removing all trees and replanting with seeds or seedlings.

CLINE: A region across which organisms or geology changes gradually.

CLOACA: A single opening through which digestive, urinary, and reproductive products pass to the exterior of squamates.

COALESCE: Join together.

COASTAL PLAIN: A region of the lowlands of Alabama characterized by relatively flat terrain, deep sandy soils, and meandering rivers and creeks.

COLUBRID: Any member of the squamate family Colubridae.

COLUBRINE: Any member of the squamate subfamily Colubrinae.

CYCLOID: Having a half-moon shaped border.

DEWLAP: A flap of skin under the chin of some squamates that can be extended during social displays.

DIBAMIDAE: A family of limbless, snake-like squamates of Mexico and Southeast Asia.

DIPSADINE: Any member of the squamate subfamily Dipsadinae, a lineage of snakes many of which are rear fanged.

DIURNAL: Active during daylight hours.

DORSOLATERALS: Scales along the sides and toward the top of the body.

DORSUM: In four-legged vertebrates, that portion of the body that faces away from the soil surface; the back.

ECDYSIS: The process of replacing the epidermis; shedding of skin.

ENVENOMATION: The process of injecting toxic compounds produced by one organism into another organism.

EXTERNAL EAR OPENING: An opening at the side of the head through which sound waves pass from the exterior, through the middle ear, to the inner ear.

EXTERNAL NARES: Openings at the front of the snout through which chemicals pass on their way to the nasal cavity.

FALL LINE: A geographic feature separating mountainous topography of the Appalachian Mountains from the relatively flat topography of the Coastal Plain. The Fall Line typically has rocky streams with short waterfalls that transition to slow, meandering streams of the Coastal Plain.

FLATWOODS: A habitat of the Coastal Plain dominated by pines and grasses and flooded seasonally by heavy rains.

FOSSORIAL: Living under the surface of the soil, typically by burrowing into the soil.

GRANULAR: Describes scales that create small, rounded projections.

GYMNOPHTHALMIDAE: A family of New World tropical squamates related to the family Teiidae.

HELIOTHERMIC: Gathering heat by intercepting the sun's radiant energy (basking).

HEMIPENIS: An evertible sac-like structure unique to male squamates and used to transfer sperm to the cloaca of a female. These occur in pairs in males, one on the right and one on the left side, and are referred to collectively as hemipenes.

HERPETOFAUNA: The assemblage of amphibian and reptile species at a particular locality or region.

HERPETOLOGY: The study of amphibians and reptiles.

IGUANIAN: Any member of the lineage Iguania, squamates with relatively short bodies, long legs, and strong dependence on vision.

IMMACULATE: Without spots.

INTERGRADIENT: Exhibiting gene flow between two or more subspecies.

INTERNASAL: Scales on the top of the snout of squamates that separate the right and left external nares.

IRIS: A sheet of tissue in the eye of vertebrates that creates an opening (pupil) through which light passes to the retina.

KEELED: Possessing a central ridge.

KERATINIZATION: The process of depositing the protein keratin into the skin of squamates.

LABIAL: Any of the scales that line the lateral edge of the upper or lower jaws.

LACERTOIDEA: A monophyletic lineage of squamates comprised of the families Teiidae, Gymnophthalmidae, and Lacertidae, plus the group Amphisbaenia.

LAMELLAE: Plate-like projections on the surface of the skin.

LEIOCEPHALIDAE: A family of Caribbean squamates referred to as curly-tailed lizards because of their habit of curling their tail over their back.

LEIOSAURINAE: A subfamily of South American squamates of arid regions.

LOREAL: A scale in squamates separating the preocular scale from the postnasal.

MAXILLA: The largest tooth-bearing bone of the upper jaw of most squamates.

MELANISTIC: Characterized by dark appearance because of excessive production of the pigment melanin.

MESIC: A habitat that is intermediate in water content between dry (xeric) and wet (hydric) habitats.

MONOPHYLETIC: A lineage containing an ancestor and all descendants of that ancestor.

MOSASAUR: A fossil radiation of large, predatory, marine squamates.

NATRICINES: Any member of the squamate subfamily Natricinae; frequently referred to as water snakes.

**NEONATE:** A newborn individual; hatchling.

**NOCTURNAL:** Active at night.

**OCELLI:** Circular dark markings bordered by light markings that together look like eyes.

**ODONATES:** A member of the group Odonata; the dragonflies and damselflies.

**OVERSTORY:** That portion of the vegetation that forms a canopy created by trees.

**OVIPAROUS:** Reproduction via egg laying.

**OVULATION:** The process of releasing unfertilized eggs from the ovary in female squamates.

**PAPILLAE:** Short, finger-like projections on the surface of skin.

**PARIETAL:** One of a pair of large scales located on the top of the head of squamates and located immediately posterior to the frontal scales.

**PARTHENOGENIC:** Capable of producing offspring from unfertilized eggs.

**PHYLLODACTYLIDAE:** A family of the squamate lineage Gekkota characterized by the presence of lamellae on the venter of the digits, creating toe pads.

**PHYLOGEOGRAPHY:** The study of patterns of gene flow among populations of the same species across broad geographic scales.

**PHYSIOGRAPHIC:** Non-random patterns of soils and vegetation across broad geographic scales.

**POLYCHROTIDAE:** A family of the squamate lineage Iguania; the family includes neotropical arboreal forms referred to as bush or monkey lizards.

**POSTANAL SCALE:** A scale located immediately posterior to the cloacal opening. A pair of these is enlarged in males of some squamates.

**POSTMENTAL:** A single scale or pair of scales located immediately posterior to the mental scale of the lower jaw.

**POSTLABIALS:** Scales of the lateral edge of the upper jaw and located immediately posterior to the labials.

**POSTNASAL:** The scale of squamates located immediately posterior to the external naris.

**PREFRONTAL:** Any of the scales located immediately anterior to the frontal scales of squamates.

**PREOCULAR:** The scale of squamates located immediately anterior to the eye.

**PROVENTRICULUS:** A muscular region of the esophagus located immediately anterior to the stomach of some squamates.

**QUADRICHOTOMY:** Four branches emerging from a single node.

**ROSTRAL:** The anterior-most scale of the upper jaw of squamates.

**SANDHILL:** A geographic area dominated by deep sands allowing excessive draining of the soil.

**SCALATION:** The pattern of distribution of scales in squamates.

SCINCOMORPHA: A lineage of squamates containing the families Scincidae, Xantusiidae, Gerrhosauridae, and Cordylidae.

SCLEROTINIZED PAPILLAE: Projections on the surface of some squamate tongues that are hardened by cross-linking of proteins on its surface.

SCUTELLATION: The pattern of distribution of scales in squamates.

SCUTES: Large, flat, plate-like scales.

SEMIFOSSORIAL: Active on the soil surface and under its surface.

SQUAMATE: Any member of the reptile lineage Squamata.

SQUAMOSAL: A bone at the posterior region of the skull connecting the quadrate to the postorbital.

SUBCAUDAL: Any of the scales located on the ventral surface of the tail.

SUBOCULAR: A scale separating the upper labials from the scales surrounding the eye.

SUBSPECIES: A geographically and morphologically distinctive metapopulation within a species.

SULCUS SPERMATICUS: A groove on the surface of an everted hemipenis that serves to carry sperm from the cloaca of the male to the tip of the hemipenis.

SUPRALABIALS: Labials of the upper jaw.

SYNAPOMORPHY: A shared derived character state.

TEMPORAL: Scales of squamates located posterior to the postoculars and adjacent to the supralabials.

VENTER: On a four-legged land vertebrate, that portion of the body that faces the ground; the belly.

VERMICULATIONS: Worm-like markings.

VIVIPARITY: Reproduction via live birth.

VOMERONASAL ORGAN: A chemosensory organ found in the roof of the mouth of squamates.

XENOSAURIDAE: A family of Central American squamates known as knob-scaled lizards.

XERIC: A habitat that is dry, in Alabama typically because of sandy soils that drain water rapidly.

# Photo Credits

Mark A. Bailey, pages 14, 15, 16, 17, 19, 29, 84, 93, 128, 245, 268 (bottom), 332
Roger D. Birkhead, pages 94, 229 (bottom)
Matt Buckingham, pages 261, 304
Adam Cooner, page 248
Alan Cressler, pages 28 (both), 46, 268 (top), 345
Rick Dowling, page 132
Brian P. Folt, page 335
James C. Godwin, pages 74, 140, 167, 205, 215, 222, 254, 270, 278 (bottom), 314
Scott Gravette, pages 49, 61, 73, 78, 136, 160, 174, 180, 185 (both), 194, 212
Matt P. Greene, page 141
Craig Guyer, pages 20, 21, 23, 24, 25, 26, 30, 67, 298
Aubrey Heupel, page 232
Pierson Hill, pages 45, 105, 110, 154, 206, 223, 238, 277, 288, 329
James Jeffrey, page 125
John B. Jensen, page 255
Ritchie L. King, pages 139, 144, 159, 184, 192, 228, 258, 267, 323, 324
Melissa Miller, page 307
Robert H. Mount, pages 87, 219, 293, 340
Todd Pierson, pages 56, 229 (top), 281
Corey Raimond, pages 50, 208, 278 (top), 346
Nathan Shephard, page 284
Robert Smith, page 60
Eric C. Soehren, pages 91, 107, 177, 351
Geoff Sorrell, pages 163, 290
James A. Stiles, pages 81, 104, 188, 191
Kevin Stohlgren, pages 120, 201, 202, 239, 301
Kenneth Wray, pages 90, 97, 137, 147, 153, 250, 263, 287, 320, 352

# Works Cited

Alabama Department of Conservation and Natural Resources, Division of Wildlife and Freshwater Fisheries. 2005. "Conserving Alabama's Wildlife: A Comprehensive Strategy." Alabama Department of Conservation and Natural Resources, Montgomery. 332 pages.

Aldridge, R. D., and W. S. Brown. 1995. "Male Reproductive Cycle, Age at Maturity, and Cost of Reproduction in the Timber Rattlesnake (*Crotalus horridus*)." *Journal of Herpetology* 29:399–407.

Aldridge, R. D., and R. D. Semlitsch. 1992. "Female Reproductive Biology of the Southeastern Crowned Snake (*Tantilla coronata*)." *Amphibia-Reptilia* 13:209–218.

Alfaro, M. E., and S. J. Arnold. 2001. "Molecular Systematic and Evolution of *Regina* and the Thamnophiine Snakes." *Molecular Phylogenetics and Evolution* 21:408–423.

Allen, C. R., D. M. Epperson, and A. S. Garmestani. 2004. "Red Imported Fire Ant Impacts on Wildlife: A Decade of Research." *American Midland Naturalist* 152:88–103.

Averill-Murray, R. C. 2006. "Natural History of the Western Hog-Nosed Snake (*Heterodon nasicus*) with Notes on Envenomation." *Sonoran Herpetologist* 19:98–101.

Barrett, R. K., and C. Guyer. 2008. "Differential Responses of Amphibians and Reptiles in Riparian and Stream Habitats to Land Use Disturbances in Western Georgia, USA." *Biological Conservation* 141:2290–2300.

Baxley, D., G. J. Lipps Jr., and C. P. Qualls. 2011. "Multiscale Habitat Selection by Black Pine Snakes (*Pituophis melanoleucus lodingi*) in Southern Mississippi." *Herpetologica* 67:154–166.

Baxley, D., and C. P. Qualls. 2009. "Black Pine Snake (*Pituophis melanoleucus lodingi*): Spatial Ecology and Associations between Habitat Use and Prey Dynamics." *Herpetologica* 43:284–293.

Bernardino, F. S., and G. H. Dalrymple. 1992. "Seasonal Activity and Road Mortality of the Snakes of the Pay-hay-okee Wetlands of the Everglades National Park, USA." *Biological Conservation* 62:71–75.

Bishop, D. C., and A. C. Echternacht. 2003. "Winter Growth and Sex Ratio of a Northern Population of *Anolis carolinensis* (Sauria Polychrotidae)." *Copeia* 2003:906–909.

———. 2004. "Emergence Behavior and Movements of Winter-Aggregated Green Anoles (*Anolis carolinensis*) and the Thermal Characteristics of Their Crevices in Tennessee." *Herpetologica* 60:168–177.

Blackburn, D. G. 2006. "Squamate Reptiles as Model Organisms for the Evolution of Viviparity." *Herpetological Monographs* 20:131–146.

Blouin-Demers, G., H. L. Gibbs, and P. J. Weatherhead. 2005. "Genetic Evidence for Sexual Selection in Black Ratsnakes, *Elaphe obsoleta.*" *Animal Behaviour* 69:225–234.

Blouin-Demers, G., J. R. Row, and P. J. Weatherhead. 2004. "Phenotypic Consequences of Nest-Site Selection in Black Rat Snakes (*Elaphe obsoleta*)." *Canadian Journal of Zoology* 82:449–456.

Brandley, M. C., T. J. Guiher, R. A. Pyron, and C. T. Winne. 2010. "Does Dispersal across an Aquatic Geographic Barrier Obscure Phylogeographic Structure in the Diamond-Backed Watersnake (*Nerodia rhombifer*)?" *Molecular Phylogenetics and Evolution* 57:552–560.

Breininger, D. R., M. L. Legare, and R. B. Smith. 2004. "Eastern Indigo Snakes (*Drymarchon couperi*) in Florida: Influence of Edge Effects on Population Viability." Pp. 299–311. In *Species Conservation and Management: Case Studies*, ed. H. R. Akçakaya, M. A. Burgman, O. Kindvall, C. C. Wood, P. Sjögren-Gulve, J. S. Hatfield, and M. A. McCarthy. Vol. 1. Oxford: Oxford University Press.

Brower, A. I., and M. R. Cranfield. 2001. "*Cryptosporidium* sp.-Associated Enteritis without Gastritis in Rough Green Snakes (*Opheodrys aestivus*) and a Common Garter Snake (*Thamnophis sirtalis*)." *Journal of Zoo and Wildlife Medicine* 32:101–105.

Bryson, R. W., J. Pastorini, F. T. Burbrink, and M. R. J. Forstner. 2007. "A Phylogeny of the *Lampropeltis mexicana* Complex (Serpentes: Colubridae) Based on Mitochondrial DNA Sequences Suggests Evidence for Species-Level Polyphyly within *Lampropeltis.*" *Molecular Phylogenetics and Evolution* 43:674–684.

Burbrink, F. T. 2001. "Systematics of the Eastern Ratsnake Complex (*Elaphe obsoleta*)." *Herpetological Monographs* 15:1–53.

Burbrink, F. T., and T. J. Guiher. 2015. "Considering Gene Flow When Using Coalescent Methods to Delimit Lineages of North American Pitvipers of the Genus *Agkistrodon.*" *Zoological Journal of the Linnean Society* 173:505-526.

Burbrink, F. T., F. Fontanella, R. A. Pyron, T. J. Guiher, and C. Jimenez. 2008. "Phylogeography across a Continent: The Evolutionary and Demographic History of the North American Racer (Serpentes: Colubridae: *Coluber constrictor*)." *Molecular Phylogenetics and Evolution* 47:274–288.

Burbrink, F. T., R. Lawson, and J. B. Slowinski. 2000. "Mitochondrial DNA Phylogeography of the North American Rat Snake (*Elaphe obsoleta*): A Critique of the Subspecies Concept." *Evolution* 54:2107–2114.

Burkett-Cadena, N. D., S. P. Graham, H. K. Hassan, C. Guyer, M. D. Eubanks, C. R. Katholi, and T. R. Unnasch. 2008. "Blood Feeding Patterns of Potential Arbovirus Vectors of the Genus *Culex* Targeting Ectothermic Hosts." *American Journal of Tropical Medicine and Hygiene.* 79:809–815.

Butler, J. A., T. W. Hull, and R. Franz. 1995. "Neonate Aggregations and Maternal Attendance of Young in the Eastern Diamondback Rattlesnake, *Crotalus adamanteus.*" *Copeia* 1995:196–198.

Carranza, S., and E. N. Arnold. 2006. "Systematics, Biogeography, and Evolution of *Hemidactylus* Geckoes (Reptilia: Gekkonidae) Elucidated Using Mitochondrial DNA Sequences." *Molecular Evolution and Systematics* 38:531–545.

Castoe, T. A., and C. L. Parkinson. 2006. "Bayesian Mixed Models and the Phylogeny of Pitvipers (Viperidae: Serpentes)." *Molecular Phylogenetics and Evolution* 39:91–110.

Christman, S. P. 1982. "*Storeria dekayi*." *Catalogue of North American and Reptiles.* 306.1–306.4.

Clark, A. M., P. E. Moler, E. E. Possardt, A. H. Savitsky, W. S. Brown, and B. W. Bowen. 2002. "Phylogeography of the Timber Rattlesnake (*Crotalus horridus*) Based on mtDNA Sequences." *Journal of Herpetology* 37:145–154.

Collins, J. T. 1991. "Viewpoint: A New Taxonomic Arrangement for Some North American Amphibians and Reptiles." *Herpetological Review* 22:42–43.

Conant, R., and J. T. Collins. 1998. *A Guide to Reptiles and Amphibians: Eastern and Central North America.* New York: Houghton Mifflin.

Conrad, J. L. 2008. "Phylogeny and Systematics of Squamata (Reptilia) Based on Morphology." *Bulletin of the American Museum of Natural History* 310:1–182.

Cooper, W. E., Jr., D. C. Buth, and L. J. Vitt. 1990. "Prey Odor Discrimination by Ingestively Naïve Coachwhip Snakes (*Masticophis flagellum*)." *Chemoecology* 1:86–91.

Cooper, W. E., Jr., and S. Secor. 2007. "Strong Response to Anuran Chemical Cues by an Extreme Dietary Specialist, the Eastern Hog-Nosed Snake (*Heterodon platirhinos*)." *Canadian Journal of Zoology* 85:619–625.

Cooper, W. E., Jr., and L. J. Vitt. 1997. "Maximizing Male Reproductive Success in the Broad-Headed Skink (*Eumeces laticeps*): Preliminary Evidence for Mate Guarding, Size-Assortative Pairing, and Opportunistic Extra-Pair Mating." *Amphibia-Reptilia* 18:59–73.

Crother, B. I. (ed.) 2017. "Scientific and Standard English Common Names of Amphibians and Reptiles of North America North of Mexico, with Comments Regarding Confidence in Our Understanding." *SSAR Herpetological Circular* no. 43. 102 pages.

Crother, B. I., J. Boundy, F. T. Burbrink, and S. Ruane. 2017. "Squamata (In Part) – Snakes." Pp. 59-81. In *Scientific and Standard English Common Names of Amphibians and Reptiles of North America North of Mexico, with Comments Regarding Confidence in Our Understanding*, ed. B. I. Crother. *SSAR Herpetological Circular* no 43.

Deitloff, J. D., V. M. Johnson, and C. Guyer. 2013. "Bold Colors in a Cryptic Lineage: Do Eastern Indigo Snakes Exhibit Color Dimorphism?" PLOS One DOI: 10.1371/journal pone 0064538.

de Queiroz, K. 1998. "A General Lineage Concept of Species, Species Criteria, and the Process of Speciation: A Conceptual Unification and Terminological Recommendations." Pp. 57–75. In *Endless Forms: Species and Speciation*, ed. D. J. Howard and S. H. Berlocher. Oxford: Oxford University Press.

de Queiroz, K., and J. Gautier. 1992. "Phylogenetic Taxonomy." *Annual Review of Ecology and Systematics* 23:449–480.

DeVault, T. L., and A. R. Krochmal. 2002. "Scavenging by Snakes: An Examination of the Literature." *Herpetologica* 58:429–436.

Diemer, J. A., and D. W. Speake. 1983. "The Distribution of the Eastern Indigo Snake, *Drymarchon corais couperi*." *Georgia Journal of Herpetology* 17:256–264.

do Amaral, J. P. S. 1999. "Lip-Curling in Redbelly Snakes (*Storeria occipitomaculata*): Functional Morphology and Ecological Significance." *Journal of Zoology* 248:289–293.

Dodd, C. K. 1993. "Population Structure, Body Mass, Activity, and Orientation of an Aquatic Snake (*Seminatrix pygaea*) during a Drought." *Canadian Journal of Zoology* 71:1281–1288.

Dodd, C. K., and W. J. Barichivich. 2007. "Movements of Large Snakes (*Drymarchon, Masticophis*) in North-Central Florida." *Florida Scientist* 70:83–94.

Dorcas, M. E., J. D. Willson, R. N. Reed, R. W. Snow, M. R. Rochford, M. A. Miller, W. E. Meshaka Jr., P. T. Andreadis, F. J. Mazzotti, C. M. Romagosa, and K. M. Hart. 2012. "Severe Mammal Declines Coincide with Proliferation of Invasive Burmese Pythons in Everglades National Park." Proceedings of the National Academy of Sciences DOI: 10.1073/pnas.1115226109.

Duran, C. M. 1998. "Status of the Black Pine Snake (*Pituophis melanoleucus lodingi* Blanchard)." Unpublished Report to USFWS, Jackson, MS.

Duval, D., R. Hershkowitz, and J. Trupiano-Duval. 1980. "Responses of Five-Lined Skinks (*Eumeces fasciatus*) and Ground Skinks (*Scincella lateralis*) to Conspecific and Interspecific Cues." *Journal of Herpetology* 14:121–127.

Erbsen, W. 1995. "Southern Mountain Fiddle." Mel Bay Publishing, Pacific, MO.

Etheridge, K. E., L. C. Wit, J. C. Sellers, and S. E. Trauth. 1986. "Seasonal Changes in Reproductive Condition and Energy Stores in *Cnemidophorus sexlineatus*." *Journal of Herpetology* 20:554–559.

Faust, T. M., and S. M. Blomquist. 2011. "Size and Growth in Two Populations of Black Kingsnakes, *Lampropeltis nigra*, in East Tennessee." *Southeastern Naturalist* 10:409–422.

Fitch, H. R. 1970. "Reproductive Cycles in Lizards and Snakes." University of Kansas, Museum of Natural History, Miscellaneous Publications 52:1–247.

———. 2006. "Ecological Succession on a Natural Area in Northeastern Kansas from 1948 to 2006." *Herpetological Conservation and Biology* 1:1–5.

Fitch, H. R., and P. L. von Achen. 1977. "Spatial Relationships and Seasonality in the Skinks *Eumeces fasciatus* and *Scincella laterale* in Northeastern Kansas." *Herpetologica* 33:303–313.

Fontanella, F. M., C. R. Feldman, M. E. Siddal, and F. T. Burbrink. 2008. "Phylogeography of *Diadophis punctatus*: Extensive Lineage Diversity and Repeated Patterns of Historical Demography in a Trans-continental Snake." *Molecular Phylogenetics and Evolution* 46:1049–1070.

Ford, N. B. 1982. "Courtship Behavior of the Queen Snake, *Regina septemvittata*." *Herpetological Review* 13:72.

Franz, R. 1984. "The Florida Gopher Frog and the Florida Pine Snake as Burrow Associates of the Gopher Tortoise in Northern Florida." Pp 16–20. In *The*

*Gopher Tortoise and Its Community*, ed. D. R. Jackson and R. J. Bryant. Proceedings of the 5th Annual Meeting of the Gopher Tortoise Council, Florida State Museum, Gainesville.

Fredriksson, G. M. 2005. "Predation on Sun Bears by Reticulated Pythons in East Kalimantan, Indonesian Borneo." *Raffles Bulletin of Zoology* 53:165–168.

Freidenfelds, N. A., T. R. Robbins, and T. Langkilde. 2012. "Evading Invaders: The Effectiveness of a Behavioral Response Acquired through Lifetime Exposure." *Behavioral Ecology* 23:659–664.

Frost, D. R., and D. M. Hillis. 1990. "Species in Concept and Practice: Herpetological Applications." *Herpetologica* 46:86–104.

Gans, C. 1975. "Tetrapod Limlessness: Evolution and Functional Corollaries." *American Zoologist* 15:455–467.

Gautier, J. A., M. Kearney, J. A. Maisano, O. Rieppel, and A. D. B. Behlke. 2012. "Assembling the Squamate Tree of Life: Perspectives from the Phenotype and the Fossil Record." *Bulletin of the Peabody Museum of Natural History* 53:3–308.

Gautier, J. A., A. G. Kluge, and T. Rowe. 1988. "Amniote Phylogeny and the Importance of Fossils." *Cladistics* 4:105–209.

Gerald, G. W., M. A. Bailey, J. N. Holmes. 2006. "Movement and Activity Range Sizes of Northern Pinesnakes (*Pituophis melanoleucus melanoleucus*) in Middle Tennessee." *Journal of Herpetology* 40:503–510.

Gerber, G. P., and A. C. Echternacht. 2000. " Evidence for Asymmetric Intraguild Predation between Native and Introduced *Anolis* Lizards." *Oecologia* 124:599–607.

Gibbons, J. W. 1972. "Reproduction, Growth, and Sexual Dimorphism in the Canebrake Rattlesnake (*Crotalus horridus atricaudatus*)." *Copeia* 1972:222–226.

Gibbons, J. W., J. W. Coker, and T. M. Murphy Jr. 1977. "Selected Aspects of the Life History of the Rainbow Snake (*Farancia erythrogramma*)." *Herpetologica* 33:276–281.

Gibbons, J. W., and M. Dorcas. 2002. "Defensive Behaviors of Cottonmouths (*Agkistrodon piscivorus*) toward Humans." *Copeia* 2002:195–198.

———. 2004. *North American Watersnakes: A Natural History*. Norman: University of Oklahoma Press.

Gibbs, H. L., S. J. Corey, G. Blouin-Demers, K. A. Prior, and P. J. Weatherhead. 2006. "Hybridization between mt-DNA-Defined Phylogeographic Lineages of Black Ratsnakes (*Pantherophis* sp.)." *Molecular Ecology* 15:3755–3767.

Giery, S. T., and R. S. Ostfeld. 2007. "The Role of Lizards in the Ecology of Lyme Disease in Two Endemic Zones of the Northeastern United States." *Journal of Parasitology* 93:511–517.

Glaudas, X., T. M. Farrell, and P. G. May. 2005. "Defensive Behavior of Free-Ranging Pygmy Rattlesnakes (*Sistrurus miliarius*)." *Copeia* 2005:196–200.

Glenn, J. L., R. C. Straight, and T. B. Wolt. 1994. "Regional Variation in the Presence of Canebrake Toxin in *Crotalus horridus* Venom." *Comparative Biochemistry and Physiology*. Part C 107:337–346.

Glor, R. E., J. B. Losos, and A. Larson. 2005. "Out of Cuba: Overwater Dispersal and Speciation among Lizards of the *Anolis carolinensis* Subgroup." *Molecular Ecology* 14:2419–2432.

Godwin, J. C. 2004a. "Eastern Indigo Snake *Drymarchon couperi* (Holbrook)." Pp. 156–157. In Mirarchi et al., *Alabama Wildlife*, vol. 4.

———. 2004b. "Northern Pine Snake *Pituophis melanoleucus melanoleucus* (Daudin)." Pp. 167. In Mirarchi et al., *Alabama Wildlife*, vol. 4.

Goldberg, S. R. 2002. "Reproduction in the Coachwhip, *Masticophis flagellum* (Serpentes: Colubridae), from Arizona." *Texas Journal of Science* 54:143–150.

Graham, S. P., R. L. Earley, S. K. Hoss, G. W. Schuett, and M. S. Grober. 2008. "The Reproductive Biology of Male Cottonmouths (*Agkistrodon piscivorus*): Do Plasma Steroid Hormones Predict the Mating Season?" *General and Comparative Endocrinology* 159:226–236.

Greenberg, C. H., D. G. Neary, and L. D. Harris. 1994. "Effect of High-Intensity Wildfire and Silviculture Treatments on Reptile Communities in Sand-Pine Scrub." *Conservation Biology* 8:1047–1057.

Griffith, G. E., J. M. Omernik, J. A. Comstock, S. Lawrence, G. Martin, A. Goddard, V. J. Hulcher, and T. Foster. 2001. "Ecoregions of Alabama and Georgia." Two-sided color poster with map, descriptive text, summary tables, and photographs. US Geological Survey, Reston, VA. Scale 1:1,700,000.

Guiher, T. J. and F. T. Burbrink. 2008. "Demographic and Phylogeographic Histories of Two Venomous North American Snakes of the Genus *Agkistrodon*." *Molecular Phylogenetics and Evolution* 48:543–553.

Guyer, C., and M. A. Bailey. 1993. "Amphibians and Reptiles of Longleaf Pine Communities." Pp. 139–158. In *The Longleaf Pine Ecosystem: Ecology, Restoration, and Management*, ed. S. M. Hermann. Proceedings of the Tall Timbers Fire Ecology Conference, No. 18. Tallahassee, FL.

Guyer, C., M. A. Bailey, J. Holmes, J. Stiles, and S. Stiles. 2007. "Herpetofaunal Response to Longleaf Pine Ecosystem Restoration, Conecuh National Forest, Alabama." www.outdooralabama.com/research-mgmt/State%20Wildlife%20Grants/Conecuh_NF_Final_Herp_Report_April_2007.pdf, accessed January 10, 2010.

Hall, P. M., and A. J. Meier. 1993. "Reproduction and Behavior of Western Mud Snakes (*Farancia abacura reinwardtii*) in American Alligator Nests." *Copeia* 1993:219–222.

Hansknecht, K. A. 2008. "Lingual Luring by Mangrove Saltmarsh Snakes (*Nerodia clarkii compressicauda*)." *Journal of Herpetology* 42:9–15.

Harper, R. M. 1943. "Forests of Alabama." Alabama Geological Survey, Monograph 10, Revised Edition.

Hart, B. 2002. "Status Survey of the Eastern Indigo Snake (*Drymarchon couperi* Holbrook), Black Pine Snake (*Pituophis melanoleucus lodingi* Blanchard), and Southern Hognose Snake (*Heterodon simus* Linnaeus) in Alabama." Unpublished report to Alabama Department of Conservation and Natural Resources.

Hein, A. M., and C. Guyer. 2009. "Body Temperatures of Over-wintering

Cottonmouth Snakes: Hibernaculum Use and Inter-individual Variation." *Journal of the Alabama Academy of Sciences* 80:33–45.

Hodges, W. L., and K. R. Zamudio. 2004. "Horned Lizard (*Phrynosoma*) Phylogeny Inferred from Mitochondrial Genes and Morphological Characters: Understanding Conflicts Using Multiple Approaches." *Molecular Phylogenetics and Evolution* 31:961–971.

Holman, J. A. 1995. "Pleistocene Amphibians and Reptiles in North America." *Oxford Monographs on Geology and Geophysics*, no. 32. New York: Oxford University Press.

Hoss, S. K., C. Guyer, L. L. Smith, and G. W. Schuett. 2010. "Multiscale Influences of Landscape Composition and Configuration on the Spatial Ecology of Eastern Diamond-Backed Rattlesnakes (*Crotalus adamanteus*)." *Journal of Herpetology* 44:110–123.

Hoss, S. K., G. W. Schuett, R. L. Earley, and L. L. Smith. 2011. "Reproduction in Male *Crotalus adamanteus* Beauvois (Eastern Diamond-Backed Rattlesnake): Relationship of Plasma Testosterone to Testis and Kidney Dimensions and the Mating Season." *Southeastern Naturalist* 109:95–108.

Howes, B. J., B. Lindsay, and S. C. Lougheed. 2006. "Range-Wide Phylogeography of a Temperate Lizard, the Five-Lined Skink (*Eumeces fasciatus*)." *Molecular Phylogenetics and Evolution* 40:183–194.

Hughes, M. H., and D. H. Nelson. 2004. "Rainbow Snake *Farancia erytrogramma erytrogramma* (Palisot de Beauvois)." Pp. 165. In Mirarchi et al., *Alabama Wildlife*, vol. 4.

Hyslop, N. L. 2007. "Movements, Habitat Use, and Survival of the Threatened Eastern Indigo Snake (*Drymarchon couperi*) in Georgia." Unpublished PhD dissertation, University of Georgia, Athens.

Hyslop, N. L., R. J. Cooper, and J. M. Meyers. 2009. "Shifts in Shelter and Microhabitat Use of *Drymarchon couperi* (Eastern Indigo Snake) in Georgia." *Copeia* 2009:458–464.

Hyslop, N. L., J. M. Meyers, R. J. Cooper, and T. M. Norton. 2009. "Survival of Radio-Implanted *Drymarchon couperi* (Eastern Indigo Snake) in Relation to Body Size and Sex." *Herpetologica* 65:199–206.

IUCN. 2014. "IUCN Red List of Threatened Species." Version 2014.2. www.iucnredlist.org, downloaded November 11, 2014.

Jackson, D. R., and R. Franz. 1981. "Ecology of the Eastern Coral Snake (*Micrurus fulvius*) in Northern Peninsular Florida." *Herpetologica* 37:213–228.

Jackson, N. D., and C. C. Austin. 2010. "The Combined Effects of Rivers and Refugia Generate Extreme Cryptic Fragmentation within the Common Ground Skink (*Scincella lateralis*)." *Evolution* 64:409–428.

———. 2012. "Inferring the Evolutionary History of Divergence Despite Gene Flow in a Lizard Species, *Scincella lateralis* (Scincidae), Composed of Cryptic Lineages." *Biological Journal of the Linnean Society* 107:192–209.

Jansen, K. P., H. R. Mushinsky, and S. A. Karl. 2008. "Population Genetics of the

Mangrove Salt Marsh Snake, *Nerodia clarkii compressicauda*, in a Linear, Fragmented Habitat." *Conservation Genetics* 9:401–410.

Jensen, J. B. 2004a. "Mimic Glass Lizard *Ophisaurus mimicus* Palmer." Pp. 164. In Mirarchi et al., *Alabama Wildlife*, vol. 4.

———. 2004b. "Southern Hognose Snake *Heterodon simus* (Linneus)." Pp. 156–157. In Mirarchi et al., *Alabama Wildlife*, vol. 4.

Jenssen, T. A., N. Greenberg, and K. A. Hovde. 1995. "Behavioral Profile of Free-Ranging Male Lizards, *Anolis carolinensis*, across Breeding and Post-breeding Seasons." *Herpetological Monographs* 9:41–62.

Johnson, C. R., and W. G. Voight. 1978. "Observations on Thermoregulation in the Western Slender Glass Lizard, *Ophisaurus attenuatus attenuatus* (Sauria: Anguidae)." *Zoological Journal of the Linnean Society* 63:305–307.

Johnson, R. W., R. R. Fleet, M. B. Keck, and D. C. Rudolph. 2007. "Spatial Ecology of the Coachwhip, *Masticophis flagellum* (Squamata: Colubridae), in Eastern Texas." *Southeastern Naturalist* 6:111–124.

Jones, K. B., and W. G. Whitford. 1989. "Feeding Behavior of Free-Roaming *Masticophis flagellum*: An Efficient Ambush Predator." *Southwestern Naturalist* 34:460–467.

Jordan, R. 1970. "Death-Feigning in a Captive Red-Bellied Snake, *Storeria occipito-maculata* (Storer)." *Herpetologica* 26:466–468.

Kaufman, G. S., H. T. Smith, R. M. Engeman, W. E. Meshaka Jr., and E. M. Cowan. 2007. "*Ophisaurus ventralis* (Eastern Glass Lizard). Fire-Induced Mortality." *Herpetological Review* 38:460–461.

Kautz, R. S., and J. A. Cox. 2001. "Strategic Habitats for Biodiversity Conservation in Florida." *Conservation Biology* 15:55–77.

Kimbell, L. M., III, D. L. Miller, W. Chavez, and N. Altman. 1999. "Molecular Analysis of the 18S rRNA Gene of *Cryptosporidium serpentis* in a Wild-Caught Corn Snake (*Elaphe guttata guttata*) and a Five-Species Restriction Fragment Length Polymorphism-Based Assay That Can Additionally Discern *C. parvum* from *C. wrairi*." *Applied Environmental Microbiology* 65:5345–5349.

Kofron, C. P. 1979. "Female Reproductive Biology of the Brown Snake, *Storeria dekayi*, in Louisiana." *Copeia* 1979:465–466.

Koons, D. N., R. D. Birkhead, S. M. Boback, M. I. Williams, and A. P. Greene. 2009. "The Effect of Body Size on Cottonmouth (*Agkistrodon piscivorus*) Survival, Recapture Probability, and Behavior in an Alabama Swamp." *Herpetological Conservation and Biology* 4:221–235.

Krysko, K. L., M. C. Granatosky, L. P. Nuñez, D. J. Smith. 2016. "A Cryptic New Species of Indigo Snake (Genus *Drymarchon*) from the Florida Platform of the United States." *Zootaxa* 4138:549–649.

Krysko, K. L., L. P. Nuñez, C. A. Lippi, D. J. Smith, M. C. Granatosky. 2016. "Pliocene-Pleistocene Lineage Diversifications in the Eastern Indigo Snake (*Drymarchon couperi*) in the Southeastern United States." *Molecular Phylogenetics and Evolution* 98:111–122.

Krysko, K. L., L. P. Nuñez, C. E. Newman, and B. W. Bowen. 2017. "Phylogenetics

of Kingsnakes, *Lampropeltis getula* Complex (Serpentes: Colubridae), in Eastern North America." *Journal of Heredity* 2017:1-13.

Kuntz, A. 2000. "The Fiddler's Companion: A Descriptive Index of North American and British Isles Music for the Folk Violin." www.ceolas.org/tunes/fc/
intro.html, accessed March 31, 2012.

Lailvaux, S. P., A. Herrel, B. vanHooydonck, J. J. Meyer, and D. J. Ischick. 2004
"Performance Capacity, Fighting Tactics and the Evolution of Life-Stage Male
Morphs in the Green Anole Lizard (*Anolis carolinensis*)." *Proceedings of the
Royal Society of London* B 271:2501–2508.

Lailvaux, S. P., and D. J. Irschick. 2007. "Effects of Temperature and Sex on Jump
Performance and Biomechanics in the Lizard *Anolis carolinensis*." *Functional
Ecology* 21:534–543.

Langford, G. E. 1987. "Native American Legends." Little Rock, AR: August House.

Langkilde, T. 2009. "Invasive Fire Ants Alter Behavior and Morphology of Native
Lizards." *Ecology* 90:208–217.

Langkilde, T., and N. A. Freidenfelds. 2010. "Consequences of Envenomation: Red
Imported Fire Ants Have Delayed Effects on Survival but Not Growth of Native
Fence Lizards." *Wildlife Research* 37:566–573.

Lawson, R. 1987. "Molecular Studies of Thamnophiine Snakes: The Phylogeny of
the Genus *Nerodia*." *Journal of Herpetology* 21:140–157.

Lawson, R., A. J. Meier, P. G. Frank, and P. E. Moler. 1991. "Allozyme Variation
and Systematics of the *Nerodia fasciata—Nerodia clarkii* Complex of Water
Snakes (Serpentes: Colubridae)." *Copeia* 1991:638–659.

Leaché, A. D., and T. W. Reeder. 2002. "Molecular Systematics of the Eastern
Fence Lizard (*Sceloporus undulatus*): Comparison of Parsimony, Likelihood,
and Bayesian Approaches." *Systematic Biology* 51:44–68.

Lee, M. S. Y. 2005a. "Molecular Evidence and Marine Snake Origins." *Biology
Letters* 22:227–230.

———. 2005b. "Squamate Phylogeny, Taxon Sampling and Data Congruence."
*Organisms Diversity and Evolution* 5:25–45.

Linehan, J. M., L. L. Smith, and D. A. Steen. 2010. "Ecology of the Eastern
Kingsnake (*Lampropeltis getula getula*) in a Longleaf Pine (*Pinus palustris*)
Forest in Southwestern Georgia." *Herpetological Conservation and Biology*
5:94–101.

Lovern, M. B., M. M. Holmes, and J. Wade. 2004. "The Green Anole (*Anolis carolinensis*): A Reptilian Model for Laboratory Studies of Reproductive Morphology and Behavior." *Institute for Laboratory Animal Research Journal* 45:54–64.

Makowsky, R., J. C. Marshall Jr., J. McVay, P. T. Chippendale, and L. J. Rissler.
2010. "Phylogeographic Analysis and Environmental Niche Modeling of the
Plainbellied Water Snake (*Nerodia erythrogaster*) Reveals Low Levels of Genetic and Ecological Differentiation." *Molecular Phylogenetics and Evolution*
55:985–995.

Marion, K. R. 2004. "Prairie Kingsnake *Lampropeltis calligaster calligaster* (Harlan)." Pp. 63–64. In Mirarchi et al., *Alabama Wildlife*, vol. 4.

Marion, K. R., and G. Bosworth. 1982. "*Hemidactylus turcicus* (Mediterranean Gecko)." *Herpetological Review* 13:52.

McConkey, E. H. 1954. "A Systematic Study of the North American Lizards of the Genus *Ophisaurus*." *American Midland Naturalist* 51:133–169.

McKnight, T. L., and D. Hess. 2000. "Physical Geography: A Landscape Appreciation." Upper Saddle River, NJ: Prentiss-Hall.

McVay, J. D., and B. Carstens. 2013. "Testing Monophyly without Well-Supported Gene Trees: Evidence from Multi-locus Nuclear Data Conflicts with Existing Taxonomy in the Snake Tribe Thamnophiini." *Molecular Phylogenetics and Evolution* 68:425–431.

Means, D. B. 2004a. "Coal Skink *Eumeces anthracinus* (Linneus)." Pp. 58–59. In Mirarchi et al., *Alabama Wildlife*, vol. 4.

———. 2004b. "Eastern Diamondback Rattlesnake *Crotalus adamanteus* Palisot de Beauvois." Pp. 169–170. In Mirarchi et al., *Alabama Wildlife*, vol. 4.

———. 2004c. "Eastern Kingsnake *Lampropeltis getula getula* (Linneus). Pp. 166. In Mirarchi et al., *Alabama Wildlife*, vol. 4.

———. 2004d. "Florida Pine Snake *Pituophis melanoleucus mugitus* Barbour." Pp. 168. In Mirarchi et al., *Alabama Wildlife*, vol. 4.

———. 2009. "Effects of Rattlesnake Roundups on the Eastern Diamondback Rattlesnake (*Crotalus adamanteus*)." *Herpetological Conservation and Biology* 4:132–141.

Means, D. B., and H. W. Campbell. 1981. "Effects of Prescribed Fire on Amphibians and Reptiles." Pp. 89–96. In *Prescribed Fire and Wildlife in Southern Forests*, ed. G. W. Wood. Georgetown, SC: Belle Baruch Forest Science Institute, Clemson University.

Mebert, K. 2008. "Good Species Despite Massive Hybridization: Genetic Research on the Contact Zone between the Water Snakes *Nerodia sipedon* and *N. fasciata* in the Carolinas, USA." *Molecular Ecology* 17:1918–1929.

Meshaka, W. E., Jr., B. P. Butterfield, and J. B. Hauge. 2004. *Exotic Amphibians and Reptiles of Florida*. Malabar, FL: Krieger Publishing.

Michaud, E. J., and A. C. Echternacht. 1995. "Geographic Variation in the Life History of the Lizard *Anolis carolinensis* and Support for the Pelvic Constraint Model." *Journal of Herpetology* 29:86–97.

Middendorf, G. A., III, and W. C. Sherbrooke. 1992. "Canid Elicitation of Blood-Squirting in a Horned Lizard (*Phrynosoma cornutum*)." *Copeia* 1992:519–527.

Miller, G. S. 2008. "Home Range Size, Habitat Associations and Refuge Use of the Florida Pine Snake, *Pituophis melanoleucus mugitus*, in Southwest Georgia, USA." Unpublished master's thesis, University of Florida, Gainesville.

Mills, H. H., and J. M. Kaye. 2001. "Drainage History of the Tennessee River: Review and New Metamorphic Quartz Gravel Locations." *Southeastern Geology* 40:75–97.

Mineka, S., and M. Cook. 1988. "Social Learning and the Acquisition of Snake

Fear in Monkeys." Pp. 51–73. In *Social Learning: Psychological and Biological Perspectives*, ed. T. R. Zentall and B. G. Galef Jr. Hillsdale, NJ: Lawrence Earlbaum Associates.

Mirarchi, R. E., M. A. Bailey, J. T. Garner, T. M. Haggerty, T. L. Best, M. F. Mettee, and P. E. O'Neil, eds. 2004. *Alabama Wildlife*. Vol. 4, *Conservation and Management Recommendations for Imperiled Wildlife*. Tuscaloosa: University of Alabama Press.

Mirarchi, R. E., M. A. Bailey, T. M. Haggerty, and T. L. Best, eds. 2004. *Alabama Wildlife*. Vol. 3, *Imperiled Amphibians, Reptiles, Birds, and Mammals*. Tuscaloosa: University of Alabama Press.

Mount, R. H. 1975. *Reptiles and Amphibians of Alabama*. Auburn: Alabama Agricultural Experiment Station, Auburn University.

———. 1980. "Survey for the Presence or Absence of Threatened or Endangered Amphibians and Reptiles, Conecuh National Forest, Alabama." Unpublished report to US Forest Service.

———. 1986. "Black Pine Snake." Pp. 35–36. In *Vertebrate Animals of Alabama in Need of Special Attention*, ed. R. H. Mount. Auburn: Alabama Agricultural Experiment Station, Auburn University.

Mount, R. H., S. E. Trauth, and W. H. Mason. 1981. "Predation by the Red Imported Fire Ant, *Solenopsis invicta* (Hymenoptera: Formicidae) on Eggs of the Lizard *Cnemidophorus sexlineatus* (Squamata: Teiidae)." *Journal of the Alabama Academy of Sciences* 52:66–70.

Mullin, S. J., R. J. Cooper, and W. H. N. Gutzke. 1998. "The Foraging Ecology of the Gray Rat Snake (*Elaphe obsolete spiloides*). III. Searching for Different Prey Types in Structurally Varied Habitats." *Canadian Journal of Zoology* 776:548–555.

Murphy, R. W., J. Fu, A. Lathrop, J. V. Feltham, and V. Kovac. 2002. "Phylogeny of the Rattlesnakes (*Crotalus* and *Sistrurus*) Inferred from Sequences of Five Mitochondrial DNA Genes." Pp. 69–92. In *Biology of the Vipers*, ed. G. W. Schuett, M. Hoggren, M. E. Douglas, and H. W. Greene. Salt Lake City, UT: Eagle Mountain Publishing.

Mushinsky, H. R. 1979. "Mating Behavior of the Common Water Snake, *Nerodia sipedon sipedon*, in Eastern Pennsylvania (Reptilia, Serpentes, Colubridae)." *Herpetologica* 13:127–129.

———. 1985. "Fire and the Florida Sandhill Herpetofaunal Community: With Special Attention to Responses of *Cnemidophorus sexlineatus*." *Herpetologica* 41:333–342.

Mushinsky, H. R., J. J. Hebrard, and D. S. Vodopich. 1982. "Ontogeny of Water Snake Foraging Ecology." *Ecology* 63:1624–1629.

Mushinsky, H. R., J. J. Hebrard, and M. G. Walley. 1980. "The Role of Temperature on the Behavioral and Ecological Associations of Sympatric Water Snakes." *Copeia* 1980:744–754.

Myers, C. W. 1967. "The Pine Woods Snake *Rhadinaea flavilata* (Cope)." *Bulletin of the Florida State Museum* 11:47–97.

———. 1974. "The Systematics of *Rhadinaea* (Colubridae), a Genus of New World Snakes." *Bulletin of the American Museum of Natural History* 153:1–262.

Nelson, D. H. 2004. "Eastern Coral Snake *Micrurus fulvius* (Linneus)." P. 169. In Mirarchi et al., *Alabama Wildlife*, vol. 4.

Nelson, D. H., and M. A. Bailey. 2004. "Black Pinesnake *Pituophis melanoleucus lodingi* Blanchard." Pp. 157–158. In Mirarchi et al., *Alabama Wildlife*, vol. 4.

Nguyen, T. Q., W. Bohme, T. T. Nguyen, Q. K. Le, K. R. Pahl, T. Haus, and T. Ziegler. 2011. "Review of the Genus *Dopasia* Gray, 1853 (Squamata: Anguidae) in the Indochina Subregion." *Zootaxa* 2894:58–68.

Nicholson, K. E., B. I. Crother, C. Guyer, and J. M. Savage. 2012. "It Is Time for a New Classification of *Anolis* (Squamata: Dactyloidae)." *Zootaxa* 3477:1–108.

Palmer, W. M. 1978. "Communal Egg-Laying and Hatchlings of the Rough Green Snakes, *Opheodrys aestivus* (Linnaeus) (Reptilia: Serpentes: Colubridae)." *Journal of Herpetology* 10:257–259.

———. 1987. "A New Species of Glass Lizard (Anguidae: *Ophisaurus*) from the Southeastern United States." *Herpetologica* 43:415–423.

Palmer, W. M., and A. L. Braswell. 1995. *Reptiles of North Carolina*. Chapel Hill: University of North Carolina Press.

Pisani, G. R. 2009. "*Virginia valeriae* and *Storeria dekayi* in a Northeast Kansas Grassland Community: Ecology and Conservation Implications." *Journal of Kansas Herpetology* 32:20–36.

Plummer, M. V. 1981. "Habitat Utilization, Diet, and Movements of a Temperate, Arboreal Species (*Opheodrys aestivus*)." *Journal of Herpetology* 14:425–432.

———. 1985. "Demography of Green Snakes (*Opheodrys aestivus*)." *Herpetologica* 41:373–381.

———. 1990. "Nesting Movements, Nesting Behavior, and Nest Sites of Green Snakes (*Opheodrys aestivus*)." *Herpetologica* 46:190–195.

———. 1993. "Thermal Ecology of Arboreal Green Snakes (*Opheodrys aestivus*)." *Journal of Herpetology* 27:254–260.

———. 1997. "Population Ecology of Green Snakes (*Opheodrys aestivus*) Revisited." *Herpetological Monographs* 11:102–123.

Plummer, M. V., and N. E. Mills. 2000. "Spatial Ecology and Survivorship of Resident and Translocated Hognose Snakes (*Heterodon platirhinos*)." *Journal of Herpetology* 34:565–575.

Pyron, R. A., and F. T. Burbrink. 2009a. "Lineage Diversification of a Widespread Species: Roles for Niche Divergence and Conservatism in the Common Kingsnake, *Lampropeltis getula*." *Molecular Ecology* 16:3443–3457.

———. 2009b. "Neogene Diversification and Taxonomic Stability in the Snake Tribe Lampropeltini (Serpentes: Colubridae)." *Molecular Phylogenetics and Evolution* 52:524–529.

Pyron, R. A., F. T. Burbrink, G. R. Colli, A. N. Montes de Oca, L. J. Vitt, C. A. Kuczynski, and J. J. Wiens. 2011. "The Phylogeny of Advanced Snakes (Colubroidea), with Discovery of a New Subfamily and Comparison of Support

Methods with Likelihood Trees." *Molecular Phylogenetics and Evolution* 58:329–342.

Pyron, R. A., F. T. Burbrink, and J. J. Wiens. 2013. "A Phylogeny and Revised Classification of Squamata, Including 4,161 Species of Lizards and Snakes." *BMC Evolutionary Biology* 13:93.

Pyron, R. A., and C. Camp. 2007. "Courtship and Mating Behaviours of Two Syntopic Species of Skinks (*Plestiodon anthracinus* and *P. fasciatus*)." *Amphibia-Reptilia* 28:263–268.

Qualls, C. P., and R. G. Jaeger. 1991. "Dear Enemy Recognition in *Anolis carolinensis*." *Journal of Herpetology* 25:361–363.

Reeder, T. D., C. J. Cole, and H. W. Dessauer. 2002. "Phylogenetic Relationships of Whiptail Lizards of the Genus *Cnemidophorus* (Squamata: Teiidae): A Test of Monophyly, Reevaluation of Karyotypic Evolution, and Review of Hybrid Origins." *American Museum Novitates* 3365:1–61.

Reeder, T. D., and J. J. Wiens. 1996. "Evolution of the Lizard Family Phrynosomatidae as Inferred from Diverse Kinds of Data." *Herpetological Monographs* 10:43–84.

Richardson, M. L., P. J. Weatherhead, and J. D. Brawn. 2006. "Habitat Use and Activity of Prairie Kingsnakes (*Lampropeltis calligaster calligaster*) in Illinois." *Journal of Herpetology* 40:423–428.

Richmond, J. Q. 2006. "Evolutionary Basis of Parallelism in North American Scincid Lizards." *Evolution and Development* 8:477–490.

Robbins, T. R., N. R. Freidenfelds, and T. Langkilde. 2013. "Native Predator Eats Invasive Toxic Prey: Evidence for Increased Incidence of Consumption Rather Than Aversion-Learning." *Biological Invasions* 15:407–415.

Rodda, G. H., C. S. Jarnevich, and R. N. Reed. 2009. "What Parts of the US Mainland Are Climatically Suitable for Invasive Alien Pythons Spreading from Everglades National Park?" *Biological Invasions* 11:241–252.

Rodríguez-Robles, J. A., and J. M. de Jesús-Escobar. 2000. "Molecular Systematics of New World Gopher, Bull, and Pinesnakes (Colubridae: *Pituophis*), Transcontinental Species Complex." *Molecular Phylogenetics and Evolution* 14:35–50.

Romagosa, C. A., C. Guyer, and M. C. Wooten. 2009. "Contribution of the Live-Vertebrate Trade toward Taxonomic Homogenization." *Conservation Biology* 23:1001–1007.

Ruby, D. E. 1984. "Male Breeding Success and Differential Access to Females in *Anolis carolinensis*." *Herpetologica* 40:272–280.

Rudolph, D. C., R. R. Schaefer, S. J. Burgdorf, M. Duran, and R. N. Conner. 2007. "Pine Snake (*Pituophis ruthveni* and *P. melanoleucus lodingi*) Hibernacula." *Journal of Herpetology* 41:560–565.

Russell, K. R., and H. G. Hanlin. 1999. "Aspects of the Ecology of Worm Snakes (*Carphophis amoenus*) Associated with Small Isolated Wetlands in South Carolina." *Journal of Herpetology* 33:339–344.

Sanderson, W. E. 1993. "Additional Evidence for the Specific Status of *Nerodia cyclopion* and *Nerodia floridana* (Reptilia: Colubridae)." *Brimleyana* 19:83–92.

Schmitz, A., P. Mausfeld, and D. Embert. 2004. "Molecular Studies on the Genus *Eumeces* Wiegmann 1834: Phylogenetic Relationships and Taxonomic Implications." *Hamadryad* 28:73–89.

Schoener, T. W. 1968. "The *Anolis* Lizards of Bimini: Resource Partitioning in a Complex Fauna." *Ecology* 49:704–726.

Schotz, A., and M. Barbour. 2009. "Ecological Assessment and Terrestrial Vertebrate Surveys for Black Belt Prairies in Alabama." www.outdooralabama.com/research-mgmt/State%20Wildlife%20Grants/Prairies%20Ecological%20Assessment%20Final%20Report.pdf, accessed January 10, 2010.

Schwaner, T., and R. H. Mount. 1976. "Systematic and Ecological Relationships between the Water Snakes *Natrix sipedon pleuralis* and *N. fasciata* in Alabama and the Florida Panhandle." *Occasional Papers of the Museum of Natural History of the University of Kansas*, 45:1–44.

Secor, S. M. 1995. "Ecological Aspects of Foraging Mode for the Snakes *Crotalus cerastes* and *Masticophis flagellum*." *Herpetological Monographs* 9:169–186.

Secor, S. M., and K. A. Nagy. 1994. "Bioenergetic Correlates of Foraging Mode for the Snakes *Crotalus cerastes* and *Coluber constrictor*." *Ecology* 75:1600–1614.

Seigel, D. S., D. M. Sever, J. L. Rheubert, and K. M. Gribbins. 2009. "Reproductive Biology of *Agkistrodon piscivorus* Lacépède (Squamata: Serpentes: Viperidae: Crotalinae)." *Herpetological Monographs* 23:74–107.

Seigel, R. A., J. W. Gibbons, and T. K. Lynch. 1995. "Temporal Changes in Reptile Populations: Effects of a Severe Drought on Aquatic Snakes." *Herpetologica* 51:424–434.

Semlitsch, R. D., K. L. Brown, and J. P. Caldwell. 1981. "Habitat Utilization, Season Activity, and Population Size Structure of the Southeastern Crowned Snake, *Tantilla coronata*." *Herpetologica* 37:40–46.

Semlitsch, R. D., and G. B. Moran. 1984. "Ecology of the Redbelly Snake (*Storeria occipitomaculata*) Using Mesic Habitats in South Carolina." *American Midland Naturalist* 111:33–40.

Semlitsch, R. D., J. H. K. Pechmann, and J. W. Gibbons. 1988. "Annual Emergence of Juvenile Mud Snakes (*Farancia abacura*) at Aquatic Habitats." *Copeia* 1988:243–245.

Sever, D. M., and W. C. Hamlett. 2001. "Female Sperm Storage in Reptiles." *Journal of Experimental Zoology* 292:187–199.

Seyle, C. W. 1980. "The Systematic Relationship between the Water Snakes *Nerodia sipedon* and *N. fasciata* in Georgia." Unpublished master's thesis, Auburn University, Auburn, AL.

Shine, R., R. N. Reed, S. Shetty, M. Lemaster, and R. T. Mason. 2002. "Reproductive Isolating Mechanisms between Two Sympatric Sibling Species of Sea Snakes." *Evolution* 56:1655–1662.

Slowinski, J. B. 1995. "A Phylogenetic Analysis of the New World Coral Snakes (Elapidae: *Leptomicrurus*, *Micruroides*, and *Micrurus*) Based on Allozymic and Morphological Characters." *Journal of Herpetology* 29:325–338.

Smith, C. R. 1987. "Ecology of Juvenile and Gravid Female Eastern Indigo Snakes in North Florida." Unpublished master's thesis, Auburn University, Auburn, AL.

Smith, H. M. 2005. "A Replacement Name for Most Members of the Genus *Eumeces* in North America." *Journal of Kansas Herpetology* 14:15–16.

Smith, L. L., and C. K. Dodd. 2003. "Wildlife Mortality on US Highway 441 across Paynes Prairie, Alachua County, Florida." *Florida Scientist* 66:128–140.

Soltis, D. E., A. B. Morris, J. S. McLachlan, P. S. Manos, and P. S. Soltis. 2006. "Comparative Phylogeography of Unglaciated Eastern North America." *Molecular Ecology* 15:4261–4293.

Spangler, J. A., and R. H. Mount. 1969. "The Taxonomic Status of the Natricine Snake *Regina septemvittata mabila* (Neill)." *Herpetologica* 25:113–119.

Speake, D. W. 1986. "Eastern Indigo Snake." Pp. 24–25 In *Vertebrate Animals of Alabama in Need of Special Attention*, ed. R. H. Mount. Auburn, AL: Alabama Agricultural Experiment Station, Auburn University.

Speake, D. W., J. A. McGlincy, and C. Smith. 1987. "Captive Breeding and Experimental Reintroduction of the Eastern Indigo Snake.' Pp. 84–88. In *Proceedings of the Third Southeast Nongame and Endangered Wildlife Symposium*, ed. R. R. Odum, K. Riddleberger, and J. Ozier. Georgia Department of Natural Resources, Game and Fish Division Technical Bulletin WL 4, Social Circle, GA.

Speake, D. W., and R. H. Mount. 1973. "Some Possible Ecological Effects of 'Rattlesnake Roundups' in the Southeastern Coastal Plain." Pp. 267–277. In *Proceedings of the 27th Annual Conference of the Southeastern Association of Game and Fish Commissioners*. Contributions to the Alabama Cooperative Wildlife Research Unit, Auburn, AL.

Steen, D. A., S. N. Becker, and L. L. Smith. 2005. "*Heterodon platirhinos* (Eastern Hog-nosed Snake)." Reproduction. *Herpetological Review* 36:457.

Steen, D. A. J. M. Linehan, and L. L. Smith. 2010. "Multiscale Habitat Selection and Refuge Use of Common Kingsnakes, *Lampropeltis getula*, in Southwestern Georgia." *Copeia* 2010:227–231.

Steen, D. A., C. J. W. McClure, L. L. Smith, B. J. Halstead, C. K. Dodd Jr., W. B. Sutton, J. R. Lee, D. L. Baxley, W. J. Humphries, and C. Guyer. 2013. "The Effect of Coachwhip Presence on Body Size of North American Racers Suggests Competition between These Sympatric Snakes." *Journal of Zoology* 289:86–93.

Steen, D. A., C. J. W. McClure, W. B. Sutton, D. C. Rudolph, J. B. Pierce, J. R. Lee, L. L. Smith, B. B. Gregory, D. L. Baxley, D. J. Stevenson, and C. Guyer. 2014. "Copperheads Are Common When Kingsnakes Are Not: Relationships between the Abundance of a Predator and One of Their Prey." *Herpetologica* 70:69–76.

Steen, D. A., A. E. R. McGee, S. M. Hermann, J. A. Stiles, S. H. Stiles, and C. Guyer. 2010. "Effects of Forest Management on Amphibians and Reptiles: Generalist Species Obscure Trends among Native Forest Associates." *Open Environmental Sciences* 4:24–30.

Steen, D. A., G. J. Miller, S. C. Sterrett, and L. L. Smith. 2007. "*Masticophis flagellum flagellum* (Eastern Coachwhip)." *Herpetological Review* 38:90.

Steen, D. A., L. L. Smith, L. M. Conner, J. C. Brock, and S. K. Hoss. 2005. "Habitat Use of Sympatric Rattlesnake Species within the Gulf Coastal Plain." *Journal of Wildlife Management* 71:759–764.

Steen, D. A., J. A. Stiles, S. H. Stiles, C. Guyer, J. B. Pierce, D. C. Rudolph, and L. L. Smith. 2011. "*Regina rigida* (Glossy Crayfish Snake). Terrestrial Movement." *Herpetological Review* 42:101.

Steen, D. A., D. J. Stevenson, J. C. Beane, J. D Willson, M. J. Aresco, J. C. Godwin, S. P. Graham, L. L. Smith, J. M. Howze, D. C. Rudolph, J. B. Pierce, J. R. Lee, B. B. Gregory, J. Jensen, S. H. Stiles, J. A. Stiles, N. H. Nazdrowicz, and C. Guyer. 2013. "Terrestrial Movements of the Red-Bellied Mudsnake (*Farancia abacura*) and Rainbow Snake (*F. erytrogramma*)." *Herpetological Review* 44:208–213.

Stevenson, D. J., K. M. Enge, L. D. Carlile, K. J. Dyer, T. M. Norton, N. L. Hyslop, and R. A. Kiltie. 2009. "An Eastern Indigo Snake (*Drymarchon couperi*) Mark-Recapture Study in Southeastern Georgia." *Herpetological Conservation and Biology* 4:30–42.

Strickland, J. L., C. L. Parkinson, J. K. McCoy, and L. K. Ammerman. 2014. "Phylogeography of *Agkistrodon piscivorus* with emphasis on the western limit of its range." *Copeia* 2014:639–649.

Todd, B. D., and K. M. Andrews. 2008. "Response of a Reptile Guild to Forest Harvesting." *Conservation Biology* 22:753–761.

Todd, B. D., J. D. Willson, C. T. Winne, R. D. Semlitsch, and J. W. Gibbons. 2008. "Ecology of the Southeastern Crowned Snake, *Tantilla coronata*." *Copeia* 2008:388–394.

Townsend, T. M., A. Larson, E. Louis, and J. R. Macey. 2004. "Molecular Phylogenetics of Squamata: The Position of Snakes, Amphisbaenians, and Dibamids, and the Root of the Squamate Tree." *Systematic Biology* 53:735–757.

Trauth, S. E. 1981. "Nesting Habitat and Reproductive Characteristics of the Lizard *Cnemidophorus sexlineatus* (Lacertilia: Teiidae)." *American Midland Naturalist* 109:289–299.

———. 1984. "Seasonal Incidence and Reproduction in the Western Slender Glass Lizard, *Ophisaurus attenuatus attenuatus* (Reptilia: Anguidae), in Arkansas." *Southwestern Naturalist* 29:271–275.

———. 1994. "Reproductive Cycles in Two Arkansas Skinks in the Genus *Eumeces* (Sauria: Scincidae)." *Proceedings of the Arkansas Academy of Sciences* 48:210–218.

Tryon, B. W., and J. B. Murphy. 1982. "Miscellaneous Notes on the Reproductive Biology of Reptiles. 5. Thirteen Varieties of the Genus *Lampropeltis*, Species *mexicana*, *triangulum*, and *zonata*." *Transactions of the Kansas Academy of Sciences* 85:96–119.

Tu, M., and V. H. Hutchinson. 1994. "Influence of Pregnancy on Thermoregulation of Water Snakes (*Nerodia rhombifera*)." *Journal of Thermal Biology* 19:255–259.

Tuberville, T. D., J. R. Brodie, J. B. Jensen, L. LaClaire, and J. W. Gibbons. 2000. "Apparent Decline of the Southern Hog-Nosed Snake, *Heterodon simus*." *Journal of the Elisha Mitchell Science Society* 116:19–40.

Tyson, A., D. A. Steen, and L. L. Smith. 2008. "*Nerodia taxispilota* (Brown Watersnake). Diet." *Herpetological Review* 39:472.

Uetz, P. 2010. "The Original Descriptions of Reptiles." *Zootaxa* 2334:59–68.

Uller, T., and M. Olsson. 2008. "Multiple Paternity in Reptiles: Patterns and Processes." *Molecular Ecology* 17:2566–2580.

US Fish and Wildlife Service. 1978. "Endangered and Threatened Wildlife and Plants: Listing of the Eastern Indigo Snake as a Threatened Species." *Federal Register* 43:4026–4028.

Utiger, U., N. Helfenberger, B. Schatti, C. Schmidt, M. Ruf, and V. Ziswiler. 2002. "Molecular Systematics and Phylogeny of Old and New World Ratsnakes, *Elaphe* Auct., and Related Genera (Reptilia, Squamata, Colubridae)." *Russian Journal of Herpetology* 9:105–124.

Utiger, U., B. Schatti, and N. Helfenberger. 2005. "The Oriental Colubrine Genus, *Coelognathus*, Fitzinger 1843 and Classification of Old and New World Racers and Ratsnakes (Reptilia, Squamata, Colubridae, Colubrinae)." *Russian Journal of Herpetology* 12:39–60.

VanZant, J. L., and M. C. Wooten. 2007. "Old Mice, Young Islands and Competing Biogeographical Hypotheses." *Molecular Ecology* 16:5070–5083.

Vidal, N., and S. B. Hedges. 2009. "The Molecular Evolutionary Tree of Lizards, Snakes, and Amphisbaenians." *Comptes Rendus Biologies* 332:129–139.

Vitt, L. J., S. S. Sartorius, T. C. S. Ávila-Pires, P. A. Zani, and M. C. Espósito. 2005. "Small in a Big World: Ecology of Leaf-Litter Geckos in New World Tropical Forests." *Herpetological Monographs* 19:137–152.

Waldron, J. L., S. H. Bennett, S. M. Welch, M. E. Dorcas, J. D. Lanham, and W. Kalinowksy. 2006. "Habitat Specificity and Home-Range Size as Attributes of Species Vulnerability to Extinction: A Case Study Using Sympatric Rattlesnakes." *Animal Conservation* 9:414–420.

Wallach, V., K. L. Williams, and J. Boundy. 2014. *Snakes of the World: A Catalog of Living and Extinct Species*. Boca Raton, FL: CRC Press.

Wharton, C. H. 1960. "Birth and Behavior of a Brood of Cottonmouths, *Agkistrodon piscivorus piscivorus* with Notes on Tail-Luring." *Herpetologica* 16:125–129.

White, G., C. Ottendorfer, S. P. Graham, and T. R. Unnasch. 2011. "Competency of Reptiles and Amphibians for Eastern Equine Encephalitis Virus." *American Journal of Tropical Medicine and Hygiene* 85:421–425.

Wiens, J. J., C. A. Kuczynski, T. Townsend, T. W. Reeder, D. G. Mulcahy, and J. W. Sites. 2010. "Combining Phylogenomics and Fossils in Higher-Level Squamate

Reptile Phylogeny: Molecular Data Change the Placement of Fossil Taxa." *Systematic Biology* 59:674–688.

Wiens, J. J., and J. L. Slingluff. 2001. "How Lizards Turn in to Snakes: A Phylogenetic Analysis of Body-Form Evolution in Anguid Lizards." *Evolution* 55:2303–2318.

Williams, K. L., 1978. *Systematics and Natural History of the North American Milk Snake*, Lampropeltis triangulum. Milwaukee: Milwaukee Public Museum.

Wilson, M. A., and A. C. Echternacht. 1987. "Geographic Variation in the Critical Thermal Minimum of the Green Anole, *Anolis carolinensis* (Sauria: Iguanidae), along a Latitudinal Gradient." *Comparative Biochemistry and Physiology Part A: Physiology* 87:757–760.

Winne, C. T., M. E. Dorcas, and S. M. Poppy. 2005. "Population Structure, Body Size, and Seasonal Activity of Black Swamp Snakes (*Seminatrix pygaea*)." *Southeastern Naturalist* 4:1–14.

Winne, C. T., T. J. Ryan, Y. Leiden, and M. E. Dorcas. 2001. "Evaporative Water Loss in Two Natricine Snakes, *Nerodia fasciata* and *Seminatrix pygaea*." *Journal of Herpetology* 35:129–133.

Winne, C. T., J. D. Willson, and J. W. Gibbons. 2006. "Income Breeding Allows an Aquatic Snake *Seminatrix pygaea* to Reproduce Normally Following Prolonged Drought-Induced Aestivation." *Journal of Animal Ecology* 75:1352–1360.

Winne, C. T., J. D. Willson, B. D. Todd, K. M. Andrews, and J. W. Gibbons. 2007. "Enigmatic Decline of a Protected Population of Eastern Kingsnakes, *Lampropeltis getula*, in South Carolina." *Copeia* 2007:507–519.

Witz, B. W. 2001. "Aspects of the Thermal Biology of the Six-Lined Racerunner *Cnemidophorus sexlineatus* (Squamata: Teiidae) in West-Central Florida." *Journal of Thermal Biology* 26:529–535.

Young, B. A., and A. Aguiar. 2002. "Response of Western Diamondback Rattlesnakes (*Crotalus atrox*) to Airborne Sounds." *Journal of Experimental Zoology* 205:3087–3092.

Zaher, H., F. G. Grazziotin, J. E. Cadle, R. W. Murphy, J. C. de Moura-Leite. 2009. "Molecular Phylogeny of Advanced Snakes (Serpentes: Caenophidia) with an Emphasis on South American Xenodontines: A Revised Classification and Descriptions of New Taxa." *Papéis Avulsos de Zoologia* 49:115–153.

# About the Authors

CRAIG GUYER was born on August 6, 1952, in Los Angeles, California, to parents who migrated to several towns across southern California before settling on a ranch on the outskirts of Oceanside. Although now decimated by high-density housing, during his childhood this area was remote, covered with coastal sage scrub vegetation, and home to healthy populations of amphibians and reptiles. So, when he reached third grade, a stage when many children develop a fascination with wildlife, his backyard provided daily opportunities for exploration, an interest that he has never outgrown. Craig attended Humboldt State University in Northern California, graduating with a basic degree in 1975. He then moved to Idaho State University, in 1978 completing a master's degree that examined homing behavior in sagebrush lizards and short-horned lizards. He entered the lab of Dr. Jay M. Savage, then at the University of Southern California but who later chaired the Department of Biological Sciences at the University of Miami, for doctoral work. This career step introduced Craig to the diverse herpetofauna of Costa Rica, where he continues to maintain active research. He accepted a position at Auburn University in 1987, replacing the retired Robert H. Mount as curator of herpetology. Over the years, the Guyer lab has maintained the premiere collection of Alabama amphibians and reptiles, provided research vital to the conservation of Alabama's amphibians and reptiles, and taught undergraduate students about global patterns of vertebrate biodiversity. He is the 2013 recipient of the Meritorious Teaching Award given by the Society for the Study of Amphibians and Reptiles and was the 2013–16 Scharnagel Professor of Biological Sciences in the College of Science and Mathematics at Auburn University. He retired in 2016.

MARK A. BAILEY was born on July 20, 1961, in Birmingham, Alabama. His fascination with reptiles and amphibians began around age nine when he was given Hobart M. Smith's *Snakes as Pets*, and his formative years were spent roaming the woods of the upper Black Warrior River watershed, bringing home more "pets" than a conservation-minded herpetologist should probably confess. He was fortunate to have parents who tolerated and supported his interests. He attended public and private schools in Huffman, Palmerdale, and Pinson, Alabama. After graduating from high school in 1979, he entered Auburn University where he completed a BS degree in Biology in 1984 and an MS degree in Zoology in 1988, with Robert H. Mount as his major professor. From 1986 to 1988, he was assistant curator of the Auburn University Museum's herpetology collection. Immediately following graduation, he worked as a biological technician at Conecuh

National Forest, and in 1989, he was hired as zoologist for the Nature Conservancy's new Alabama Natural Heritage Program. There, he conducted statewide reptile and amphibian field surveys, organized conservation plans, and initiated the Alabama Herp Atlas Project. In 1998, he left the Heritage Program and with his wife, Karan, founded Conservation Southeast, a consulting firm they continue to operate from Andalusia. He has served on the advisory board of the Auburn University School of Forestry and Wildlife Science and is a past president of the Alabama Chapter of the Wildlife Society. He has served as the Alabama state representative to the Gopher Tortoise Council, and he continues to serve as a director of the Alabama Wildlife Federation. He received the Governor's Conservation Achievement Award for Wildlife Biologist of the Year in 2007. Mark has been involved in Partners for Amphibian and Reptile Conservation (PARC) and the Alabama chapter (ALAPARC) since their inception and was lead author on the 2006 PARC publication, *Management Guidelines for Amphibians and Reptiles of the Southeastern United States*.

ROBERT H. MOUNT was born December 25, 1931, in Lewisburg, Tennessee. Following the death of his mother four years later, he was cared for by his father and great-aunt until his aunt died in 1937. He then went to live with his maternal aunt and uncle in Waynesboro, Tennessee. His father visited him weekly during that period and taught him to appreciate the natural world.

Waynesboro, with its crystal clear creeks and surrounding forests teeming with wildlife, was an idyllic place perfectly suited for a budding naturalist. Robert's father remarried when Robert was ten, and he, with his father and adoring stepmother, settled in Jackson, Tennessee, until moving to Albany, Georgia, where Robert attended high school. Following his graduation, he enrolled at Alabama Polytechnic Institute, now Auburn University, where he received BS and MS degrees. He then served in the army in the United States and Korea as a medical entomologist for two years, after which he enrolled at the University of Florida as a PhD student under the direction of the late Dr. Archie Carr. There, he majored in zoology and following his graduation taught at Montevallo College in central Alabama, where he began his research on the herpetofauna of the state. Dr. Mount accepted a position at Auburn University in 1966, where he continued his research, ably assisted by colleagues Drs. Dan Speake, James Dobie, and George Folkerts and his many graduate students, culminating in the 1975 publication of *The Reptiles and Amphibians of Alabama*. Mount held offices in a number of professional and environmental organizations and received several awards, including a Dudley Beaumont Fellowship, J. Kelly Mosley Award, Gopher Tortoise Award, Outstanding Environmental Writer Award by the Alabama Environmental Council, and a Lifetime Environmental Achievement Award by the Southern Appalachian Mountain Coalition. He retired in 1987 but continued to influence conservation within Alabama through weekly newspaper columns, which he continued until his passing on September 10, 2017.

# Index